T0313216

Fuel Processing and Energy Utilization

Fuel Processing and Energy Utilization

Edited by

Sonil Nanda

Prakash Kumar Sarangi

Dai-Viet N. Vo

CRC Press is an imprint of the
Taylor & Francis Group, an **informa** business
A CHAPMAN & HALL BOOK

CRC Press
Taylor & Francis Group
6000 Broken Sound Parkway NW, Suite 300
Boca Raton, FL 33487-2742

Printed on acid-free paper

International Standard Book Number-13: 978-1-1385-9320-6 (Hardback)

Visit the Taylor & Francis Web site at
http://www.taylorandfrancis.com

and the CRC Press Web site at
http://www.crcpress.com

Contents

Preface vii
Editors xi
Contributors xiii

Chapter 1 Fossil Fuels versus Biofuels: Perspectives on Greenhouse Gas Emissions, Energy Consumptions, and Projections 1

Kang Kang, Mingqiang Zhu, Guotao Sun, and Xiaohui Guo

Chapter 2 An Overview of Fossil Fuel and Biomass-Based Integrated Energy Systems: Co-firing, Co-combustion, Co-pyrolysis, Co-liquefaction, and Co-gasification 15

Ejaz Ahmad, Ayush Vani, and Kamal K. Pant

Chapter 3 Catalytic Conversion of Lignocellulosic Biomass into Fuels and Value-Added Chemicals 31

Shireen Quereshi, Suman Dutta, and Tarun Kumar Naiya

Chapter 4 Production and Characterization of Biodiesel through Catalytic Routes 53

Yun Hin Taufiq-Yap and Nasar Mansir

Chapter 5 Recent Advances in Hydrogen Production through Bi-Reforming of Biogas 71

Tan Ji Siang, Doan Pham Minh, Sharanjit Singh, Herma Dina Setiabudi, and Dai-Viet N. Vo

Chapter 6 Effects of Mesoporous Supports and Metals on Steam Reforming of Alcohols 93

Richard Y. Abrokwah, William Dade, Sri Lanka Owen, Vishwanath Deshmane, Mahbubur Rahman, and Debasish Kuila

Chapter 7 Current Developments in the Production of Liquid Transportation Fuels through the Fischer-Tropsch Synthesis 109

Venu Babu Borugadda and Ajay K. Dalai

Chapter 8 Production of Biolubricant Basestocks from Structurally Modified Plant Seed Oils and Their Derivatives 123

Venu Babu Borugadda, Vaibhav V. Goud, and Ajay K. Dalai

Chapter 9 Recent Advances in Consolidated Bioprocessing for Microbe-Assisted Biofuel Production 141

Prakash Kumar Sarangi and Sonil Nanda

Chapter 10 Cultivation and Conversion of Algae for Wastewater Treatment and Biofuel Production 159

Priyanka Yadav, Sivamohan N. Reddy, and Sonil Nanda

Chapter 11 Life-Cycle Assessment of Biofuels Produced from Lignocellulosic Biomass and Algae..... 177

Naveenji Arun and Ajay K. Dalai

Chapter 12 Synthetic Crude Processing: Impacts of Fine Particles on Hydrotreating of Bitumen-Derived Gas Oil..... 187

Rachita Rana, Sonil Nanda, Ajay K. Dalai, Janusz A. Kozinski, and John Adjaye

Index..... 207

Preface

Since the industrial revolution, fossil fuels in the form of crude oil, gasoline, diesel, coal, and natural gas have dominated the global energy sector. The extensive use of fossil fuels is continually leading to the cumulative emission of greenhouse gases, which result in many adverse environmental conditions such as air pollution, acid rain, global warming, and ozone layer depletion. The role of energy conversion and utilization has become significantly important because of the fluctuations in availability of energy resources and the volatility in the fuel prices. Moreover, the crude oil industries based on fossil fuel resources have been facing many technical impediments such as high carbon footprint, shortage of resources, unaccountable greenhouse gas emissions, and subsequent environmental damage. The fuel refineries could be more competitive by increasing the energy efficiency and eliminating wastes that degrade the environment and natural ecosystems.

The consumption of fossil fuels and the emission of greenhouse gases are closely associated with each other. In recent years, there has been a paradigm shift in the global interest from fossil-based energy sources to green fuels. Green fuels, especially biofuels, are produced from wide-ranging organic waste resources that have numerous environmental and socio-economic benefits. Biofuels have tremendous scope in supplementing the aggregating fuel demands, mitigating CO_2 emissions, ensuring energy security and economic sustainability, as well as preventing ecological degradation. Furthermore, when utilized in connection with fossil fuels through several co-processing technologies, biofuels could potentially bring many advantages to existing energy infrastructures. This book provides current information on the production and utilization of fossil fuels and biofuels; co-processing technologies for fossil energy and bioenergy; synthetic crude oil processing; waste-to-energy and chemical generation; conversion of biomass through consolidated thermochemical, hydrothermal and biochemical pathways; reforming technologies; techno-economic analysis; as well as life-cycle assessment studies.

This book, consisting of 12 chapters, is dedicated to the new developments and prospects in energy conversion and utilization as well as fuel processing and upgrading. **Chapter 1** by Kang et al. provides an all-inclusive overview of the comparison between fossil fuels and biofuels from an environmental perspective. The chapter gives special attention to the rapidly increasing worldwide energy demand and its environmental impacts related to greenhouse gas emissions. The chapter also discusses the many potential advantages of biofuels and biorefining technologies over fossil fuel refineries.

Chapter 2 by Ahmad et al. introduces several fossil fuel and biofuel integrated energy systems such as co-firing, co-combustion, co-pyrolysis, co-liquefaction, and co-gasification. The chapter justifies the need of co-processing technologies that could bring many economic and environmental benefits such as mitigation of greenhouse gas emissions, high-energy efficiency, and circular economy.

Chapter 3 by Quereshi et al. throws light on the catalytic conversion of lignocellulosic biomass into value-added chemicals and fuel products. The chapter highlights the generation of 5-hydroxymethylfurfural, 5-ethoxymethylfurfural, levulinic acid, and ethyl levulinate from waste lignocellulosic materials as the platform building block compounds to produce many commodity chemicals and fuels. Several Brønsted and Lewis acid-catalyzed reactions such as depolymerization, isomerization, dehydration, and rehydration involved in the production of such value-added compounds are discussed.

Chapter 4 by Taufiq-Yap and Mansir presents the catalyst-assisted production and characterization of biodiesel from various feedstocks including edible oils, non-edible oils, microalgal oil, and waste cooking oil. The chapter highlights several catalysts used in the production of biodiesel such as homogeneous and heterogeneous catalysts as well as heterogeneous acid and heterogeneous solid base catalysts.

Chapter 5 by Siang et al. recounts the recent advances in hydrogen production by bi-reforming of biogas. The chapter systematically discusses the thermodynamic aspects, mechanisms, and kinetics of bi-reforming methane including the influence of process parameters such as gas hourly space velocity, reaction temperature, and feed composition. The catalysts involved in bi-reforming methane and the effects of catalyst supports and promoters are also described.

Chapter 6 by Abrokwah et al. evaluates the performance of mesoporous supports and metals for hydrogen production by steam reforming of alcohols. The study describes the synthesis and characterization of various high surface area catalytic systems by a one-pot hydrothermal method for steam reforming of methanol and glycerol. The activity and stability tests for the synthesized catalysts during steam reforming of alcohols are methodically reported.

Chapter 7 by Borugadda and Dalai gives an overview of the current developments in the production of liquid transportation fuels by Fischer-Tropsch synthesis. The chapter describes the advantages of integrated routes for biomass-to-gas and gas-to-liquid conversion technologies. The technological advancements, reaction chemistry, catalysis and reactor engineering involved in Fischer-Tropsch process are also discussed.

Chapter 8 by Borugadda et al. describes the production of biolubricant basestocks from structurally modified plant seed oils and their derivatives. The chapter elucidates the chemical composition, structure, and properties of plant seed oils, as well as several essential reactions involved in biolubricant generation such as epoxidation, hydroxylation, di-ester, tri-ester, and tetra-ester formation by esterification and anhydrides addition.

Chapter 9 by Sarangi and Nanda is a synopsis of the recent advances in consolidated bioprocessing for microbe-assisted biofuel production. The chapter describes the involvement of several solventogenic bacteria, filamentous fungi, and yeasts for bioconversion of lignocellulosic biomass and organic wastes for bioethanol and biobutanol production. This chapter also highlights microbial strain development, metabolic engineering, and bioprocess technologies in alcohol-based biofuel production.

Chapter 10 by Yadav et al. reviews the cultivation and conversion of algae for wastewater treatment and biofuel production. The chapter summarizes the aspects of algal metabolism and technologies involved in the cultivation and harvesting of algae. The role of algae in carbon sequestration and wastewater treatment is described along with biofuel production through its conversion by hydrothermal liquefaction.

Chapter 11 by Arun and Dalai is based on life-cycle assessment of biofuels produced from lignocellulosic biomass and algae. The chapter describes different methodologies for life-cycle assessment of biofuels. The impacts of fertilizers used for the cultivation of biofuel feedstocks on the environmental credibility as well as the influence of by-products and co-products of biomass conversion process on the environment are also discussed.

Chapter 12 by Rana et al. is the final chapter of the book, which is a review of synthetic crude processing and the impacts of fine particles on hydrotreating bitumen-derived gas oil. The chapter is a technical appraisal on the upgrading of bitumen, catalytic hydrotreating, and the mechanisms of fine particle deposition. This chapter focuses on significant concepts in the hydrotreating of gas oil with special emphasis on the physicochemical properties and behavior of fine particles present in bitumen-derived gas oil during the industrial hydrotreating process.

This book is a unification of chapters relating to the cutting-edge applications of green technologies that could reinvigorate the conventional oil industries and consolidated biorefineries by positioning them within a competitive energy market. This book also evaluates the potentials of integrated fossil fuel and biofuel refineries in attaining a relatively low carbon footprint, circular economy, and environmental sustainability.

The editors thank all the authors who contributed their scholarly materials to develop this book. Our heartfelt thanks to CRC Press for providing the opportunity to publish this edited volume. We appreciate the sincere efforts by the CRC Press publishing team led by Ms. Renu Upadhyay and Ms. Shikha Garg for their editorial assistance in preparing this book.

Sonil Nanda
Department of Chemical and Biochemical Engineering
University of Western Ontario
London, Ontario, Canada

Prakash Kumar Sarangi
Directorate of Research
Central Agricultural University
Imphal, Manipur, India

Dai-Viet N. Vo
Faculty of Chemical and Natural Resources Engineering
Universiti Malaysia Pahang
Kuantan, Pahang, Malaysia

Editors

Sonil Nanda is a research fellow at the University of Western Ontario in London, Ontario, Canada. He earned his PhD in Biology from York University, Toronto, Ontario, Canada; MSc in applied microbiology from Vellore Institute of Technology (VIT University), Tamil Nadu, India; and BSc in microbiology from Orissa University of Agriculture and Technology, Bhubaneshwar, Odisha, India. Dr. Nanda's research areas are related to the production of advanced biofuels and biochemicals through thermochemical and biochemical conversion technologies such as gasification, pyrolysis, carbonization, and fermentation. He has expertise in hydrothermal gasification of a wide variety of organic wastes and biomass including agricultural and forestry residues, industrial effluents, municipal solid wastes, cattle manure, sewage sludge, and food wastes to produce hydrogen fuel. His parallel interests are also in the generation of hydrothermal flames for treatment of hazardous wastes, agronomic applications of biochar, phytoremediation of heavy metal contaminated soils, as well as carbon capture and sequestration. Dr. Nanda has published more than 65 peer-reviewed journal articles, 15 book chapters, and has presented at many international conferences. He serves as a fellow member of the Society for Applied Biotechnology in India, as well as a life member of the Indian Institute of Chemical Engineers, Association of Microbiologists of India, Indian Science Congress Association, and the Biotech Research Society of India. He is also an active member of several chemical engineering societies across North America such as the American Institute of Chemical Engineers, the Chemical Institute of Canada, and the Combustion Institute-Canadian Section.

Prakash Kumar Sarangi is a scientist with specialization in food microbiology at Central Agricultural University in Imphal, India. He earned his PhD in microbial biotechnology specialization from Botany department, Ravenshaw University, Cuttack, India; MTech in applied botany from Indian Institute of Technology Kharagpur, India; and MSc in botany from Ravenshaw University, Cuttack, India. Dr. Sarangi's current research is focused on bioprocess engineering, renewable energy, biochemicals, biomaterials, fermentation technology, and post-harvest engineering and technology. He has expertise in bioconversion of crop residues and agro-wastes into value-added phenolic compounds. He has more than 10 years of teaching and research experience in biochemical engineering, microbial biotechnology, downstream processing, food microbiology, and molecular biology. He has served as a reviewer for many international journals, published more than 40 research articles in peer-reviewed journals, and authored more than 15 book chapters. He has published approximately 50 national and international conference papers. He is associated with many scientific societies as a fellow member (Society for Applied Biotechnology) and life member (Biotech Research Society of India; Society for Biotechnologists of India; Association of Microbiologists of India; Orissa Botanical Society; Medicinal and Aromatic Plants Association of India; Indian Science Congress Association; Forum of Scientists, Engineers & Technologists; and International Association of Academicians and Researchers).

Dai-Viet N. Vo earned his PhD in chemical engineering from the University of New South Wales in Sydney, Australia, in 2011. He has worked as a postdoctoral fellow at the University of New South Wales in Sydney and Texas A&M University at Qatar, Doha. He is currently a senior lecturer at the Faculty of Chemical & Natural Resources Engineering in Universiti Malaysia Pahang in Kuantan, Malaysia. His research areas are the production of green synthetic fuels via Fischer-Tropsch synthesis using biomass-derived syngas from reforming processes. He is also an expert in advanced material synthesis and catalyst characterization. During his early career, he has worked as the principal investigator and co-investigator for 19 different funded research projects related to sustainable and alternative energy. He has published 2 books, 3 book chapters, 50 peer-reviewed journal articles, and more than 60 conference proceedings. He has served on the technical and publication committees of numerous international conferences in chemical engineering and as the guest editor for some special issues in the *International Journal of Hydrogen Energy* (Elsevier) and *Catalysts* (MDPI). He is also a regular reviewer for many prestigious international journals.

Contributors

Richard Y. Abrokwah
Department of Energy and Environmental Systems
North Carolina A&T State University
Greensboro, North Carolina

John Adjaye
Syncrude Edmonton Research Centre
Edmonton, Canada

Ejaz Ahmad
Department of Chemical Engineering
Indian Institute of Technology Delhi
New Delhi, India

Naveenji Arun
Department of Chemical and Biological Engineering
University of Saskatchewan
Saskatoon, Canada

Venu Babu Borugadda
Department of Chemical and Biological Engineering
University of Saskatchewan
Saskatoon, Canada

William Dade
Department of Chemistry
North Carolina A&T State University
Greensboro, North Carolina

Ajay K. Dalai
Department of Chemical and Biological Engineering
University of Saskatchewan
Saskatoon, Canada

Vishwanath Deshmane
Department of Chemistry
North Carolina A&T State University
Greensboro, North Carolina

Suman Dutta
Department of Chemical Engineering
Indian Institute of Technology Dhanbad (Indian School of Mines)
Dhanbad, India

Vaibhav V. Goud
Department of Chemical Engineering
Indian Institute of Technology Guwahati
Guwahati, India

Xiaohui Guo
College of Mechanical and Electronic Engineering
Northwest A&F University
Yangling, China

Kang Kang
College of Mechanical and Electronic Engineering
Northwest A&F University
Yangling, China

Janusz A. Kozinski
Department of Chemical Engineering
University of Waterloo
Waterloo, Canada

Debasish Kuila
Department of Chemistry
North Carolina A&T State University
Greensboro, North Carolina

Nasar Mansir
Catalysis Science and Technology Research Centre
Universiti Putra Malaysia
Serdang, Malaysia

Doan Pham Minh
Université de Toulouse
IMT Mines Albi
Albi, France

Tarun Kumar Naiya
Department of Petroleum Engineering
Indian Institute of Technology Dhanbad (Indian School of Mines)
Dhanbad, India

Sonil Nanda
Department of Chemical and Biochemical Engineering
University of Western Ontario
London, Canada

Sri Lanka Owen
Department of Chemistry
North Carolina A&T State University
Greensboro, North Carolina

Kamal K. Pant
Department of Chemical Engineering
Indian Institute of Technology Delhi
New Delhi, India

Shireen Quereshi
Department of Chemical Engineering
Indian Institute of Technology Dhanbad (Indian School of Mines)
Dhanbad, India

Mahbubur Rahman
Department of Chemistry
North Carolina A&T State University
Greensboro, North Carolina

Rachita Rana
Department of Chemical and Biological Engineering
University of Saskatchewan
Saskatoon, Canada

Sivamohan N. Reddy
Department of Chemical Engineering
Indian Institute of Technology Roorkee
Roorkee, India

Prakash Kumar Sarangi
Directorate of Research
Central Agricultural University
Imphal, India

Herma Dina Setiabudi
Faculty of Chemical and Natural Resources Engineering
University Malaysia Pahang
Kuantan, Malaysia

Tan Ji Siang
Faculty of Chemical and Natural Resources Engineering
University Malaysia Pahang
Kuantan, Malaysia

Sharanjit Singh
Faculty of Chemical and Natural Resources Engineering
University Malaysia Pahang
Kuantan, Malaysia

Guotao Sun
College of Mechanical and Electronic Engineering
Northwest A&F University
Yangling, China

Yun Hin Taufiq-Yap
Catalysis Science and Technology Research Centre
Universiti Putra Malaysia
Serdang, Malaysia

Ayush Vani
Department of Textile Technology
Indian Institute of Technology Delhi
New Delhi, India

Dai-Viet N. Vo
Faculty of Chemical and Natural Resources Engineering
Universiti Malaysia Pahang
Kuantan, Malaysia

Priyanka Yadav
Department of Chemical Engineering
Indian Institute of Technology Roorkee
Roorkee, India

Mingqiang Zhu
College of Mechanical and Electronic Engineering
Northwest A&F University
Yangling, China

1 Fossil Fuels versus Biofuels

Perspectives on Greenhouse Gas Emissions, Energy Consumptions, and Projections

Kang Kang, Mingqiang Zhu, Guotao Sun, and Xiaohui Guo

CONTENTS

1.1 Introduction 1
1.2 Fossil Fuel: A Historical Overview 2
1.3 Environmental Issues Related to Fossil Fuels 4
1.4 Greenhouse Gas Emissions from Fossil Fuels 5
1.5 Research on Biofuels: Drivers and Benefits 6
1.5.1 Current Status of Biofuels 7
1.5.2 Basic Concept of Biorefinery 9
1.6 Potentials in Biofuel Development 9
1.6.1 Obtaining Sustainable Feedstocks at Low Cost 9
1.6.2 Developing Biomass Processing Technologies with High Efficiency 10
1.7 Conclusions 11
References 11

1.1 INTRODUCTION

Fossil fuels have been explored and used extensively for many centuries, and consumption became particularly intensive after the industrial revolution, which occurred in the late eighteenth and nineteenth centuries. Currently, the major contributors of fossil energy include coal, oil, and natural gas (Armaroli and Balzani 2011). The chemical energy stored in the hydrocarbons of fossil fuels has been utilized in various ways (e.g., in the direct combustion for generating heat or power, in the conversion to other value-added chemicals, or in further conversion to industrially relevant products by physiochemical or biological routes). Despite its increasing demand, many are criticizing the fossil fuel-based energy system due to the concerns over depletion of resources, pollution problems, rising fuel prices, and increasing political concerns (Luque et al. 2008; Nanda et al. 2015). One of the most severe issues is the emission of greenhouse gases related to fossil fuel burning in developed and developing countries (Chavez-Rodriguez and Nebra 2010). The industries have explored and implemented many types of alternatives including hydropower, solar, wind, geothermal, wave, and biomass that could potentially mitigate the greenhouse gas emissions with low carbon footprints (Panwar et al. 2011; Nanda et al. 2017b).

Among all the alternatives, biofuels derived from biomass have attracted great interests worldwide due to their carbon-neutral nature (Basu 2010). By far, biofuels have been developed for many generations and generated various forms of biofuels including bioethanol, biobutanol, biodiesel, bio-oil, bio-char, biomass fuel pellets, and biobased-syngas (Nanda et al. 2014b; Remón et al. 2016; Kasmuri et al. 2017; Kang et al. 2018). Currently, biofuels cannot entirely replace fossil fuel due to certain limitations such as poor fuel properties, higher production cost, low yield, need for fuel upgrading technologies, and engine modifications. However, biofuels have shown great potential for sustainability. Moreover, findings show that recycling of biomass and other bio-based waste materials for energy production boosts rural economy, employment opportunities, energy security, and carbon neutrality (Nanda et al. 2015; Voloshin et al. 2016; Rodionova et al. 2017; Stephen and Periyasamy 2018). Many countries have started subsidizing the production and utilization of biofuels, whereas researchers are still debating on the net-benefits of switching from fossil to bio-based energy systems.

To get a clear idea of the future direction of the fuel research, it is neccessary to compare the existing challenges and opportunities between fossil fuels and biofuels. With all the trends mentioned, it is obvious that, for a considerate period, fossil energy system will continue to remain the main source of energy that we rely on but will be augmented by many other renewables sources, including biofuels in the near future. Based on this trend, this chapter provides a general overview of the comparison between fossil fuels and biofuels. Special attention is given to the environmental impacts such as greenhouse gas emissions, energy consumptions, but the details about different conversion processes and fuel upgrading technologies are not the focus of this chapter. The projections on energy utilization are also made based on the knowledge obtained from the literature.

1.2 FOSSIL FUEL: A HISTORICAL OVERVIEW

The fossil fuels including coal, lignite, oil shales, tar, asphalt, petroleum, and natural gas have been used as the main sources of fuels since the industrial revolution. The name fossil fuels arise from their production from fossilized remains of living organisms deep under the Earth's crust for thousands of years as hydrocarbons. On the other hand, biofuels are a result of photosynthesis by the green plants that converts and stores solar energy in the form of carbohydrates, a chemical energy. However, human civilization was solely dependent on the energy from sunlight and dried plant biomass until the thirteenth century, when coal was discovered. Since the nineteenth century, the extraction and utilization of petroleum and natural gas significantly broadened the spectra of fossil energy resources.

The production and utilization routes of fossil fuels have been significantly advanced by the development in related research areas in geochemistry as well as in chemical, mechanical, and civil engineering. With the world's increasing population, prosperity, and economic growth following the industrial revolution, the pursuit for cost effectiveness and reliability in supplying fossil fuels has compromised environmental sustainability. In the twenty-first century, the exploitation of fossil fuels has been at a dramatic pace. Therefore, the sustainability should be given high priority when developing any energy-related technologies or implementing energy-related policies (Chu and Majumdar 2012). Fossil fuel-based energy systems play a dominant role in the energy supply sector because most of the energy supply is from crude oil (e.g., petroleum, diesel, and kerosene), natural gas, and coal. The worldwide liquid fuels consumption is shown in Figure 1.1. From the figure, it is obvious that in recent years the total consumption is steadily increasing (USEIA 2018c).

With the fast pace of industrialization and urbanization in developing counties such as China and India, unless a dramatic breakthrough occurs in the energy sector, the global energy demand will be gradually increasing in the near future. And most of the energy will still be from the fossil fuel resources since they are still more reliable, cheaper, and more efficient, and easier to produce, store, and transport, (Biresselioglu and Yelkenci 2016). Above all, based on the current situation, the basic strategy is improving the utilization efficiency of fossil fuels while searching for alternative green fuels.

The basic concept of conventional fossil energy recovery is maximizing the energy recovery using different technologies including purification through physical or chemical separation such as fractional distillation, pressure swing adsorption, membrane separation, solvent extraction, and so on. The chemical conversion technologies available for upgrading the fossil fuels or utilizing their by-products are catalytic cracking, liquefaction, reforming (using steam or CO_2), gasification,

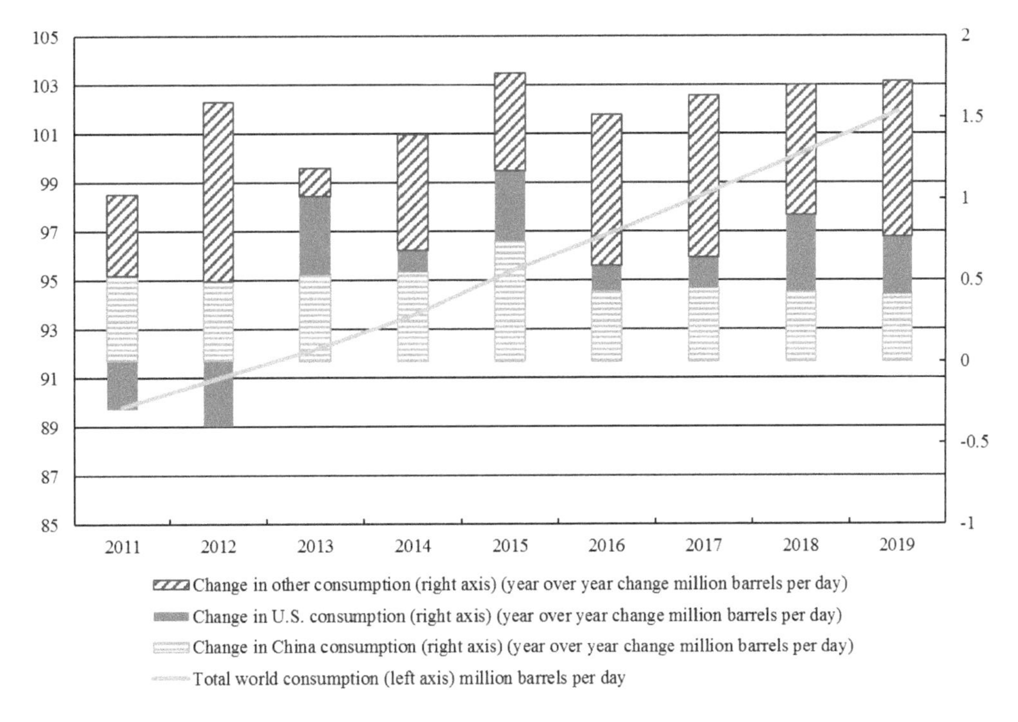

FIGURE 1.1 Worldwide liquid fuels consumption. (Data adapted from USEIA, Short-term energy outlook, March 2018, United States Energy Information Administration, https://www.eia.gov/outlooks/steo/USEIA, 2018c.)

Fischer-Tropsch (FT) synthesis, and so on (Anthony and Howard 1976; Babich and Moulijn 2003; Thomas and Dawe 2003).

As shown in Figure 1.2, the processing or conversion routes of coal, natural gas, and crude oil are different, but they share some critical technologies, which point to the same final products such as conversion into H_2 by reforming and upgrading of syngas into other fuels through Fischer-Tropsch synthesis. In addition, the transformation of fossil fuel to the synthetic final products often requires a combination of two or more processes and considerable amounts of energy input. Traditionally, the energy and carbon balance is compensated by other forms of fossil energy, which in turn leads to further consumption of fossil energy and drives the demand for developing highly efficient recovery technologies. Therefore, the development of greener processes that could generate cleaner fuels seems to be direction of the future.

Many academic researchers and industrial leaders are dedicated to developing technologies that could boost the efficiency of fossil energy utilization, thus yielding novel conversion and upgrading routes featuring high-efficiency, cost-effectiveness and eco-friendless. An interesting example is the chemical looping processes such as chemical looping combustion (CLC) and chemical looping reforming (CLR), which cover a wide range for the utilization of solid, gaseous, and liquid fuels (Tang et al. 2015). The CLC process does not involve any direct mixing of air and fuel, but a solid oxygen carrier (normally a metal oxide) supplies the oxygen through a redox cycle. The oxygen carrier is first reduced to supply the oxygen and then oxidized with air. By the exclusion of the gas separation step, CLC can gain dramatic energy savings compared to conventional combustion technologies (Adánez et al. 2018). Another example would be the development of hybrid, battery, and fuel-cell electric vehicles, which are being developed to mitigate the dependency on conventional vehicular fuels and the adverse effects of related greenhouse gas emissions. Although the poor energy density of batteries and capacitors is hindering their practical utilization, their reducing costs and innovation in technology is also creating future potentials (Pollet et al. 2012). Moreover, the

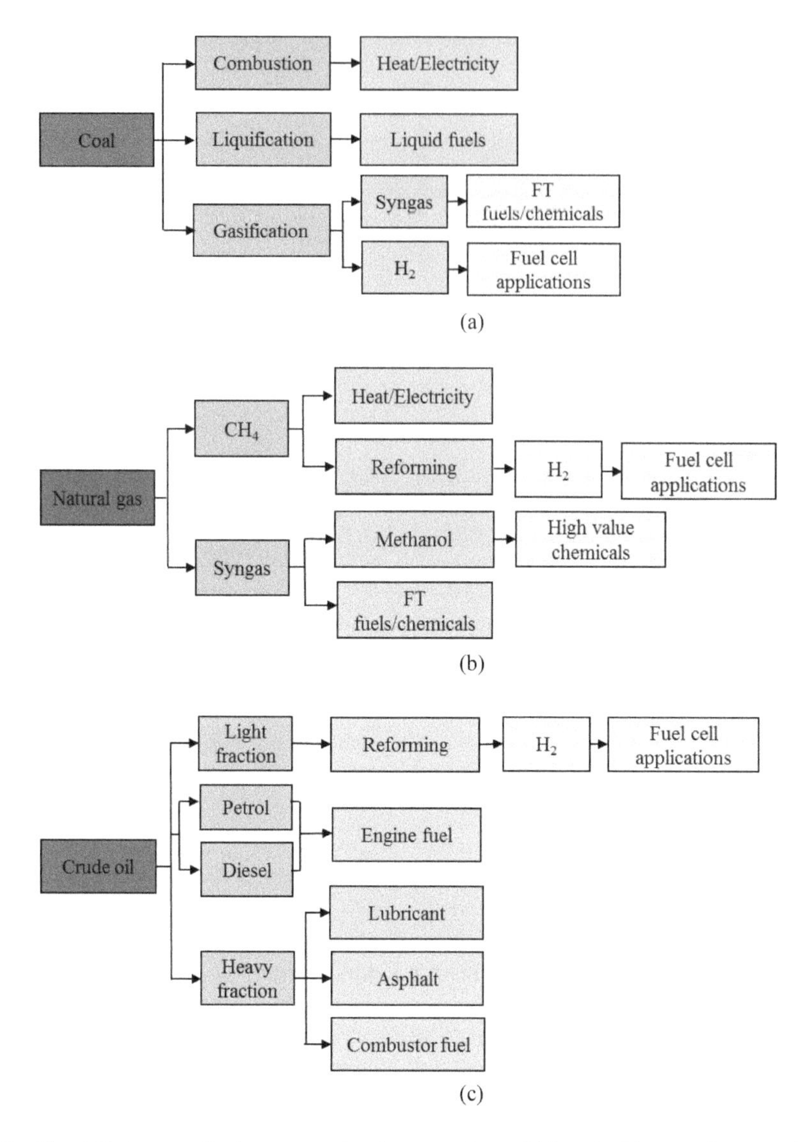

FIGURE 1.2 Simplified routes for the conventional conversions of (a) coal, (b) natural gas, and (c) crude oil.

idea of coupling renewable energy (e.g., solar, wind, water, geothermal, and biomass) with the fossil energy-based system seems to be a promising way to reduce energy consumption in the foreseeable future. In addition, the advances in novel materials such as catalysts, electrodes, reactor coatings, and so on may lead to breakthroughs in fossil fuel energy technologies.

1.3 ENVIRONMENTAL ISSUES RELATED TO FOSSIL FUELS

Other than the energy crisis, fossil fuel utilization also causes many environmental problems, which involves issues in the atmosphere, hydrosphere, biosphere, and lithosphere. The hydrocarbon composition of the fossil fuels when decomposed or transformed releases many unwanted chemicals, emissions, and particulate matter, which potentially lead to environmental problems. Also, the existing technologies for the utilization of fossil fuels, although improving, are still not capable of preventing the formation and release of these contaminants into the environment and available prevention technologies often are too expensive to be fully implemented at a commercial level.

The most commonly studied environmental issues which could be correlated to fossil fuel utilization is air pollution. A good example is photochemical smog, which is formed mainly due to emissions of volatile organic compounds (VOCs) and NO_x gases from motor vehicles and industrial sources (Rye 1995). The issue of haze-fog pollution in China, although formed by interrelated reasons, seems to be closely related to vehicular emissions from the extensive use of fossil energy and fuel combustion for industries and for domestic heating (Fu and Chen 2017).

Moreover, the impact of fossil fuel utilization also extends to water bodies. The discharge of effluents from power stations, fuel refineries, or chemical processing industries contain many hazardous and volatile components, which may mix with the surface water and/or groundwater and disperse widely within a short time (Reddy et al. 2016). In addition, the effluents, coolants, and recycled water used for cooling purposes and released by a power plant or fuel refinery may also contain dissolved chemicals which could cause contamination if not treated properly before disposal (Macqueen 1980).

The direct impact of the fossil energy system on the lithosphere mainly is mining where the fossil fuel precursors are buried. During the mining process, geological structures of the sites undergo different extents of destruction and might not recover. In addition, the migration and transformation behavior of some fossil fuel-derived contaminants could cause them potentially to participate in different types of environmental issues. In addition, some uneven weather conditions are indirectly linked to the greenhouse gas emissions and global warming caused by the increasing usage of fossil fuels such as acid rain, increase in the atmospheric temperature, droughts, storms, increase in the sea level, and so on. Other than the environmental issues mentioned, an unavoidable concern correlated to the fossil energy is the enormous quantities of CO_2, CH_4, SO_x, and NO_x emissions, which are the major sources of greenhouse gas emissions that accelerate global warming. In recent years, CO_2 emissions from fossil fuel sources have raised serious concerns about the effects on the environment and human health.

1.4 GREENHOUSE GAS EMISSIONS FROM FOSSIL FUELS

Different types of fossil fuels emit different amounts of CO_2 into the atmosphere when burned. The analysis of emissions across the fossil fuels could be achieved by comparing the amount of CO_2 emitted per unit of energy generated or heat content. Generally, the amount of CO_2 released by different types of fossil fuels follows the order of coal > diesel > gasoline > natural gas (USEIA 2018a). The amount of CO_2 produced during the combustion of a fuel is dependent mainly on the content of carbon and hydrogen in the fuel. Since natural gas (methane) mainly consists of only carbon and hydrogen, it has relatively higher energy content than other fossil fuels and releases the lowest content of CO_2 when combusted. Therefore, natural gas is considered to be a cleaner fuel than other types of fossil fuels. However, methane is considered 30 times more potent as a heat-trapping gas (Kelly 2014). The disadvantage of natural gas lies in its uneven distribution worldwide and its limited reserve. Facing the current carbon scenario and the stringent regulations on sulfur level in fuels, the valorization of the natural gas reserves into liquid fuels through the Fischer-Tropsch process has raised significant interest in recent years (Perego et al. 2009).

Coal is more abundant than natural gas and is still the dominant source of fossil energy for many countries, yet it is also considered the highest CO_2 emitter. In addition, the combustion of coal is usually the dominant source of CO_2 emission for developing countries like China and India, which still rely on coal as a major source for electricity and steam (Govindaraju and Tang 2013; Chen et al. 2016). Hence, controlling of the CO_2 emissions from coal is critical in the long run. Normally, the source of coal or the manner of coal utilization cause differences in the CO_2 emission because the low-rank coals emit higher CO_2 than the high-rank coals. Significant reduction in CO_2 emissions from coal could be achieved by implementing alternative technologies such as the integrated gasification combined cycle (IGCC) or deploying carbon capture and storage (CCS) facilities (Skodras et al. 2015; Nanda et al. 2016b, 2016c).

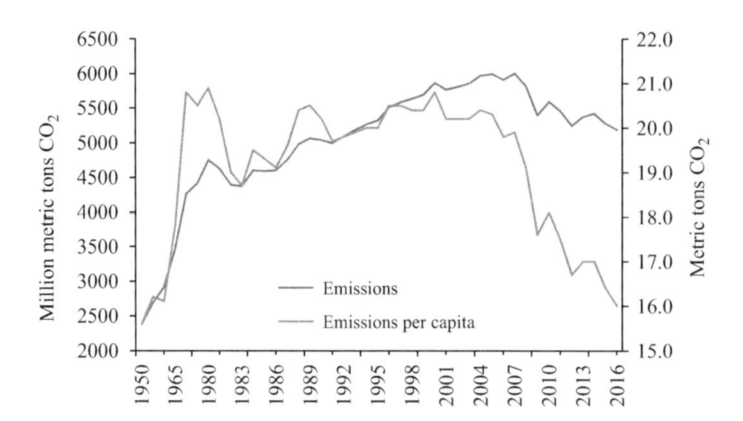

FIGURE 1.3 CO_2 emissions from energy consumption over the years. (Data adapted from USEIA, Monthly energy review, March 2018, United States Energy Information Administration, https://www.eia.gov/totalenergy/data/monthly/pdf/mer.pdf, 2018b.)

The liquid fossil fuels such as diesel and gasoline show intermediate levels of CO_2 emission, which is lower than coal but higher than natural gas. In addition, diesel-powered vehicles show higher fuel economy and lower CO_2 emissions than gasoline-powered vehicles (Sullivan et al. 2004). Other than modification of vehicular engines, which has never stopped evolving, a critical strategy for reducing the CO_2 emission from liquid fossil fuel utilization is blending with biofuels such as biodiesel, bioethanol, and biobutanol (Balat and Balat 2009; Mata et al. 2010; Nanda et al. 2014a, 2017a).

Less CO_2 emission could be achieved by optimizing the fossil fuel utilization network such as using more natural gas, developing clean coal technologies, and enhancing the efficiency of liquid fuel production. Compared to biofuels, the CO_2 released during the burning of fossil fuels is considered to increase their atmospheric levels because the emitted CO_2 could not be recycled within a short period of time unlike for biofuels.

The CO_2 emissions caused by the energy consumption in the United States over decades is plotted in Figure 1.3 (USEIA 2018b). As shown in the figure, from 1950 to 2016, the CO_2 emission caused by energy consumption increased dramatically until maximum emission in 2007. After 2007, emissions flucuated with a slight reduction. The declining trend in emissions per capita might possibly be due to public awareness to reduce carbon emissions and advances in energy-saving technologies. However, many countries are still consuming increasing amount of energy every year, and these uneven contributions to the CO_2 emission bring serious challenges to the global community to find effective and equitable solutions.

1.5 RESEARCH ON BIOFUELS: DRIVERS AND BENEFITS

It becomes increasingly important to seek a more sustainable method for societal development due to the increasing global energy demands, the limiting fossil fuel reserves, the need to reduce greenhouse gas emissions, and the fluctuating crude oil prices. Therefore, a cost-competitive and reliable solution is needed to reduce fossil fuel dependency and to achieve potential reductions in greenhouse gas. Consequently, it is necessary to search for a cleaner, more secure, and affordable fuels for transportation to drive a low-carbon technology. In this regard, biofuels which are often produced from plants or other organic wastes, have emerged to make significant contributions to a carbon neutral fuel economy (Naik et al. 2010).

One of the major advantages of biofuel over fossil fuels is the potential reduction of CO_2 emission since the amount emitted burning could be recycled by plants through the photosynthesis process and fixed in the newly generated biomass. The emissions of CO_2 from the utilization of different biofuels are shown in Figure 1.4 (USEIA 2018b). The total emission of CO_2 from biofuel utilization

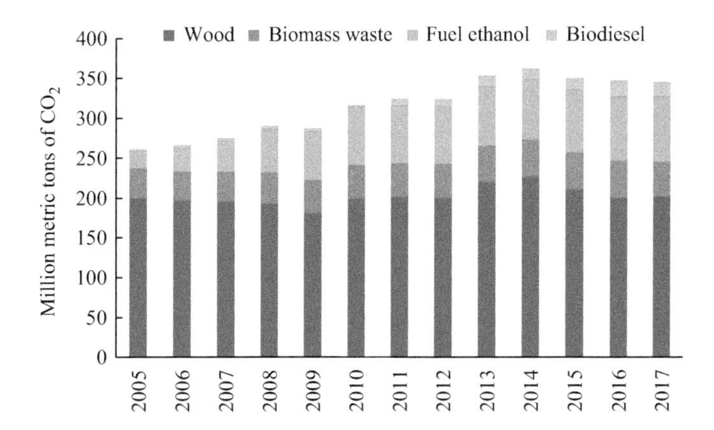

FIGURE 1.4 Emissions of CO_2 from biofuel and bioenergy utilization by different fuels. (Data source from USEIA, Monthly energy review March 2018, United States Energy Information Administration, https://www.eia.gov/totalenergy/data/monthly/pdf/mer.pdf, 2018b.)

fluctuates, although the emissions increased in 2014 and then decreased gradually. In terms of different biofuels, the CO_2 emission from burning wood is at a the highest followed by bioethanol, solid biomass wastes, and biodiesel.

1.5.1 Current Status of Biofuels

Biofuels can be produced from a wide range of biomass sources through many promising technologies to fulfill the targets of using bio-based renewable resources and mitigating greenhouse gas emissions (Gaurav et al. 2017). Many thermochemical (e.g., pyrolysis, gasification, liquefaction, torrefaction, and combustion) and biochemical technologies (e.g., fermentation, anaerobic digestion, and microbial fuel cells) can convert biomass and other organic wastes to produce diverse biofuels such as bio-oil, biodiesel, bioethanol, biobutanol, hydrogen, syngas, bio-char, green electricity, and heat (Nanda et al. 2014a, 2015, 2016a, 2017b). Based on the type of the sources of biomass used, the biofuels could be categorized into first-, second-, and third-generation biofuels (Alam et al. 2015).

As shown in Figure 1.5, the first-generation biofuels such as bioethanol and biodiesel are produced directly from conventional food crops including corn, maize, sugarcane, rapeseed, and soybean. The production of first-generation biofuels is mainly through mature processes, which have been widely used commercially due to its cost-effectiveness, time-efficiency, and techno-economic feasibility. The alcohol fermented from sugars and starch currently is used in many countries as additives to gasoline. Another first-generation biofuel (i.e., biodiesel) is produced from vegetable oil or animal fat through transesterification. Biodiesel can be an alternative to petroleum diesel. However, the most contentious issue related to first-generation biofuels is that the biomass precursors used for biofuel production often compete with the food crops and arable lands for their cultivation.

To avoid the issue of food-versus-fuel, second-generation biofuels are produced from non-food sources such as dedicated energy crops and waste biomass residues. The second-generation biofuels are produced mostly from non-edible lignocellulosic biomass such as agricultural biomass (e.g., straws, grasses, husk, shell, seed carps, etc.) and forest residues (e.g., woody biomass, sawdust, dead trees, etc.) (Nanda et al. 2013). The processes to generate second-generation biofuels usually require a physicochemical, biochemical, or hydrothermal pretreatment to release the trapped sugars from the biomass for conversion to biofuels. This process requires more cost, energy, and materials compared to first-generation biofuels. Although the second-generation biofuels overcome the criticism of first-generation biofuels, they are limited in being cost competitive to existing fossil fuels. However, the technology for second-generation biofuels production is under development, so they still have the potential for reduced processing costs and improved production efficiency with technological advances.

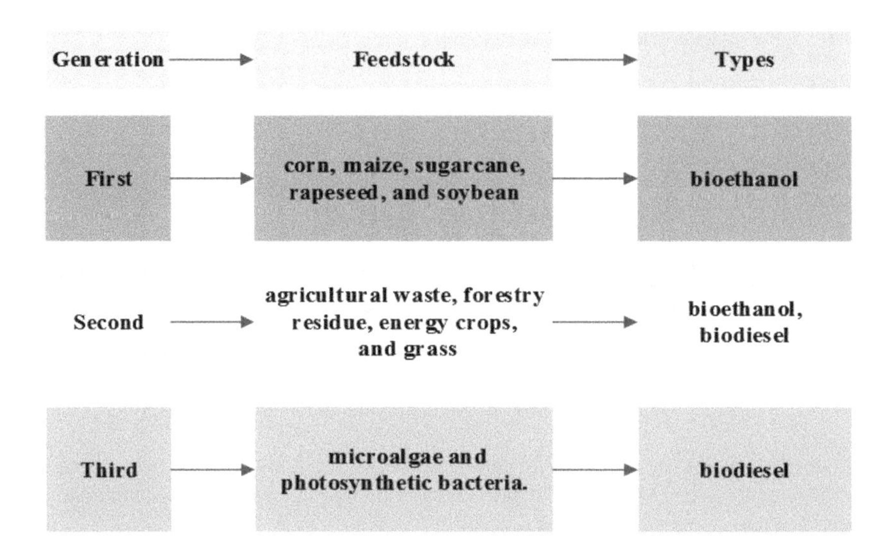

FIGURE 1.5 Different generations of biofuels.

Third-generation biofuels mostly use algae and photosynthetic bacteria as the feedstocks. Mature algal species are subjected to enzymatic, physicochemical, or hydrothermal extraction processes to obtain oil, which is then processed to produce biodiesel or refined into other petroleum-based fuels like bioethanol, biopropanol, and biohydrogen. The major advantage of algae is that it can grow in photobioreactors and open ponds that do not compete with arable agricultural lands as in the case of first-generation biofuels. However, further research is required still to make algal cultivation more economically and environmentally sustainable and cost-competitive with petroleum, diesel, and other fossil fuels.

The trends in global biofuel production by different regions in recent years are shown in Figure 1.6. As the data indicates, the global production of biofuels has been increasing over the last ten years; however, North, South, and Central America and Europe are significantly ahead of other regions in the total amount produced, indicating that there is still a gap between the developed and developing areas in adopting biofuel production and utilization.

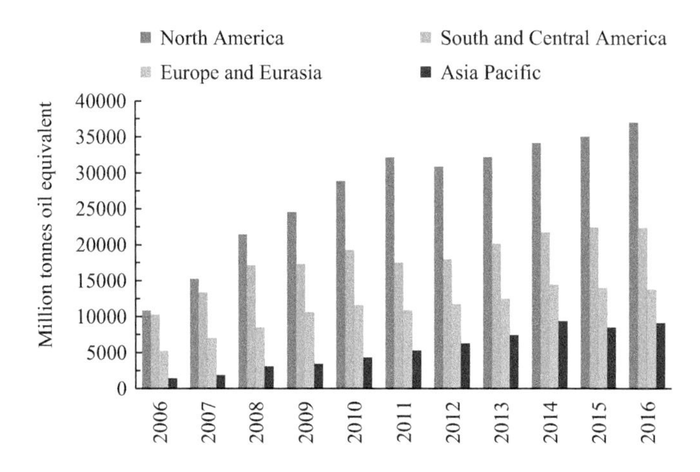

FIGURE 1.6 Global biofuel production by different regions. (Data source from British Petroleum, BP Statistical Review of World Energy—June 2016, https://www.bp.com/content/dam/bp/pdf/energy-economics/statistical-review-2016/bp-statistical-review-of-world-energy-2016-full-report.pdf, 2016.)

1.5.2 Basic Concept of Biorefinery

It remains uncertain how important the biofuels will be in the future as a sustainable alternative to fossil fuels since many factors such as biomass productivity, future science advances, environmental policies, subsidies for biofuel production, and public support for replacing fossil fuels with biofuels might affect the outcome. However, if research and development continue to improve the biofuel production efficiency, then the cost investment in biorefineries could be minimized to make the biofuels competitive with fossil fuels in the energy market. The most promising way to minimize the economics of the overall process in biofuel production is to make full use of the different components of biomass and by-products from biorefineries.

The concept of biorefinery can be compared to a petroleum refinery. The recoverable products in a biorefinery include high-value products (e.g., biofuels) and medium-value products (e.g., fibers, bioplastic precursors, biochemicals, animal feed, etc.), although the toxicity and quality of the products determines their end use (Kamm and Kamm 2004; Halasz et al. 2005). Therefore, biofuels are the main drivers for the developments of a biorefinery, but other relevant by-products are expected to be developed as technology becomes more and more sophisticated over time. This development could also help in a circular economy (i.e., utilization of all the end-products and by-products of a biorefining process).

One of the major challenges in converting lignocellulosic biomass as a feedstock into biofuels comes from the recalcitrance of lignin to various chemical and biological reactions (Hu et al. 2018). Many novel technologies have been developed for lignin depolymerization and valorization such as catalytic gasification, enzymatic depolymerization, alkaline treatment, and chemi-mechanical pretreatments, but the conversion selectivity and efficiency still require optimization (Fougere et al. 2016; Chen and Wan 2017; Kang et al. 2017; Ma et al. 2018).

1.6 POTENTIALS IN BIOFUEL DEVELOPMENT

As a major source of renewable fuels, biofuel has been researched and developed intensively in recent decades for energy security. Biofuels have the potential to be carbon neutral or even carbon negative (with the application of biochar for carbon capture), which makes them superior to fossil fuels for environmental friendliness and sustainability (Nanda et al. 2016b). Given the current technical status of biofuel production, its economic incompetence with fossil fuels is still a major concern hindering its commercialization. Furthermore, whether biofuels are a low-carbon energy source depends on their lifecycles because a carbon debt might be formed if food crops were used as the biomass feedstock and the fossil fuels used in the process should be considered.

1.6.1 Obtaining Sustainable Feedstocks at Low Cost

To overcome the challenges of economic and environmental crisis, low-cost biomass is generally favored for producing biofuels such as lignocellulosic biomass (e.g., cereal straw, husk, sugarcane bagasse, and forest residues), dedicated energy crops (e.g., short rotation coppice, switchgrass, timothy grass, alfalfa, and algae), and organic wastes (e.g., animal manure, organic fractions in municipal solid wastes, and sewage sludge) (Wright 2006; Sims et al. 2010; Koutinas et al. 2016). Most of the potential feedstocks from plant residue such as lignocellulosic biomass have many promising abilities to produce biofuel through thermochemical and biochemical technologies as mentioned earlier. Moreover, the dedicated energy crops do not compete with food crops for arable land, available water, and nutrients because they are cultivated in marginal or degraded lands (Fargione et al. 2008).

As an aquatic crop, algae appear to be one of the most promising feedstocks for biofuel production (Adenle et al. 2013). Research on microalgae using organic solid wastes or wastewaters derived from human activities has also been conducted (Craggs et al. 2011; Pittman et al. 2011). The biomass yield and composition of different algal varieties are affected depending on the nutrient replenishment as well as sources of carbon, energy, and sunlight. Transgenic plants are also being developed with the aim to increase the lipid content or modify the lignin structure within the plant cell walls for improving sugar yield from energy crops (Chen et al. 2006; Liu et al. 2011). Most of the current assessments on the lifecycle, techno-economic feasibility, and supply chain have been reported based on lab-scale data (Quinn et al. 2011; Ramachandra et al. 2013). The potential of energy crops for biofuel production and lifecycle assessment are yet to be verified in a large-scale field experiments (Lam and Lee 2012; Adenle et al. 2013). Moreover, to reduce the cost of obtaining a suitable biofuel feedstock, the distribution, production, varieties, speciation, transport, storage, and initial processing of biomass also should be considered.

For a commercial-scale biofuel plant to meet all year-round demands in a large geographical region, several different kinds of biomass might be prepared as feedstocks. In addition, biomass feedstocks with diverse physical characteristics and chemical compositions would require the flexibility of process adjustment for pretreatment. Therefore, cost-effective feedstock supplies also are related to the size and coverage of the biofuel refineries.

1.6.2 Developing Biomass Processing Technologies with High Efficiency

Most of the feedstocks for biofuel production require pretreatments to reduce the biomass recalcitrance for more effective conversion during downstream processing. The current pretreatment technologies are not only energy-intensive but also time-consuming. Among the emerging pretreatment processes, integrated pretreatment processes via physicochemical and biochemical methods seem to be more effective and promising (Zheng et al. 2014; An et al. 2015; Kroon et al. 2015; Zhang et al. 2016; Bhutto et al. 2017). For example, the ammonia fiber explosion (AFEX) process can provide significant environmental benefits since it only uses anhydrous ammonia at moderate temperature and high pressures within a short time with less waste effluent production (Lau and Dale 2009; Flores-Gómez et al. 2018). The combination of biological (fungal) pretreatment and physicochemical methods integrates the merits of low-energy demands through the biological steps with shortened processing times and improved products yield (Zhang et al. 2016). Immobilizing the enzymes and the applications of catalytic nanoparticles are also good examples of bio-physical pretreatment (Menetrez 2012; Rai et al. 2017). Due to differences in the structure and composition of feedstocks, the adopted conversion pathways and optimum pretreatment can lead to obtaining intended products.

Thermochemical and biological conversion technologies are two main pathways to obtain valuable energy products and by-products (biochemicals and biomaterials). In thermochemical conversion routes, the temperature plays a significant role in the yield and composition of the product. Understanding the patterns of temperature distribution among different feedstock particles with a confined thermochemical conversion system is crucial (Taba et al. 2012). A better understanding of the kinetics of heat and mass transfer could be helpful for obtaining the designed products with fewer impurities and undesired products. In addition, the performance of a metal catalyst or biocatalyst is also critical for the product upgrading. Chemical selectivity, susceptibility to impurities, and lifetimes are important factors for a catalyst (Chew and Bhatia 2008; Suopajärvi et al. 2013; Thegarid et al. 2014). Developing new and efficient catalysts for biomass conversion with good merits in these aspects could help improve the cost-effectiveness of a biorefining process.

In biological conversion routes, the substrate competitiveness, product inhibition, and conversion selectivity between C_5 and C_6 sugars using fungal and bacterial species are some usual technical challenges (Nanda et al. 2014a). However, the use of genetically modified strains and engineered microorganisms as well as metabolic engineering can help overcome these issues (Ragauskas et al. 2006; Olofsson et al. 2010; Nanda et al. 2017a). The various inhibitory substances produced during the pretreatment process such as furfurals, hydroxymethylfurfural (HMF), and resin acids can inhibit the microbial growth and fermentation process. Removal of these inhibitors through adsorbents, neutralization, and dilution could potentially reduce the toxicity of these inhibitors but complete elimination of these inhibitors in a continuous bioprocess is still a major challenge in biological conversion technologies.

Given many processes for the conversion of biomass to biofuels, there remains great potential for process integration. For example, the syngas produced from biomass gasification can be converted to ethanol through syngas fermentation using *Clostridium* bacteria. This suggests that integrating the thermochemical and biological pathways could lead to the improvement in cost-competitiveness of the overall biorefining process (Mohammadi et al. 2011; Liew et al. 2014). Besides, maximizing the value of the co-products (e.g., heat, electricity, and biochemicals) within a single biorefinery process is also highly preferred (Sims et al. 2010).

1.7 CONCLUSIONS

The intent of this chapter is to provide an overview of the current trends in fossil fuel refineries and biorefineries. The fossil fuel-based energy system still will play a dominant role in the energy supply sector in the near future. However, with the advantages of carbon neutrality, ensuring energy security, and many other socio-environmental benefits such as reinvigorating rural economy and creating community employment opportunities, the potential of biofuels should not be overlooked.

Critical issues should be tackled in terms of reduction in the cost of the feedstock, the pretreatment and conversion process, and increasing the profitability of the final products. Non-food based plant residues such as agricultural and forestry residues should be given high priority for biofuel production. Consolidation and integration of thermochemical and biological technologies could lead to the development of low-cost, green, and high-efficiency processes, which combines the benefits from different conversion routes. Special attention also should be given to subjects such as genetic modifications, enzymatic decomposition, catalysis, and chemical reaction engineering. In addition to the main fuel product, other value-added and specialized chemicals and by-products obtained from the biorefineries should be explored to find additional consumer markets to enhance the attention from petrochemical industries to biofuel industries. In addition, when developing biofuels, the negative impacts on the environment should also be minimized by reducing and efficiently recycling the waste streams.

REFERENCES

Adánez, J., A. Abad, T. Mendiara, P. Gayán, L. F. de Diego, and F. García-Labiano. 2018. Chemical looping combustion of solid fuels. *Progress in Energy and Combustion Science* 65:6–66.

Adenle, A. A, G. E. Haslam, and L. Lee. 2013. Global assessment of research and development for algae biofuel production and its potential role for sustainable development in developing countries. *Energy Policy* 61:182–195.

Alam, F., S. Mobin, and H. Chowdhury. 2015. Third generation biofuel from algae. *Procedia Engineering* 105:763–768.

An, Y. X., M. H. Zong, H. Wu, and N. Li. 2015. Pretreatment of lignocellulosic biomass with renewable cholinium ionic liquids: Biomass fractionation, enzymatic digestion and ionic liquid reuse. *Bioresource Technology* 192:165–171.

Anthony, D. B., and J. B. Howard. 1976. Coal devolatilization and hydrogasification. *AIChE Journal* 22:625–656.

Armaroli, N., and V. Balzani. 2011. Towards an electricity-powered world. *Energy & Environmental Science* 4:3193–3222.

Babich, I. V., and J. A. Moulijn. 2003. Science and technology of novel processes for deep desulfurization of oil refinery streams: A review. *Fuel* 82:607–631.

Balat, M., and H. Balat. 2009. Recent trends in global production and utilization of bio-ethanol fuel. *Applied Energy* 86:2273–2282.

Basu, P. 2010. *Biomass Gasification and Pyrolysis: Practical Design and Theory*. Academic Press, Burlington, MA.

Bhutto, A. W., K. Qureshi, K. Harijan, R. Abro, T. Abbas, A. A. Bazmi, S. Karim, and G. Yu 2017. Insight into progress in pre-treatment of lignocellulosic biomass. *Energy* 122:724–745.

Biresselioglu, M. E., and T. Yelkenci. 2016. Scrutinizing the causality relationships between prices, production, and consumption of fossil fuels: A panel data approach. *Energy* 102:44–53.

British Petroleum. 2016. BP Statistical Review of World Energy—June 2016. https://www.bp.com/content/dam/bp/pdf/energy-economics/statistical-review-2016/bp-statistical-review-of-world-energy-2016-full-report.pdf (accessed on April 20, 2018).

Chavez-Rodriguez, M. F., and S. A. Nebra. 2010. Assessing GHG emissions, ecological footprint, and water linkage for different fuels. *Environmental Science & Technology* 44:9252–9257.

Chen, F., M. S. S. Reddy, S. Temple, L. Jackson, G. Shadle, and R. A. Dixon. 2006. Multi-site genetic modulation of monolignol biosynthesis suggests new routes for formation of syringyl lignin and wall-bound ferulic acid in alfalfa (*Medicago sativa* L.). *Plant Journal* 48:113–124.

Chen, J., S. Cheng, M. Song, and J. Wang. 2016. Interregional differences of coal carbon dioxide emissions in China. *Energy Policy* 96:1–13.

Chen, Z., and C. Wan. 2017. Biological valorization strategies for converting lignin into fuels and chemicals. *Renewable and Sustainable Energy Reviews* 73:610–621.

Chew, T. L., and S. Bhatia. 2008. Catalytic processes towards the production of biofuels in a palm oil and oil palm biomass-based biorefinery. *Bioresource Technology* 99:7911–7922.

Chu, S., and A. Majumdar. 2012. Opportunities and challenges for a sustainable energy future. *Nature* 488:294–303.

Craggs, R. J., S. Heubeck, T. J. Lundquist, and J. R. Benemann. 2011. Algal biofuels from wastewater treatment high rate algal ponds. *Water Science and Technology* 63:660–665.

Fargione, J., J. Hill, D. Tilman, S. Polasky, and P. Hawthorne. 2008. Land clearing and the biofuel carbon debt. *Science* 319:1235–1238.

Flores-Gómez, C. A., E. M. E. Silva, C. Zhong, B. E. Dale, L. C. Sousa, and V. Balan. 2018. Conversion of lignocellulosic agave residues into liquid biofuels using an AFEX™-based biorefinery. *Biotechnology for Biofuels* 11:7.

Fougere D., S. Nanda, K. Clarke, J. A. Kozinski, and K. Li. 2016. Effect of acidic pretreatment on the chemistry and distribution of lignin in aspen wood and wheat straw substrates. *Biomass and Bioenergy* 91:56–68.

Fu, H., and J. Chen. 2017. Formation, features and controlling strategies of severe haze-fog pollutions in China. *Science of the Total Environment* 578:121–138.

Gaurav, N., S. Sivasankari, G. S. Kiran, A. Ninawe, and J. Selvin. 2017. Utilization of bioresources for sustainable biofuels: A Review. *Renewable and Sustainable Energy Reviews* 73:205–214.

Govindaraju, V. G. R. C., and C. F. Tang. 2013. The dynamic links between CO_2 emissions, economic growth and coal consumption in China and India. *Applied Energy* 104:310–318.

Halasz, L., G. Povoden, and M. Narodoslawsky 2005. Sustainable processes synthesis for renewable resources. *Resources, Conservation, and Recycling* 44:293–307.

Hu, J., Q. Zhang, and D. J. Lee. 2018. Kraft lignin biorefinery: A perspective. *Bioresource Technology* 247:1181–1183.

Kamm, B., and M. Kamm 2004. Principles of biorefineries. *Applied Microbiology & Biotechnology* 64:137–145.

Kang, K., L. Qiu, M. Zhu, G. Sun, Y. Wang, and R. Sun. 2018. Co-densification of agroforestry residue with bio-oil for improved fuel pellets. *Energy & Fuels* 32:598–606.

Kang, K., R. Azargohar, A. K. Dalai, and H. Wang. 2017. Hydrogen generation via supercritical water gasification of lignin using Ni-Co/Mg-Al catalysts. *International Journal of Energy Research* 41:1835–1846.

Kasmuri, N. H., S. K. Kamarudin, S. R. S. Abdullah, H. A. Hasan, and A. M. D. Som. 2017. Process system engineering aspect of bio-alcohol fuel production from biomass via pyrolysis: An overview. *Renewable and Sustainable Energy Reviews* 79:914–923.

Kelly, M. 2014. A more potent greenhouse gas than CO_2, methane emissions will leap as Earth warms. Princeton Research. https://blogs.princeton.edu/research/2014/03/26/a-more-potent-greenhouse-gas-than-co2-methane-emissions-will-leap-as-earth-warms-nature/ (accessed on April 15, 2018).

Koutinas, A., M. Kanellaki, A. Bekatorou, P. Kandylis, K. Pissaridi, A. Dima, K. Boura, K. Lappa, P. Tsafrakidou, and P.Y. Stergiou. 2016. Economic evaluation of technology for a new generation biofuel production using wastes. *Bioresource Technology* 200:178–185.

Kroon, M. C., M. F. Casal, and A. van den Bruinhorst. 2015. Pretreatment of lignocellulosic biomass and recovery of substituents using natural deep eutectic solvents/compound mixtures with low transition temperatures. Patents Application number: WO2013153203A1.

Lam, M. K., and K. T. Lee. 2012. Microalgae biofuels: A critical review of issues, problems and the way forward. *Biotechnology Advances* 30:673–690.
Lau, M. W., and B. E. Dale. 2009. Cellulosic ethanol production from AFEX-treated corn stover using *Saccharomyces cerevisiae* 424A (LNH-ST). *Proceedings of the National Academy of Sciences* 106:1368–1373.
Liew, W. H., M. H. Hassim, and D. K. S. Ng. 2014. Review of evolution, technology and sustainability assessments of biofuel production. *Journal of Cleaner Production* 71:11–29.
Liu, X., J. Sheng, and R. Curtiss III. 2011. Fatty acid production in genetically modified cyanobacteria. *Proceedings of the National Academy of Sciences* 108:6899–6904.
Luque, R., L. Herrero-Davila, J. M. Campelo, J. H. Clark, J. M. Hidalgo, D. Luna, J. M. Marinas, and A. A. Romero. 2008. Biofuels: A technological perspective. *Energy & Environmental Science* 1:542–564.
Ma, R., M. Guo, and X. Zhang. 2018. Recent advances in oxidative valorization of lignin. *Catalysis Today* 302:50–60.
Macqueen, J. F. 1980. Concentration of contaminants discharged with power station cooling water. *Advances in Water Resources* 3:165–172.
Mata, T. M., A. A. Martins, and N. S. Caetano. 2010. Microalgae for biodiesel production and other applications: A review. *Renewable and Sustainable Energy Reviews* 14:217–232.
Menetrez, M. Y. 2012. An overview of algae biofuel production and potential environmental impact. *Environmental Science & Technology* 46:7073–7085.
Mohammadi, M., G. D. Najafpour, H. Younesi, P. Lahijani, M. H. Uzir, and A. R. Mohamed. 2011. Bioconversion of synthesis gas to second-generation biofuels: A review. *Renewable and Sustainable Energy Reviews* 15:4255–4273.
Naik, S. N., V. V. Goud, P. K. Rout, and A. K. Dalai. 2010. Production of first- and second-generation biofuels: A comprehensive review. *Renewable and Sustainable Energy Reviews* 14:578–597.
Nanda S., A. K. Dalai, and J. A. Kozinski. 2014a. Butanol and ethanol production from lignocellulosic feedstock: Biomass pretreatment and bioconversion. *Energy Science and Engineering* 2:138–148.
Nanda S., D. Golemi-Kotra, J. C. McDermott, A. K. Dalai, I. Gökalp, and J. A. Kozinski. 2017a. Fermentative production of butanol: Perspectives on synthetic biology. *New Biotechnology* 37:210–221.
Nanda S., J. A. Kozinski, and A. K. Dalai. 2016a. Lignocellulosic biomass: A review of conversion technologies and fuel products. *Current Biochemical Engineering* 3:24–36.
Nanda S., P. Mohanty, K. K. Pant, S. Naik, J. A. Kozinski, and A. K. Dalai. 2013. Characterization of North American lignocellulosic biomass and biochars in terms of their candidacy for alternate renewable fuels. *Bioenergy Research* 6:663–677.
Nanda, S., A. K. Dalai, F. Berruti, and J. A. Kozinski. 2016b. Biochar as an exceptional bioresource for energy, agronomy, carbon sequestration, activated carbon, and specialty materials. *Waste and Biomass Valorization* 7:201–235.
Nanda, S., J. Mohammad, S. N. Reddy, J. A. Kozinski, and A. K. Dalai. 2014b. Pathways of lignocellulosic biomass conversion to renewable fuels. *Biomass Conversion and Biorefinery* 4:157–191.
Nanda, S., R. Azargohar, A. K. Dalai, and J. A. Kozinski. 2015. An assessment on the sustainability of lignocellulosic biomass for biorefining. *Renewable and Sustainable Energy Reviews* 50:925–941.
Nanda, S., R. Rana, Y. Zheng, J. A. Kozinski, and A. K. Dalai. 2017b. Insights on pathways for hydrogen generation from ethanol. *Sustainable Energy and Fuels* 1:1232–1245.
Nanda, S., S. N. Reddy, S. K. Mitra, and J. A. Kozinski. 2016c. The progressive routes for carbon capture and sequestration. *Energy Science and Engineering* 4:99–122.
Olofsson, K., M. Wiman, and G. Lidén. 2010. Controlled feeding of cellulases improves conversion of xylose in simultaneous saccharification and co-fermentation for bioethanol production. *Journal of Biotechnology* 145:168–175.
Panwar, N. L., S. C. Kaushik, and S. Kothari. 2011. Role of renewable energy sources in environmental protection: A review. *Renewable and Sustainable Energy Reviews* 15:1513–1524.
Perego, C., R. Bortolo, and R. Zennaro. 2009. Gas to liquids technologies for natural gas reserves valorization: The Eni experience. *Catalysis Today* 142:9–16.
Pittman, J. K., A. P. Dean, and O. Osundeko. 2011. The potential of sustainable algal biofuel production using wastewater resources. *Bioresource Technology* 102:17–25.
Pollet, B. G., I. Staffell, and J. L. Shang. 2012. Current status of hybrid, battery and fuel cell electric vehicles: From electrochemistry to market prospects. *Electrochimica Acta* 84:235–249.
Quinn, J., L. D. Winter, and T. Bradley. 2011. Microalgae bulk growth model with application to industrial scale systems. *Bioresource Technology* 102:5083–5092.

Ragauskas, A. J., C. K. Williams, B. H. Davison, G. Britovsek, J. Cairney, C. A. Eckert, W. J. Frederick, J. P. Hallett, D. J. Leak, and C. L. Liotta. 2006. The path forward for biofuels and biomaterials. *Science* 311:484–489.

Rai, M., A. P. Ingle, S. Gaikwad, K. J. Dussán, and S. S. da Silva. 2017. Role of nanoparticles in enzymatic hydrolysis of lignocellulose in ethanol. In *Nanotechnology for Bioenergy and Biofuel Production*, ed. M. Rai and S. S. da Silva, 153–171. Cham, Switzerland: Springer.

Ramachandra, T. V., M. D. Madhab, S. Shilpi, and N. V. Joshi. 2013. Algal biofuel from urban wastewater in India: Scope and challenges. *Renewable and Sustainable Energy Reviews* 21:767–777.

Reddy S. N., S. Nanda S., and J. A. Kozinski. 2016. Supercritical water gasification of glycerol and methanol mixtures as model waste residues from biodiesel refinery. *Chemical Engineering Research and Design* 113:17–27.

Remón, J., P. Arcelus-Arrillaga, L. García, and J. Arauzo. 2016. Production of gaseous and liquid bio-fuels from the upgrading of lignocellulosic bio-oil in sub- and supercritical water: Effect of operating conditions on the process. *Energy Conversion and Management* 119:14–36.

Rodionova, M. V., R. S. Poudyal, I. Tiwari, R. A. Voloshin, S. K. Zharmukhamedov, H. G. Nam, B. K. Zayadan, B. D. Bruce, H. J. M. Hou, and S. I. Allakhverdiev. 2017. Biofuel production: Challenges and opportunities. *International Journal of Hydrogen Energy* 42:8450–8461.

Rye, P. J. 1995. Modelling photochemical smog in the Perth region. *Mathematical and Computer Modelling* 21:111–117.

Sims, R. E. H., W. Mabee, J. N. Saddler, and M. Taylor. 2010. An overview of second generation biofuel technologies. *Bioresource Technology* 101:1570–1580.

Skodras, G., G. Nenes, and N. Zafeiriou. 2015. Low rank coal—CO_2 gasification: Experimental study, analysis of the kinetic parameters by Weibull distribution and compensation effect. *Applied Thermal Engineering* 74:111–118.

Stephen, J. L., and B. Periyasamy. 2018. Innovative developments in biofuels production from organic waste materials: A review. *Fuel* 214:623–633.

Sullivan, J. L., R. E. Baker, B. A. Boyer, R. H. Hammerle, T. E. Kenney, L. Muniz, and T. J. Wallington. 2004. CO_2 emission benefit of diesel (versus gasoline) powered vehicles. *Environmental Science & Technology* 38:3217–3223.

Suopajärvi, H., E. Pongrácz, and T. Fabritius. 2013. The potential of using biomass-based reducing agents in the blast furnace: A review of thermochemical conversion technologies and assessments related to sustainability. *Renewable and Sustainable Energy Reviews* 25:511–528.

Taba, L. E., M. F. Irfan, W. A. M. W. Daud, and M. H. Chakrabarti. 2012. The effect of temperature on various parameters in coal, biomass, and co-gasification: A review. *Renewable and Sustainable Energy Reviews* 16:5584–5596.

Tang, M., L. Xu, and M. Fan. 2015. Progress in oxygen carrier development of methane-based chemical-looping reforming: A review. *Applied Energy* 151:143–156.

Thegarid, N., G. Fogassy, Y. Schuurman, C. Mirodatos, S. Stefanidis, E. F. Iliopoulou, K. Kalogiannis, and A. A. Lappas. 2014. Second-generation biofuels by co-processing catalytic pyrolysis oil in FCC units. *Applied Catalysis B: Environmental* 145:161–166.

Thomas, S., and R. A. Dawe. 2003. Review of ways to transport natural gas energy from countries which do not need the gas for domestic use. *Energy* 28:1461–1477.

USEIA. 2018a. United States Energy Information Administration. How much carbon dioxide is produced per kilowatthour when generating electricity with fossil fuels? https://www.eia.gov/tools/faqs/faq.php?id=74&t=11 (accessed on March 15, 2018).

USEIA. 2018b. Monthly energy eview March 2018. United States Energy Information Administration. https://www.eia.gov/totalenergy/data/monthly/pdf/mer.pdf (accessed on March 15, 2018).

USEIA. 2018c. Short-term energy outlook, March 2018. United States Energy Information Administration. https://www.eia.gov/outlooks/steo/ (accessed on March 15, 2018).

Voloshin, R. A., M. V. Rodionova, S. K. Zharmukhamedov, T. N. Veziroglu, and S. I. Allakhverdiev. 2016. Review: Biofuel production from plant and algal biomass. *International Journal of Hydrogen Energy* 41:17257–17273.

Wright, L. 2006. Worldwide commercial development of bioenergy with a focus on energy crop-based projects. *Biomass and Bioenergy* 30:706–714.

Zhang, K., Z. Pei, and D. Wang. 2016. Organic solvent pretreatment of lignocellulosic biomass for biofuels and biochemicals: A review. *Bioresource Technology* 199:21–33.

Zheng, Y., J. Zhao, F. Xu, and Y. Li. 2014. Pretreatment of lignocellulosic biomass for enhanced biogas production. *Progress in Energy and Combustion Science* 42:35–53.

2 An Overview of Fossil Fuel and Biomass-Based Integrated Energy Systems

Co-firing, Co-combustion, Co-pyrolysis, Co-liquefaction, and Co-gasification

Ejaz Ahmad, Ayush Vani, and Kamal K. Pant

CONTENTS

2.1 Introduction 15
2.2 The Need for Co-processing Technologies 17
2.3 Co-combustion and Co-firing 18
2.4 Co-pyrolysis 20
2.5 Co-liquefaction 24
2.6 Co-gasification 26
2.7 Conclusions 27
References 27

2.1 INTRODUCTION

Fossil fuel reserves have remained the primary source of energy and chemicals since the last century. However, excessive exploitation of these fossil fuel reserves, consequent depletion, and ongoing socio-political changes worldwide have encouraged the scientific community to find alternate sources to meet the energy and chemicals demand. One such alternative resource is renewable waste biomasses available in all demographic regions in the form of agricultural waste, forest residues, household wastes, and municipal solid wastes, which are often burnt to clear space and reduce volume in the developing countries. It is important to note that burning residual biomass is a waste of a valuable resource as well as the reason for severe environmental problems such as smog formation, emission of greenhouse gases, and particulate matters. Thus, it is essential to develop sustainable technologies for efficient utilization of residual biomass to produce chemicals and energy. Indeed, biomass is a versatile fuel that has direct applications in heat and power generation, household cooking, and as a source of solid fuel.

Interestingly, power generation using biomass or biofuels is carbon-neutral and causes fewer greenhouse gas emissions when compared to those produced by fossil fuels such as oil, coal, and natural gas. Fewer emissions and the possibility to create a closed carbon loop for energy and chemicals production is one of the main reasons for making application of biofuel mandatory by several developed and developing nations (Ragauskas et al. 2014). According to the data recapitulated by

World Energy Council, bioenergy contributes nearly 10% to the global energy supply at present, mainly from woody biomass. Moreover, approximately 56 exa-joules (EJ) of primary energy supply is estimated to be provided by forest-based biomass at the global level (World Energy Resources 2016). The overall trade driven by biomass pellets and liquid biofuels was 27 million ton in 2015 (World Energy Resources 2016). Undoubtedly, liquid biofuels derived from biomass such as biodiesel and bioethanol have received significant attention from the transport sector as a viable and sustainable alternative to reduce dependency on conventional gasoline fuel. Nevertheless, the uncertainty in land availability and seasonal crop yield create a conundrum in the quantitative assessment of biomass availability in future. Furthermore, the efficient processing of biomass is necessary to make a healthy shift towards a sustainable and renewable energy alternative for reducing greenhouse gas emissions.

Several technologies have been developed for biomass conversion according to the desired end product (Alam et al. 2015, 2016; Ahmad et al. 2016, 2018; Gupta et al. 2017; Quereshi et al. 2017a, 2017b). In addition, the suitability of these methods is decided based on their compatibility with inherent variations in feedstock such as mass, molecular composition, energy density, calorific value, size, microbial activity, and moisture content. For example, low moisture-containing biomass (<50% moisture) is suitable for thermal conversion, whereas biological conversion processes are more favorable for high moisture-containing feedstocks. Similarly, high calorific value and the proportion of fixed carbon and volatile matter content provide a basis for the biomass application in combustion and gasification processes. In contrast, higher oxygen contents relative to carbon is not favorable for the energy value of the fuel. Similarly, other parameters such as higher ash or solid residue content, alkali metal content, and higher lignin-to-cellulose ratio may cause a higher handling and processing cost due to tar and char formation. In general, biomass conversion processes can be broadly classified into three major categories: thermochemical (combustion, gasification and pyrolysis), biochemical (anaerobic digestion and fermentation), and physical (mechanical extraction). The combustion, gasification, and anaerobic digestion are preferred methods for heat and power generation, whereas fermentation and mechanical extraction methods are preferred for producing transportation fuels and chemicals.

Biomass offers immense advantages in the form of bioenergy over conventional fossil fuels. It may be its renewability and carbon neutrality that makes it a source of cleaner fuel. The applications of biofuels have a positive ecological impact and boost rural economy by involving farmers in energy crops production (Nanda et al. 2015). To make a transition towards a low carbon-intensive economy and fulfil our future energy demands at the global scale (as predicted 150–400 EJ/year of bioenergy production) will require all available biomass resources including energy crops, organic waste, and other residues (van Vuuren et al. 2010a, 2010b). Although it is possible to solve this problem by energy crops cultivation, it may require excessive land use and high-water demand for irrigation, which may result in deforestation and biodiversity losses. Furthermore, food security could be compromised by large-scale cultivation of energy crops on arable lands and seasonal supply. The use of degraded and marginal lands for energy crop plantation may be a suitable alternative (Nanda et al. 2015). However, the land privatization could have social costs related to the displacement of rural communities whose livelihoods are dependent on small agricultural lands.

The extensive clearing of carbon-rich ecosystems for facilitating a biomass plantation can call forth the significant loss of CO_2 from soil and vegetation into the atmosphere. These long-term assessment studies together build a paradox that extensive farming of energy crops and ill-management of biomass could lead to the increase in anthropogenic greenhouse gas emissions (Chakravorty et al. 2009; Hoehstra et al. 2010; Popp et al. 2014). Using biomass alone in the global energy sector can solve some environmental problems on this planet as supported by a majority of environmental scientists. However, the contradicting situations discussed along with the higher investment costs, and dispersion and poor infrastructure of biomass supply in most of the countries may become the limitations where energy policymakers and stakeholders need to focus.

2.2 THE NEED FOR CO-PROCESSING TECHNOLOGIES

In the light of all these factors, there is an urgent need to develop a balance between existing fossil fuel utilization and bio-renewable resources exploitation. It is evident that the dependence of humanity on only one of the fuels (i.e., fossil fuels or biomass) may result in an imbalance in the environment as well as economic and socio-political situations, thereby causing an uneven distribution of wealth and resource worldwide. Therefore, the co-processing of biomass and other feedstocks such as fossil fuels may be one possible solution for continual generation of heat and power, transport fuels, and chemical products. Co-processing with biomass as a partial substitute can be practiced by introducing some technical modifications in conventional methods of energy conversion. It can be helpful in resolving the problems related to biomass quality, insufficiency of feedstocks, and costs related to the installation of separate biorefining infrastructures for biomass processing with minimal compromises in the existing boilers or other existing petrochemical refineries. The recent research is directed towards the development of such integrated units where fossil fuels and biomass can be processed together. Interestingly, there are few commercial technologies available in this regard for co-processing biomass and fossil fuels together, which are used differently from region to region and use varied cost-benefit analysis. These technologies include co-firing, co-combustion, co-pyrolysis, co-liquefaction, and co-gasification (Figure 2.1).

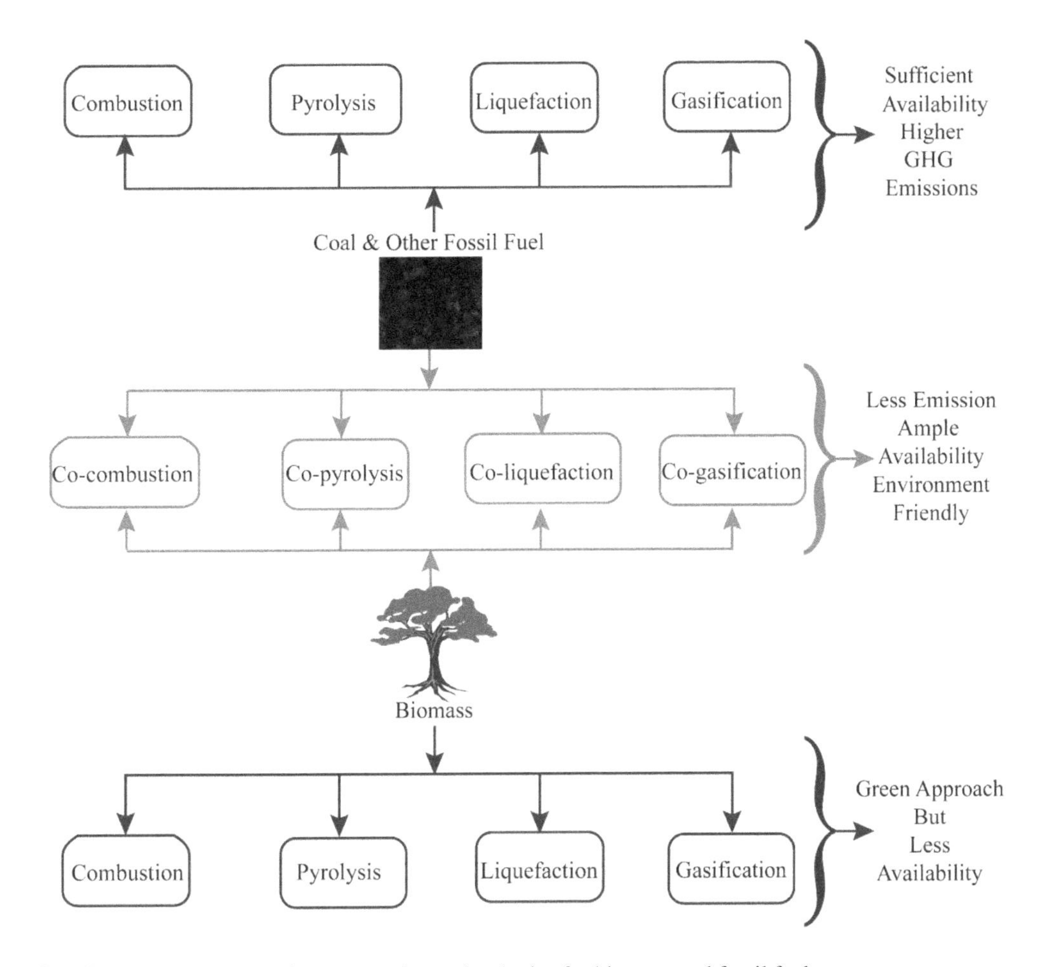

FIGURE 2.1 Advantages of co-processing technologies for biomass and fossil fuels.

There are several co-firing plants (over 230 units) of total capacity 50 megawatts electric (MWe) to 700 MWe operating at commercial scales primarily in the United States, Europe, Asia, and Australia (Energy Technology Network 2013). Several companies that have begun to introduce and expand the co-firing technology are Tokyo Electric Power Co. Inc. (Japan), Soma Kyodo Power Co. Ltd. (Japan), Hokuriku Electric Power Co. (Japan), Chubu Electric Power Co. Inc. (Japan), Tacoma Public Utilities (United States), Northern States Power (United States), Southern Company (United States), American Electric Power (United States), KEMA (Netherlands), EPON (Netherlands), EPZ (Netherlands), EZH (Netherlands), and a few other European companies and power plants (Tumuluru et al. 2011; Energy Technology Network 2013; European Biomass Industry Association 2018; Asia Biomass Office 2018). Similarly, there are several industrial units based on co-combustion, co-pyrolysis, co-liquefaction, and co-gasification technologies.

2.3 CO-COMBUSTION AND CO-FIRING

Co-firing is a cost-effective and efficient technique for utilizing biomass as a partial substitute with a base fuel in highly efficient boilers to generate power. Recently, the most emphasized practice of co-firing is blending biomass with coal in pulverized coal and cyclone boilers for producing electricity. Nevertheless, the choice of biomass for blending is limited. The co-firing process relates back to the conventional method of firing coal in the presence of air by utilizing the whole volume of the burner or the furnace in a boiler to generate steam to provide electricity using turbines. Thus, co-firing can be considered as an expansion of this existing technology where more than one fuel is fired together in the same setup at the same time with one of the fuels being coal. The co-firing of coal and biomass in existing coal-fired power plants could be a cost-effective approach to reduce anthropogenic greenhouse gas emissions in the atmosphere. However, there are also other commercial facilities available for co-firing of biomass and fossil fuels to tackle the problem of global warming. An illustration of co-firing and co-combustion is shown in Figure 2.2. Co-firing represents a family of technologies (Tillman 2000), which includes:

1. Direct co-firing using a single boiler and common or separate burner setup.
2. Indirect co-firing by gasifying biomass and subsequent gaseous fuel is burned with other fuel (e.g., coal) in same boiler setup.
3. Parallel co-firing in which separate boilers are used for burning biomass and later supplying steam generated to steam output from other sources (e.g., coal).

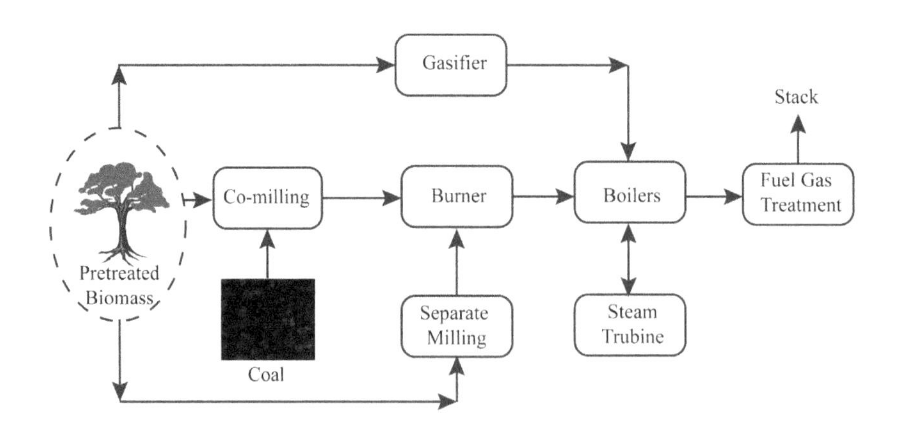

FIGURE 2.2 Overview of co-firing and co-combustion technologies.

The first approach of direct blending of biomass with coal using a conventional boiler is simplistic and cost-effective compared to other co-firing approaches. It involves direct milling (or co-milling) of biomass and its injection into pulverized coal boilers with as low as 5 weight pcerent (wt%) or less depending on the type of pulverizer. Nevertheless, this number could be increased to as high as 20 wt% in the case of cyclone boilers. In general, the most emphasized practice of co-firing is blending biomass with coal in pulverized coal and cyclone boilers for producing electricity. However, co-milling is a challenging task in certain cases that involve bark, straw, and switchgrass blending with coal because of high ash and tar generation. Although the cyclone boilers are more compatible with this type of approach, there exist some constraints to the speed of the cyclone feeder with coal as one of the fuels. Thus, the blending of a wide range of feedstocks based on their physiochemical properties has been tested to overcome such limitations. For example, low volatility and low ash content of woody biomass make it more suitable and easier to process as compared to agricultural biomass. Similarly, the grindability and particle size of the feedstock plays a crucial role in limiting the amount of biomass to be blended in co-milling. Similarly, another approach for direct co-firing involves pre-milling (or separate milling) of biomass feedstock usually in the dust or pellet form followed by feeding it separately into the same boiler. Nevertheless, the burner can be either joint or separate for the two fuels. The separate milling allows flexibility to introduce a wide range of biomass feedstocks in different quantities due to pre-mixing and separate feeding.

The second approach of indirect co-firing is a less common and relatively costly at the industrial scale due to additional requirements of gasifying biomass through a gasifier. Subsequently, the gases formed (also called as syngas) are combusted with other fuel (e.g., coal) in the same furnace boiler. The syngas formed in the first step constitutes gases like H_2, CH_4, CO, and CO_2, as well as the diverse type of hydrocarbons. The heating value of syngas varies according to the moisture content of biomass feedstock. In addition, this approach is more flexible regarding the choice of different base fuels with biomass and diverse varieties of biomass feedstocks in a higher amount. On the contrary, there are many economic and technical challenges still to overcome such as the additional cost of gasification, cleaning and filtering of obtained gas, separation of the ash of the two fuels, and ensuring complete combustion with less residence time for gas.

The third approach of parallel co-firing requires separate boilers for burning biomass. Another base fuel and the steam generated from biomass boiler are later supplied for outputs from other resources (e.g., coal). The higher amount of biomass fractions and separation of ash could be facilitated in this type of co-firing approach. Such an approach has extensive usage in pulp and paper power plants. This approach also provides the flexibility to use different types of biomass feedstocks especially those with high chlorine and alkali compositions (e.g., some agricultural biomass and sewage sludge).

Overall, co-firing of biomass with fossil fuels has some economic and environmental benefits. It reduces the dependency on fossil fuels and lessens the greenhouse gas emissions as well as improves biomass-based waste management practices. This technique has the potential to reduce oxides of nitrogen, thereby eliminating smog and ozone degradation problems. In addition, it helps to reduce sulfurous gases (e.g., SO_2), which can be attributed to the lower sulfur content in biomass. Co-firing of biomass is advantageous in many ways as it overcomes the feedstock scarcity issue, low adaptation costs in existing power plants, and lowers the investment costs for power plants. Biomass can be a low-cost alternative to the conventional coal-firing electricity generation because of its high conversion efficiency without compromising with the boiler efficiency. It may prove to be a sustainable option for refineries to manage the vast amount of woody and agricultural wastes worldwide.

The economic feasibility of biomass co-firing depends upon a number of factors, such as the cost of biomass fuel, which is governed by its supply in the energy market. The cost of biomass fuel per unit of heat for co-firing should never be higher than that of the other base fuel (e.g., coal) to make this technology economically viable. However, it is worth noting that the cost of biomass fuel varies according to quality, quantity, heating value, ash content, moisture content, and demographic region. The capital investment required for retrofitting an existing coal power plant for biomass co-firing is in the range of (in U.S. dollars) \$430–500/kW, \$760–900//kW and \$3,000–4,000/kW for co-feed, separate , and indirect co-firing plants, respectively. This cost is lower than the establishment costs of full-sized biomass-based refineries (Energy Technology Network 2013). Similarly, the electrical efficiency of a power plant practicing biomass co-firing with coal as a base fuel is between 36% and 44% depending on the type of technology adopted, a variety of biomasses used, and the plant size (Energy Technology Network 2010).

At present, the biomass pellets cost (in Euros) nearly €24–30/MWh, which is €12/MWh more than that of the coal in the world trade. However, the technical modifications like densification or pelletizing and torrefaction not only bring down the costs of biomass pellets but also make the biomass feedstock more amenable for co-firing (Hawkins Wright 2011). In torrefaction, biomass is processed to form pellets, and briquettes increases the bulk density, which makes the transportation of these feedstocks cost-effective. In this process, biomass is exposed to a temperature range between 250°C and 300°C in the absence of oxygen resulting in a feedstock much more reactive with low H/C and O/H ratios (Agar and Wihersaari 2012; Li et al. 2012). However, these technologies are yet to be implemented on a commercial scale due to several technical and economic challenges.

Many varieties of biomass (e.g., wheat straw) are not amiable to coal pulverizers due to high chlorine and alkali content, which cause corrosion on boilers surfaces. Furthermore, a mixture of biomass-derived ash and other fuel-derived ash (e.g., coal) is difficult to separate or dispose of and have the least industrial utility as compared to ashes of individual fuels. For example, the ash produced by the firing of coal alone is suitable to use as a Portland cement ingredient in construction and mining industries according to the ASTM standards, whereas ash produced from biomass firing alone can be used in fertilizers, building materials, or directly as a fuel. On the contrary, a combination of ashes derived from both the sources makes them unusable. Furthermore, the loss in boiler efficiencies, fouling, and slagging of the boiler could become a severe problem when biomass feedstocks with very high moisture content and relatively low ash fusion temperatures compared to coal are used.

Although co-firing of biomass with coal is a commonly accepted approach in industries, there has been considerable attention towards a combination of biomass with other fossil fuels. Like coal power plants, natural gas combined cycle (NGCC) plants can be retrofitted for co-firing biomass. The alternative technological options include indirect co-firing of biomass with natural gas where syngas produced in the first step is fired with natural gas in the same furnace boiler. Recently, the idea of co-firing natural gas with bio-oil is gaining considerable momentum (Agbor 2015). Kymijärvi power plant in Lahti, Finland, exemplifies the indirect co-firing technology by using a circulating fluidized bed gasification system and firing the gas with natural gas or coal (Granatstein 2002). Interestingly, indirect co-firing of biomass with natural gas is found to be relatively more advantageous than the commonly adopted biomass and coal co-firing because higher firing rates, and biomass substitution (up to 40%), can be facilitated (Agbor 2015). Secondly, the ash is formed from only one source (i.e., biomass). Therefore, there are no additional efforts required for ash separation and utilization. The ash can be used in soil amendment or for other agricultural activities. However, this technology is yet to attain commercial success possibly due to the capital-intensive gasification process.

2.4 CO-PYROLYSIS

Co-pyrolysis is another cost-effective technique where more emphasis has been given to the production of high-grade bio-oil, which can be used as a fuel or feedstock for many chemicals production. This technology facilitates the production of oil with higher calorific value, higher stability, less oxygen, less water content, and reduced corrosion problems. Moreover, co-pyrolysis leads to a higher performance and cost ratio as compared to oil produced by conventional pyrolysis (Abnisa et al. 2014b).

In general, pyrolysis is defined as the thermal decomposition of organic materials by the application of heat at variable heating rates and residence times in the absence of oxygen. Consequently, the pyrolysis of biomass transforms it into various products like oil, char, and tar along with the release of some volatile products (gases) as an irreversible change in chemical composition. Pyrolysis is also a fundamental endothermic step prior to related technologies such as combustion, liquefaction, and gasification. It is carried out at a temperature range of 350°C–550°C, which may be extended up to 700°C–800°C in the absence of a reactive atmosphere (e.g., oxygen) (Abnisa et al. 2014b). Elevated temperatures cause the breakage of chemical linkages and longer molecular chains into products like solid biochar, bio-oil, and combustible gases (Onay et al. 2007). Nevertheless, the quality and yield of products are a function of various parameters such as operating temperature, heating rate, reactor design, pressure, residence time for solids, and the type of biomass feedstock used.

Slow pyrolysis is an energy-intensive process characterized by slow heating rates, low temperatures, and high vapor residence times, which favors higher char production. Consequently, slow pyrolysis is unfavorable for producing a high quantity bio-oil due to longer vapor residence time, thereby cracking the main product (Tippayawong et al. 2008). On the contrary, fast pyrolysis is characterized by treatment of the biomass with rapid heating rates and high temperature with short residence time for vapors and fast cooling to facilitate the high production of bio-oil. This technology has gained much attention in fuel industries as a highly energy efficient system that can be established with relatively fewer investment demands (Demirbaş and Arin 2002). On the contrary, flash pyrolysis favors the production of high-yielding bio-oil under extremely reactive conditions. However, there are some qualitative limitations of the product such as low thermal stability, corrosion issues, and viscosity build up due to catalytically active char (Demirbaş and Arin 2002; Brammer et al. 2006).

The bio-oil produced from slow pyrolysis has a higher oxygen content (~35–60 wt%), which can be attributed to the presence of a higher number of oxygenated compounds resulting in lower calorific value, poor stability, corrosion issues, and lower combustion efficiency (Bridgwater et al. 1999; Parihar et al. 2007; Guillain et al. 2009; Jahirul et al. 2012). Thus, there is a need for producing qualitative bio-oil at higher performance and cost ratios for which co-pyrolysis can be a sustainable solution.

Co-pyrolysis has a similar procedure to conventional pyrolysis. However, some modifications have been made in the pre-processing steps and feedstock processing parameters. First, the drying of feedstock is necessary to reduce the moisture content ($\leq$ 10%) in the primary end-product (i.e., bio-oil). The drying of samples could also be facilitated by the heat supply through combustion of the by-products (i.e., char) to reduce the cost of the heating system. Moreover, the use of inert gases like nitrogen is a usual practice for speeding up the sweeping of vapors from the co-pyrolysis unit to the condensation unit. The reason for maintaining less hot vapor residence time is to reduce the extent of undesirable secondary chemical reactions, which could lead to the decrease in the oil yield (Demirbaş and Arin 2002). However, the inert gas flow at very high flow rates may impart an adverse effect on the oil yield. The choice of inert gas flow rate depends upon the type of reactor design such as high flow rate in case of fluid bed reactor, circulating fluid bed reactor, and entrained flow reactor, whereas no mandates are essential for vacuum and ablative reactors.

In a typical co-pyrolysis process, the temperature range is maintained at 400°C–600°C, which facilitates the production of more than 45% bio-oil in the absence of oxygen, although this temperature range may vary depending upon the nature of feedstock used (Wei et al. 2011; Abnisa et al. 2014a, 2014b). Similar to pyrolysis, the co-pyrolysis process also is affected by the type of feedstock, biomass particle size, heating rate, and reactor design. In addition, the vapors generated during the co-pyrolysis need to be cooled at a faster rate to convert them into high-yield bio-oil, otherwise they could further crack or polymerize to yield char and gases.

The quality of the primary product (i.e., biofuel) is enhanced at relatively low cost by the co-pyrolysis technique because the oil produced has lower oxygen content, high calorific value, and homogeneity. However, the magnitude of these parameters also significantly depends on the selection of feedstock for co-pyrolyzing with biomass. Interestingly, co-pyrolysis can be a useful technology to address the issue of the plastic and waste biomass management. Furthermore, co-pyrolysis of biomass with coal may be a practical step for addressing problems like lower energy density of biomass, feedstock shortage, transportation issues, and seasonal availability issues. It is reported that the co-pyrolysis of lignite and biomass causes a synergistic effect where biomass behaves as a hydrogen donor leading to low char and high liquid yield as compared to the pyrolysis of the biomass alone (Quan et al. 2014). Nevertheless, the co-feed can be coal or the synthetic materials derived from fossil fuels (e.g., waste plastics and tires), which enhance the quality of bio-oil by reducing oxygen content in the final product. Although there are several factors affecting co-feeding and co-pyrolysis, the principal effect can be observed in the biochar and liquid yield.

Table 2.1 summarizes a few feedstocks and co-feedstocks used in co-pyrolysis. Replacing lignite coal with polystyrene plastic as a co-feed with palm shell in 1:1 ratio enhanced bio-oil production by 15.5 wt% (Abnisa et al. 2014a). On the contrary, a decrease of 15.49 wt% was measured in biochar yield, which indicated that the bio-oil yield and biochar yield are contemporary to each other. Similarly, co-pyrolysis of lignite coal with safflower seed enhanced bio-oil yield by 17 wt%, whereas biochar yield reduced by 10 wt% concerning the yields obtained from individual processing of biomass (Onay et al. 2007). Moreover, the decrease in the biomass-to-coal ratio further enhanced the liquid yield. A 26.3 wt% increase in liquid yield has been reported when coal amount was increased to 30 wt% in a continuous free fall reactor at 500°C co-pyrolysis temperature (Wei et al. 2011). Equivalent liquid yield also can be obtained by replacing coal with a polymeric material such as block polypropylene under the same pyrolysis temperature (Jeon et al. 2011). Nevertheless, liquid yield from co-pyrolysis may vary according to the type of feed and co-feed. For example, co-pyrolysis of wood cellulose and polystyrene yield showed a 13.3 wt% increase in liquid yield, which was significantly lower than the 23.8 wt% liquid yield obtained from co-pyrolysis of wood chips and polypropylene together (Rutkowski and Kubacki 2006; Jeon et al. 2011). One possible reason could be attributed to the molecular composition of cellulose that undergoes cracking reactions to yield gaseous products at elevated temperatures, whereas the vapors derived from lignin in woody biomass are difficult to crack at 500°C. Therefore, it condenses to yield bio-oil. Interestingly, co-pyrolysis of potato skin and high-density polyethylene showed improvements in the liquid yield up to 16 wt% compared to the liquid yield obtained from biomass pyrolysis alone (Önal et al. 2012).

It should be noted that the co-pyrolysis of biomass and polymers might not always increase liquid yield. For example, co-pyrolysis of different types of biomass with polylactic acid, waste tires, and tire powder have not shown a significant increase in liquid yield (Cornelissen et al. 2008; Cao et al. 2009; Martínez et al. 2014). Nevertheless, positive cash flow is reported for co-pyrolysis of biomass and polymers in a techno-economic feasibility study performed by Kuppens et al. (2010). However, all type of plastics cannot be used in a co-pyrolysis process; for example, plastics like polyvinyl chloride tend to degrade the product quality due to the production of highly chlorinated, toxic, and corrosive products (Wei et al. 2011). At present, Ensyn and DynaMotive are the two North American companies that have dominantly used the fast pyrolysis technology since the 1990s and underscored the development of biomass-based clean energy systems. There are other popular companies like Renewable Oil International and Pyrovac, which are contributing to bringing biomass pyrolysis to the commercial platform (Quan et al. 2014).

TABLE 2.1
Summary of Selected Feedstocks and Co-feedstocks Used in Co-pyrolysis

Feedstock	Co-feedstock	Feedstock/ Co-feedstock ratio (wt.)	Temperature (°C)	Heating Time/ Rate	Process	Decrease in Char Yield with respect to Biomass	Increase in Liquid Yield with respect to Biomass	References
Cellulose	Polystyrene	1:1	500	5°C/min	Vertical pyrex reactor	~9.8%	~13.3%	Rutkowski and Kubacki (2006)
Legume straw	Dayan lignite	73:23	600	–	Continuous free fall reactor	~12%	~9%	Quan et al. (2014)
Legume straw	Tiefa bituminous coal	7:3	500	–	Continuous free fall reactor	54.9% with respect to coal	26.3% with respect to coal	Wei et al. (2011)
Palm Shell	Polystyrene	1:1	500	60 min	Continuous fixed-bed reactor	15.5%	15.5%	Abnisa et al. (2014a)
Pinewood chips	Waste tires	9:1	500	45 min	Auger reactor	~26.7% absolute yield	~6%	Martínez et al. (2014)
Potato skin	High-density polyethylene	1:1	500	5°C/min	Stainless steel retort	~6%	~16%	Önal et al. (2012)
Safflower seed	Lignite coal	19:1	550	7°C/min	Continuous fixed-bed reactor	~10%	~17%	Onay et al. (2007)
Sawdust powder	Tire powder	2:3	500	20°C/min	Continuous fixed-bed reactor	–	~2%	Cao et al. (2009)
Willow	Polylactic acid	1:1	450	10°C/min	Semi continuous Stainless-steel reactor	~9.9%	~2.5%	Cornelissen et al. (2008)
Wood chip	Block polypropylene	1:1	500	60 min	Continuous fixed-bed reactor	~24% absolute yield	~23.8%	Jeon et al. (2011)

2.5 CO-LIQUEFACTION

Liquefaction is the type of thermochemical biomass conversion technology that differs in its characteristics from other thermochemical conversions concerning its operating conditions such as temperature. In both gasification and pyrolysis, the biomass processing occurs above 600°C with the requirement of dry feedstock, whereas liquefaction is carried out below 400°C for feedstocks having water content. Moreover, a suitable catalyst is used to improve the oil yield and product quality. Hydrothermal liquefaction of biomass is characterized by a moderate temperature range (250°C–380°C) and high pressures (4–22 megapascal (MPa)) in the presence of water and sufficient time to break down the large polymeric compounds (Lalvani et al. 1991; Gollakota et al. 2018). Interestingly, the bio-oil produced through the liquefaction process has less tar and water content as compared to the conventional pyrolysis process. Furthermore, direct liquefaction of coal is a commercially established technology for producing liquid fuels. In this process, hydrogen donors (solvents or other additives) are introduced to facilitate the combination of hydrogen with free radicals generated by the thermal fragmentation of coal. However, there are techno-economic constraints to the commercial expansion of coal liquefaction technology such as higher cost of hydrogen donors, high greenhouse gas emissions, and the lower reactivity of coal. Therefore, co-processing of coal with biomass may provide an opportunity to redress these issues. The biomass can act as a potential hydrogen donor and reduce greenhouse gas emissions through a closed carbon cycle.

Co-liquefaction technology has been developed in which biomass mixed with other feedstock (predominantly coal in case of fossil fuel-based feedstock) are used. Under a specific range of temperature, pressure, and duration, a slurry is produced as the end-product, which is further fractioned to produce combustible gases and oils with higher yields and qualities (Tumuluru et al. 2011). However, no significant synergistic effects between coal and (untreated) biomass is observed at such low temperature possibly due to low reactivity of biomass. Therefore, the reactivity of biomass is usually enhanced by the torrefaction process before its application to liquefaction. In torrefaction, the biomass is exposed to a temperature range of 200°C–300°C, in the absence of oxygen, resulting in a biomass that is much more reactive with low H/C and O/H ratios (Jahirul et al. 2012; Agbor 2015). Similarly, there are several other factors such as temperature, pressure, solvent, holding time, pre-treatment of feedstock, blending ratio, composition of feedstock, processing, separation of liquid, and gaseous products, all of which affect liquefaction's overall efficiency (Lalvani et al. 1991).

In a typical co-liquefaction process, biomass and coal are mixed in a sealed autoclave reactor in a highly pressurized environment in the presence of a solvent and catalyst, which is later exposed to a temperature range of 300°C–450°C (Guo et al. 2011). Consequently, coal-fragmented free radicals and biomass-derived hydrogen react to form the desired products. Later, a Soxhlet extractor is used to separate the various products and the slurry mixture obtained from the co-liquefaction of coal and biomass. These products include oils, asphaltenes, preasphaltenes, gases, and solid residues. Furthermore, biomass can be pre-treated through torrefaction to bring a synergistic impact in the co-liquefaction process.

It is difficult to compare the effects of the co-liquefaction process on liquid and gaseous product yields due to the variation in process parameters and type of feedstocks. A few examples of the co-liquefaction process are given in Table 2.2. A 1:1 bagasse-to-bituminous coal ratio when subjected to co-liquefaction process at 420°C temperature for 40 minutes yielded 47.8% liquid yield (Rafiqul et al. 2000). Similarly, co-liquefaction of sawdust and Turkish lignite under similar process conditions led to a 14 and 3.9 wt% increase in liquid and gas yield, respectively, when compared to liquid and gas yields measured for liquefaction of lignite alone (Karaca et al. 2002). Likewise, a 20 wt% increase in liquid yield was measured as compared to coal liquefaction alone when bituminous coal underwent the co-liquefaction process with lignin in 1:1 ratio at 400°C for 45–60 minutes (Altieri and Coughlin 1987). The increase in liquid yield was attributed to the fact that lignin on escalated pressure (700–1000 pound per square inch (psi)) and temperature (400°C) underwent a depolymerization reaction to yield liquid fuel.

TABLE 2.2
Summary of Selected Feedstocks and Co-feedstocks Used in Co-liquefaction Process

Feedstock	Co-feedstock	Feedstock/ Co-feedstock ratio (wt.)	Temperature and Pressure	Holding Time (min)	Reactor	Increase in Liquid Yield	Increase in Gas Yield	References
Bagasse	Bituminous coal	1:1	420°C, 500 psi	40	Magnetically driven autoclave	~47.8% absolute yield	–	Rafiqul et al. (2000)
Caustic lignin	Illinois coal	3:2	375°C, 140 psi	60	Glass-lined autoclave	~48.6% with respect to coal	~26.3% with respect to coal	Lalvani et al. (1991)
Chlorella (microalgae)	Yallourn coal	1:1	350°C, 725 psi	60	Magnetically-stirred autoclave	~24% absolute yield	~2%	Ikenaga et al. (2001)
Cow manure	Bituminous coal	1:1	350°C, 1000 psi	60	Tubing-bomb reactor	~15%	–	Stiller et al. (1996)
Lignin	Bituminous coal	1:1	400°C, 700–1000 psi	45–60	Magna-drive autoclave	~20% with respect to coal	~9.98% absolute yield	Altieri and Coughlin (1987)
Rice straw	Shenfu coal	1:1	400°C, 725 psi	60	Tubing reactor	~42.5% absolute yield	~15.5% absolute yield	Hua et al. (2011)
Sawdust	Turkish lignite	1:1	400°C, 588 psi	60	Stainless steel autoclave	~14% with respect to lignite	~3.91% with respect to lignite	Karaca et al. (2002)
Sawdust	Sub-bituminous coal	1:1	400°C, 725 psi	60	Tubing reactor	~11.4%	~8.9% with respect to coal	Shui et al. (2011)
Sawdust	Yitai lignite	1:1	360°C, 507 psi	30	Magnetically-stirred autoclave	~32% absolute yield	~12% absolute yield	Guo et al. (2011)
Tea pulp	Lignite	1:3	400°C, 290 psi	60	Magnetically-stirred autoclave	~15.5% (oil and gas) with respect to lignite alone		Karaca and Koyunoglu (2017)

Interestingly, increasing the lignin content in the feed enhances both liquid and gas yield during the co-liquefaction process. For example, when caustic lignin and coal in 3:2 ratio were subjected to co-liquefaction process at 375°C temperature for 60 minutes, a 48.6 wt% increase in the liquid was measured along with 26.3 wt% increase in gaseous products as compared to liquid and gaseous yields obtained from coal processing alone (Lalvani et al. 1991). The presence of chemicals in the caustic lignin acted as catalysts for high liquid and gaseous yields.

Unlike co-pyrolysis and gasification processes, co-liquefaction does not necessarily require dry feedstock for processing. Thus, water-containing feedstock such as algal biomass can be used directly as a feedstock for this process. For example, when the microalgae *Chlorella* along with coal was subjected to the co-liquefaction process, it yielded nearly 24 wt% liquid yield (Ikenaga et al. 2001). Similarly, other high moisture-containing solid wastes such as cow manure have been used in the co-liquefaction process to produce liquid and gaseous products (Stiller et al., 1996). Overall, co-liquefaction of different types of biomass such as sawdust, tea pulp, and rice straw with different grades of coal improved liquid and gaseous products yield (Guo et al. 2011; Hua et al. 2011; Shui et al. 2011; Karaca and Koyunoglu 2017).

2.6 CO-GASIFICATION

Co-gasification is another emerging co-processing technology for biomass and fossil fuel, which has gained attention recently. It can be considered as an expansion of the traditional biomass or coal gasification process where a mixture of both feedstocks is gasified to enhance the quality of products obtained from gasification alone and to deal with the costs related to separate infrastructure development. The co-gasification of biomass and coal has many benefits in producing methane and hydrogen-rich syngas at higher conversion rates due to the reaction between the organic constituents of coal and the volatile-derived free radicals from biomass. Nevertheless, the qualitative standards of syngas and char produced from the gasification approach may vary depending upon the nature of feedstock, reactor design, catalyst, and the gasifying agent used (Farzad et al. 2016). However, unlike other co-pyrolysis and co-liquefaction where biomass serves as feedstock, coal is a primary feedstock in co-gasification process. Interestingly, it has been observed that the carbon conversion in biomass is faster than the coal-biomass blend, whereas carbon conversion is lowest for both feedstocks separately under identical experimental conditions.

A higher feed (coal) to co-feed (biomass) ratio enhances the reactivity of char when pine sawdust was subjected to gasification in a fixed bed system at 1000°C for 60 minutes with co-feed Shinwa coal (Jeong et al. 2015). The blending of coal and biomass reduces CO_2 emissions due to the catalytic activity of the natural compounds present in the biomass (Howaniec and Smoliński 2013). Moreover, an increase in the biomass ratio concerning coal enhances hydrogen generation in the co-gasification process owing to the higher elemental hydrogen content in biomass as compared to the negligible elemental hydrogen content in coal (Li et al. 2010). Interestingly, co-gasification of coal with polyethylene led to the production of higher hydrocarbons, which further validated the effect of feed type on final products from the co-gasification (Pinto et al. 2003). In general, a temperature above 840°C is considered effective for the co-gasification process to enhance the energy efficiency and carbon conversion (Pan et al., 2000). Such a temperature may be helpful in co-gasification of plastic material along with coal and biomass (Narobe et al. 2014).

It is also possible to decrease the higher hydrocarbon contents either through an increase in the gasification temperature or the oxygen content. In addition, an increase in the gasification temperature enhances hydrogen production and suppresses methane production possibly due to cracking of formed products at elevated pressures. On the contrary, gaseous products obtained from the low biomass ratio are found suitable for methanol production, whereas products obtained from the gasification of higher biomass containing feed favors dimethyl ether (DME) production (Kumabe et al. 2007). Nevertheless, it is worth noting that the co-gasification of coal and biomass may result into the decrease in energy efficiency as compared to coal alone, although hydrogen content remains higher in mixed feed gaseous products (Vélez et al. 2009). However, the co-gasification may be beneficial in the carbon footprint reduction due to lower greenhouse gas emissions.

2.7 CONCLUSIONS

At present, fossil fuels are the primary source of energy and chemicals. However, the excessive use of fossil-derived fuel has resulted in severe environmental damages especially air and water pollution. It is well known that the emissions of greenhouse gases are the apparent cause of global warming. Thus, recent efforts worldwide are directed towards the development of carbon-neutral technologies. In this regard, biomass has emerged as a potential feedstock, which provides an opportunity to create a closed carbon cycle. Plenty of thoughtful research has been done in the past two decades, thereby leading to the development of several co-processing technologies.

Biomass has limited supply and is not available in enough quantity to completely replace the fossil fuels. Although the production of energy crops may be helpful to overcome such limitations, there is a possibility that energy crop production may cause an adverse effect on the ecological balance. Furthermore, possible conflict in land utilization for food biomass and energy crops may arise. In addition, replacing fossil fuel entirely with biomass may require an entirely new infrastructure, equipment, and machinery, which in turn will become an extra economic burden to the refineries. Nevertheless, partial replacement of fossil fuel with biomass can be done without requiring any significant change in existing refinery setups. In this context, co-processing techniques such as co-firing, co-pyrolysis, co-liquefaction, and co-gasification can offer economic and environmental advantages. Co-processing of fossil fuel and biomass does not necessarily require a significant change in the existing setups, whereas it improves the end-products properties. Moreover, replacing fossil fuel with bio-renewable feedstock could partially create a closed carbon loop, thereby suppressing an increase in overall greenhouse gas emissions.

REFERENCES

Abnisa, F., W. M. A. Wan Daud, and J. N Sahu. 2014a. Pyrolysis of mixtures of palm shell and polystyrene: An optional method to produce a high-grade of pyrolysis oil. *Environmental Progress & Sustainable Energy* 33:1026–1033.

Abnisa, F., and W. M. A. Wan Daud. 2014b. A review on co-pyrolysis of biomass: An optional technique to obtain a high-grade pyrolysis oil. *Energy Conversion and Management* 87:71–85.

Agar, D., and M. Wihersaari. 2012. Bio-coal, torrefied lignocellulosic resources – Key properties for its use in co-firing with fossil coal – Their status. *Biomass and Bioenergy* 44:107–111.

Agbor, E. U. 2015. Biomass co-firing with coal and natural gas. M.Sc. Thesis. University of Alberta, Canada.

Ahmad, E., M. I. Alam, K. K. Pant, and H. A. Haider. 2016. Catalytic and mechanistic insights into the production of ethyl levulinate from biorenewable feedstocks. *Green Chemistry* 18:4804–4823.

Ahmad, E., N. Jäger, A. Apfelbacher, R. Daschner, A. Hornung, and K. K. Pant. 2018. Integrated thermo-catalytic reforming of residual sugarcane bagasse in a laboratory scale reactor. *Fuel Processing Technology* 171:277–286.

Alam, M. I., S. Gupta, A. Bohre, E. Ahmad, T. S. Khan, B. Saha, and M. A. Haider. 2016. Development of 6-amyl-α-pyrone as a potential biomass-derived platform molecule. *Green Chemistry* 18:6399–6696.

Alam, M. I., S. Gupta, E. Ahmad, and M. A. Haider. 2015. Integrated bio- and chemocatalytic processing for biorenewable chemicals and fuels. In *Sustainable Catalytic Processes*, ed. B. Saha, M. Fan, and J. Wang, 157–177. The Netherlands: Elsevier.

Altieri, P., and R. W. Coughlin. 1987. Characterization of products formed during coliquefaction of lignin and bituminous coal at 400°C. *Energy and Fuels* 1:253–256.

Asia Biomass Office. 2018. Steady expansion of co-firing power generation by woody biomass with coal. https://www.asiabiomass.jp/english/topics/1010_01.html.

Brammer, J. G., M. Lauer, and A. V. Bridgwater. 2006. Opportunities for biomass-derived "bio-oil" in European heat and power markets. *Energy Policy* 34:2871–2880.

Bridgwater, A. V., D. Meier, and D. Radlein. 1999. An overview of fast pyrolysis of biomass. *Organic Geochemistry* 30:1479–1493.

Cao, Q., L. Jin, W. Bao, and Y. Lv. 2009. Investigations into the characteristics of oils produced from co-pyrolysis of biomass and tire. *Fuel Processing Technology* 90:337–342.

Chakravorty, U., M. H. Hubert, and L. Nøstbakken. 2009. Fuel versus food. *Annual Review of Resource Economics* 1:645–663.

Cornelissen, T., J. Yperman, G. Reggers, S. Schreurs, and R. Carleer. 2008. Flash co-pyrolysis of biomass with polylactic acid. Part 1: Influence on bio-oil yield and heating value. *Fuel* 87:1031–1041.

Demirbaş, A., and G. Arin. 2002. An overview of biomass pyrolysis. *Energy Sources* 24:471–482.

Energy Technology Network. 2010. Biomass for Heat and Power. https://iea-etsap.org/E-TechDS/PDF/E05-BiomassforHP-GS-AD-gct.pdf.

Energy Technology Network. 2013. Biomass co-firing in coal power plants. https://iea-etsap.org/E-TechDS/PDF/E21IR_Bio-cofiring_PL_Jan2013_final_GSOK.pdf.

European Biomass Industry Association (EUBIA). 2018. Experiences in europe and list of biomass co-firing plants. http://www.eubia.org/cms/wiki-biomass/co-combustion-with-biomass/european-experiences-in-co-combustion/.

Farzad, S., M. A. Mandegari, and J. F. Görgens. 2016. A critical review on biomass gasification, co-gasification, and their environmental assessments. *Biofuel Research Journal* 3:483–495.

Gollakota, A. R. K., N. Kishore, and S. Gu. 2018. A review on hydrothermal liquefaction of biomass. *Renewable and Sustainable Energy Reviews* 81:1378–1392.

Granatstein, D. L. 2002. Case Study on Lahden Lampovoima Gasification Project Kymijarvi Power Station, Lahti, Finland. http://citeseerx.ist.psu.edu/viewdoc/download?doi=10.1.1.579.5883&rep=rep1&type=pdf.

Guillain, M., K. Fairouz, S. R. Mar, F. Monique, and L. Jacques. 2009. Attrition-free pyrolysis to produce bio-oil and char. *Bioresource Technology* 100:6069–6075.

Guo, Z., Z. Bai, J. Bai, Z. Wang, and W. Li. 2011. Co-liquefaction of lignite and sawdust under syngas. *Fuel Processing Technology* 92:119–125.

Gupta, D., E. Ahmad, K. K. Pant, and B. Saha. 2017. Efficient utilization of potash alum as a green catalyst for production of furfural, 5-hydroxymethylfurfural and levulinic acid from mono-sugars. *RSC Advances* 7:41973–41979.

Hawkins, W. 2011. Forest Energy monitor. https://www.hawkinswright.com/bioenergy/forest-energy-monitor.

Hoehstra, A. Y., P. W Gerbens-Leenes, and T. H. van der Meer. 2010. The water footprint of bio-energy. In *Climate Change and Water: International Perspectives on Mitigation and Adaptation*, ed. J. Smith, C. Howe, and J. Henderson, 81–95. London, UK: International Water Association (IWA) and Denver, CO, USA, American Water Works Association (AWWA) American Water Works Association.

Howaniec, N., and A. Smoliński. 2013. Steam co-gasification of coal and biomass – Synergy in reactivity of fuel blends chars. *International Journal of Hydrogen Energy* 38:16152–16160.

Hua, Z., C. A. I. Zhen-yi, S. Heng-fu, L. E. I. Zhi-ping, W. Zhi-cai, and L. I. Hai-ping. 2011. Co-liquefaction properties of Shenfu coal and rice straw. *Journal of Fuel Chemistry and Technology* 39:721–727.

Ikenaga, N., C. Ueda, T. Matsui, M. Ohtsuki, and T. Suzuki. 2001. Co-liquefaction of micro algae with coal using coal liquefaction catalysts. *Energy and Fuels* 15:350–355.

Jahirul, M. I., M. G. Rasul, A. A. Chowdhury, and N. Ashwath. 2012. Biofuels production through biomass pyrolysis- A technological review. *Energies* 5:4952–5001.

Jeon, M. J., S. J. Choi, K. S. Yoo, C. Ryu, S. H. Park, J. M. Lee, J. K. Jeon, Y. K. Park, and S. Kim. 2011. Copyrolysis of block polypropylene with waste wood chip. *Korean Journal of Chemical Engineering* 28:497–501.

Jeong, H. J., I. S. Hwang, and J. Hwang. 2015. Co-gasification of bituminous coal – pine sawdust blended char with H_2O at temperatures of 750°C–850°C. *Fuel* 156:26–29.

Karaca, F., E. Bolat, and S. Dinçer. 2002. Coprocessing of a Turkish lignite with a cellulosic waste material – 3. A statistical study on product yields and total conversion. *Fuel Processing Technology* 75:117–127.

Karaca, H., and C. Koyunoglu. 2017. Co-liquefaction of Elbistan lignite with manure biomass; part 2 – effect of biomass type, waste to lignite ratio and solid to liquid ratio. *IOP Conference Series: Earth and Environmental Science* 95:042074.

Kumabe, K., T. Hanaoka, S. Fujimoto, T. Minowa, and K. Sakanishi. 2007. Co-gasification of woody biomass and coal with air and steam. *Fuel* 86:684–689.

Kuppens, T., T. Cornelissen, R. Carleer, J. Yperman, S. Schreurs, M. Jans, and T. Thewys. 2010. Economic assessment of flash co-pyrolysis of short rotation coppice and biopolymer waste streams. *Journal of Environmental Management* 91:2736–2747.

Lalvani, S. B., C. B. Muchmore, B. Akash, P. Chivate, J. Koropchak, and C. Chavez. 1991. Lignin-augmented coal depolymerization under mild reaction conditions. *Energy and Fuels* 5:347–352.

Li, J., A. Brzdekiewicz, W. Yang, and W. Blasiak. 2012. Co-firing based on biomass torrefaction in a pulverized coal boiler with aim of 100% fuel switching. *Applied Energy* 99:344–354.

Li, K., R. Zhang, and J. Bi. 2010. Experimental study on syngas production by co-gasification of coal and biomass in a fluidized bed. *International Journal of Hydrogen Energy* 35:2722–2726.

Martínez, J. D., A. Veses, A. M. Mastral, R. Murillo, M. V. Navarro, N. Puy, A. Artigues, J. Bartrolí, and T. García. 2014. Co-pyrolysis of biomass with waste tyres: Upgrading of liquid bio-fuel. *Fuel Processing Technology* 119:263–271.

Nanda, S., R. Azargohar, A. K. Dalai, and J. A. Kozinski. 2015. An assessment on the sustainability of lignocellulosic biomass for biorefining. *Renewable and Sustainable Energy Reviews* 50:925–941.

Narobe, M., J. Golob, D. Klinar, V. Francetič, and B. Likozar. 2014. Co-gasification of biomass and plastics: Pyrolysis kinetics studies, experiments on 100 kW dual fluidized bed pilot plant and development of thermodynamic equilibrium model and balances. *Bioresource Technology* 162:21–29.

Önal, E., B. B. Uzun, and A. E. Pütün. 2012. An experimental study on bio-oil production from co-pyrolysis with potato skin and high-density polyethylene (HDPE). *Fuel Processing Technology* 104:365–370.

Onay, Ö., E. Bayram, and Ö. M. Koçkar. 2007. Copyrolysis of seyitömer-lignite and safflower seed: Influence of the blending ratio and pyrolysis temperature on product yields and oil characterization. *Energy and Fuels* 21:3049–3056.

Pan, Y. G., E. Velo, X. Roca, J. J. Manyà, and L. Puigjaner. 2000. Fluidized-bed co-gasification of residual biomass/poor coal blends for fuel gas production. *Fuel* 79:1317–1326.

Parihar, M. F., M. Kamil, H. B. Goyal, A. K. Gupta, and A. K. Bhatnagar. 2007. An experimental study on pyrolysis of biomass. *Process Safety and Environmental Protection* 85:458–465.

Pinto, F., C. Franco, R. N. André, C. Tavares, M. Dias, I. Gulyurtlu, and I. Cabrita. 2003. Effect of experimental conditions on co-gasification of coal, biomass and plastics wastes with air/steam mixtures in a fluidized bed system. *Fuel* 82:1967–1976.

Popp, J., Z. Lakner, M. Harangi-Rákos, and M. Fári. 2014. The effect of bioenergy expansion: Food, energy, and environment. *Renewable and Sustainable Energy Reviews* 32:559–578.

Quan, C., S. Xu, Y. An, and X. Liu. 2014. Co-pyrolysis of biomass and coal blend by TG and in a free fall reactor. *Journal of Thermal Analysis and Calorimetry* 117:817–823.

Quereshi, S., E. Ahmad, K. K. Pant, and S. Dutta. 2017a. Insights into the metal salt catalyzed ethyl levulinate synthesis from biorenewable feedstocks. *Catalysis Today* 291:187–194.

Quereshi, S., E. Ahmad, K. K. Pant, and S. Dutta. 2017b. Recent advances in production of biofuel and commodity chemicals from algal biomass. In *Algal Biofuels: Recent Advances and Future Prospects*, ed. S. Gupta, A. Malik, and F. Bux. Cham, Germany: Springer.

Rafiqul, I., B. Lugang, Y. Yan, and T. Li. 2000. Study on co-liquefaction of coal and bagasse by factorial experiment design method. *Fuel Processing Technology* 68:3–12.

Ragauskas, A. J., G. T. Beckham, M. J. Biddy, R. Chandra, F. Chen, M. F. Davis, B. H. Davison, R. Dixon, P. Gilna, M. Keller, P. Langan, A. K. Naskar, J. N. Saddler, T. J. Tschaplinski, G. Tuskan, and C. E. Wyman. 2014. Lignin valorization: Improving lignin processing in the biorefinery. *Science* 344:1246843.

Rutkowski, P., and A. Kubacki. 2006. Influence of polystyrene addition to cellulose on chemical structure and properties of bio-oil obtained during pyrolysis. *Energy Conversion Management* 47:716–731.

Shui, H., C. Shan, Z. Cai, Z. Wang, Z. Lei, S. Ren, and C. Pan. 2011. Co-liquefaction behavior of a sub-bituminous coal and sawdust. *Energy* 36:6645–6650.

Stiller, A. H., D. B. Dadyburjor, J. Wann, D. Tian, and J. W. Zondlo. 1996. Co-processing of agricultural and biomass waste with coal. *Fuel Processing Technology* 49:167–175.

Tillman, D. A. 2000. Biomass cofiring: The technology, the experience, the combustion consequences. *Biomass and Bioenergy* 19:365–384.

Tippayawong, N., J. Kinorn, and S. Thavornun. 2008. Yields and gaseous composition from slow pyrolysis of refuse-derived fuels. Energy sources, Part A *Energy Sources, Part A: Recovery, Utilization, and Environmental Effects* 30:1572–1580.

Tumuluru, J. S., C. T. Wright, R. D. Boardman, N. A. Yancey, and S. Sokhansanj. 2011. A review on biomass classification and composition, co-firing issues and pretreatment methods. *American Society of Agricultural and Biological Engineers*. Louisville, Kentucky. doi:10.13031/2013.37191.

van Vuuren, D. P., E. Bellevrat, A. Kitous, and M. Isaac. 2010a. Bio-energy use and low stabilization scenarios. *The Energy Journal* 31:193–221.

van Vuuren, D. P., E. Isaac, M. G. J. den Elzen, E. Stehfest, and J. van Vliet. 2010b. Low stabilization scenarios and implications for major world regions from an integrated assessment perspective. *The Energy Journal* 31:165–191.

Vélez, J. F., F. Chejne, C. F. Valdés, E. J. Emery, and C. A. Londoño. 2009. Co-gasification of colombian coal and biomass in fluidized bed: An experimental study. *Fuel* 88:424–430.

Wei, L., L. Zhang, and S. Xu. 2011. Effects of feedstock on co-pyrolysis of biomass and coal in a free-fall reactor. *Journal of Fuel Chemistry and Technology* 39:728–734.

World Energy Resources. 2016. World Energy Council. https://www.worldenergy.org/wp-content/uploads/2016/10/World-Energy-Resources-Full-report-2016.10.03.pdf.

3 Catalytic Conversion of Lignocellulosic Biomass into Fuels and Value-Added Chemicals

Shireen Quereshi, Suman Dutta, and Tarun Kumar Naiya

CONTENTS

3.1 Introduction 31
3.2 Conventional Technologies for Biomass Conversion 33
3.3 Catalysis in the Production of Value-Added Chemicals and Fuels 34
3.4 Catalytic Production of 5-Hydroxymethylfurfural from Biomass 35
3.5 Effects of Polar and Aprotic Solvents on 5-Hydroxymethylfurfural Production 40
3.6 Catalytic Production of 5-Ethoxymethylfurfural from Biomass 41
3.7 Effects of Solvent for the Production of 5-Ethoxymethylfurfural 44
3.8 Catalytic Production of Levulinic Acid from Biomass 44
3.9 Catalytic Production of Ethyl Levulinate 47
3.10 Conclusions 48
References 48

3.1 INTRODUCTION

Huge gaps in fuel demand and supply have forced the modern world to develop alternative technologies to counter the existing energy problem. Moreover, the excessive exploitation of conventional energy sources causes global warming, severe environmental problems, and geopolitical conflicts (Asif and Muneer 2007). There are other causes and effects of the imminent energy crisis; however, several efforts have been made to enhance energy-saving techniques (Abdelaziz et al. 2011). Furthermore, the demand for chemicals increases hand-in-hand with an increase in energy requirements. These chemicals are broadly classified into major categories namely basic chemicals and specialty chemicals. Basic chemicals are the petrochemicals, which are synthesized from crude oil, polymers, and inorganic chemicals, whereas specialty chemicals include fine chemicals, commodity chemicals, and pharmaceuticals. Sustainable production of the basic and specialty chemicals is essential and crucial for the social and economic development of any industrialized nation. At present, most of these chemicals and energy demands is met from conventional fossil fuel sources. On the contrary, these fossil fuels such as crude oil are not expected to last longer. Thus, an energy crisis is imminent if these issues are not addressed at this stage.

Nevertheless, several renewable sources such as solar energy, wind energy, tidal energy, and biomass have potentials to counter the upcoming energy crisis. Other promising sources of energy include nuclear energy and geothermal energy. However, making chemicals from other sources is difficult except from biomass, which can be used to produce a wide range of value-added chemicals, fuels, and materials. Biomass is a renewable source available in ample amounts in all demographic regions and follows a closed carbon cycle. Indeed, biomass is a clean and green energy source, which can easily be converted to chemicals in a cost-effective manner. When the biomass is burnt or decomposed, it releases energy in the form of heat, biofuel, steam for running turbines, and various other products depending on the mode of conversion technique applied. It burns to give energy and emits CO_2 to the atmosphere, which is further utilized in photosynthesis reaction, thereby making it a closed-loop carbon cycle (Figure 3.1). However, the major drawback associated with these technologies is the conversion efficiency, which needs to be improved. There are several sources of biomass such as agricultural crop residues, forest residue, and energy crops, which can further be subcategorized as follows:

1. Agricultural crop residues: Rice husk, sugarcane bagasse, wheat straw, corncobs, and other agricultural wastes.
2. Forest residues: Wood, dead trees, sawdust, wood chips, and other residues.
3. Energy crops: Grasses, fast-growing plants, and algae.

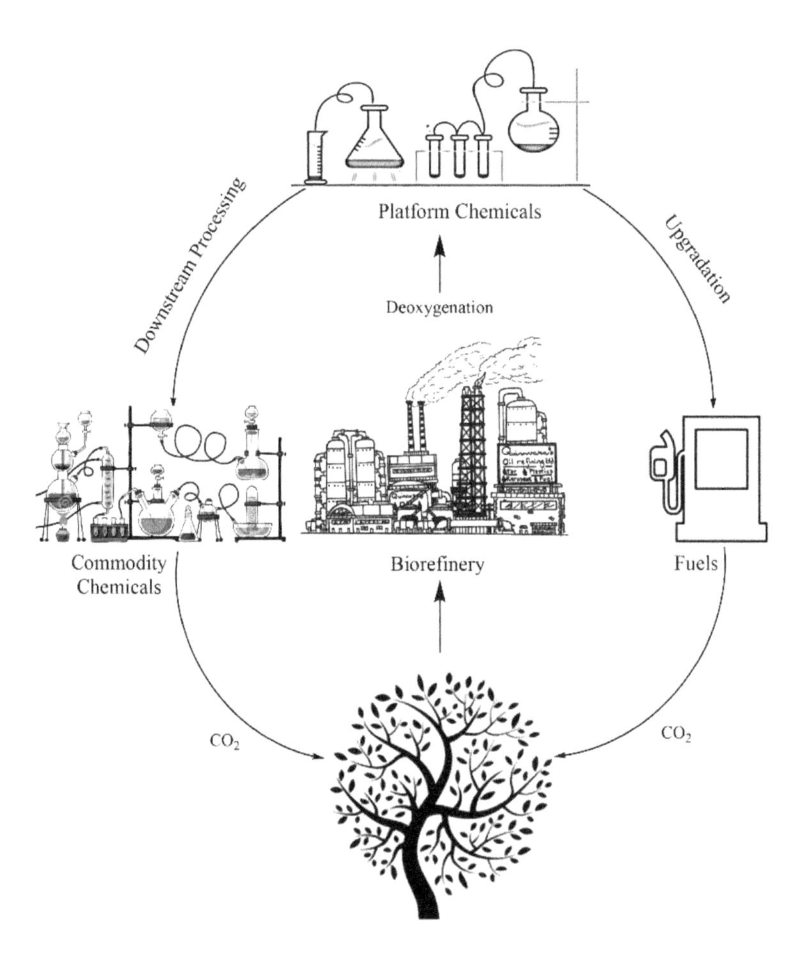

FIGURE 3.1 Closed carbon cycle of biomass.

It is worth mentioning that lignocellulosic biomass is one of the major feedstock varieties readily available worldwide. The lignocellulosic biomass mainly consists of cellulose, hemicellulose, and lignin (Hadar 2013). The polymer consists of C_6 carbon in the form of glucose unit linked with β-1,4 glycosidic bonds. This bonding provides strong structural intermolecular and intramolecular hydrogen and Van der Waals bonding. Hemicellulose is the branched polysaccharide comprised of C_6 sugars or hexoses (glucose, mannose, and galactose) and C_5 sugars or pentoses (xylose, arabinose, and rhamnose). Lignin is an amorphous complex heteropolymer in the plant cell wall that provides elasticity, mechanical strength, resistance, and thermal stability against any decomposition in the cell wall. It is stronger, more stable, and more supportable than hemicellulose and is formed by cross-linking phenylpropane units. The structure of a plant cell is such that cellulose is bound with hemicellulose and lignin (Mosier et al. 2005; Banerjee et al. 2010; Pasangulapati et al. 2012). The biomass having a high concentration of cellulose and hemicellulose is more suitable to produce fuels and chemicals, thus its variation is responsible for optimizing the selectivity and efficiency. The pretreatment of biomass is done before its conversion to break the lignin seal that provides the protective barrier for the plant cell and to decompose the cellulose and hemicellulose structure. Moreover, cellulose and hemicellulose are hydrolyzed with a dilute acid, concentrated acid, and/or microbial activity to process biomass into value-added fuel and chemical products (Hendriks and Zeeman 2009; Kumar et al. 2009).

3.2 CONVENTIONAL TECHNOLOGIES FOR BIOMASS CONVERSION

Industries mainly use sugarcane bagasse, bamboo, rice husk, sorghum, corn, switchgrass, and oil palm for converting into fuels and chemicals. The type of biomass categorizes the production of fuel into first-, second- and third-generation. First-generation fuels utilize food crops like sugarcane, wheat, and corn starch for production of bioethanol, whereas rapeseed oil is utilized for production of biodiesel. However, diversion of food crops towards industrial application for production of biofuel may cause a problem in the food supply and a rise in food prices. Thus, second-generation biofuels are mainly produced from plant wastes, wood, and non-food crop wastes, whereas algae and grasses are considered the source for third-generation biofuels as the energy crops. The various biomass conversion technologies are direct combustion, thermochemical conversion, chemical conversion, and biochemical conversion.

Direct combustion is a technique in which biomass is burnt in the presence of oxygen to produce energy in the form of electricity. It is an exothermic reaction, which generates heat energy from the chemical energy stored in biomass for further application in boilers, furnaces, heat exchangers, steam turbines, and various industrial and household applications. On the contrary, thermochemical conversion is the process that utilizes the heat and chemicals to produce energy products in absence or presence of oxygen. In such processes, the solid biomass is transformed to gases, which are further modified to oil to produce various fuels and chemicals. The thermochemical conversion consists of pyrolysis, gasification, and liquefaction techniques (Mckendry 2002). However, the co-firing is the cost-effective technology usually used in power plants where biomass and coal are burnt to produce energy or biomass is gasified to produce clean fuel and that fuel is burnt with coal for power generation.

Fossil fuels cause the emission of SO_x, NO_x, and other greenhouse gases, although they generate a large amount of power as compared to other feedstocks. Thus, combining fossil fuels and biomass enhances the co-firing technology by reducing the emission of gases as well as increasing the power generation capacity (Zhang et al. 2010). Similarly, liquefaction is a technique in which biomass is processed into biocrude oil under a moderate temperature of 300°C–400°C and a high pressure of 10–20 MPa in the presence of a reducing agent (Alonso et al. 2010). The resulting biomass is depolymerized to its monomeric units, which is further decomposed into smaller molecular weight chemicals through hydrolysis, dehydration, decarboxylation, and other reactions depending on the type of product required.

Similarly, biochemical conversion is a process in which biomass is transformed to biogas or liquid fuels in the presence of the microorganisms and enzymes. The two basic biochemical technologies are anaerobic digestion and fermentation. The anaerobic digestion is notably the bio-methanation technique in which the biodegradable waste from kitchen scraps, crop residues, sewage, and manure are converted into biogas rich in methane. Industrially, it is used for running gas turbines directly as fuel after the removal of CO_2. The residual solid products can be further utilized as fertilizer for the growth of plants, for animal feeding, and in fiberboard as a building product.

Bio-methanation is a multistep process involving biomass hydrolysis and conversion into sugars, amino acids, fatty acids, and glycerol with the help of fermentative bacteria. However, these compounds are further broken down into acetate, hydrogen, ammonia, organic acids, and CO_2 in the presence of acetogenic bacteria. The presence of acetate results in a 70% production of CH_4, whereas 30% comprises of H_2 and CO_2. This process should be maintained at an optimal temperature (5°C–70°C) and optimum pH (6.8–7.4) for the growth of acidogenic and methanogenic microorganisms (Yokoyama and Matsumura 2008). Fermentation in the presence of yeast and bacteria results in ethanol production, which is recovered through distillation, separation, and dehydration for automotive fuel. The various biomass feedstocks for production of ethanol are sugarcane, bagasse, corn, wheat, and other crops and lignocellulosic biomass.

3.3 CATALYSIS IN THE PRODUCTION OF VALUE-ADDED CHEMICALS AND FUELS

Biomass contains a relatively higher content of oxygen and a lower amount of carbon and hydrogen compared to petroleum sources, which make biorefineries unique and more versatile than petroleum refineries. Variation in the content of carbon, hydrogen, and oxygen enables the formation of a wide range of chemicals and fuels. Nevertheless, the high content of oxygen lowers the heat content and increases its polarity towards blending with other fossil fuels. The basic disadvantages associated with biorefineries are their low efficiency, which needs to be improved by deoxygenation and depolymerization of biomass. The deoxygenation is required for removal of oxygen before it is further processed into chemicals. Depolymerization of lignocellulosic biomass is an initial step for converting into its monomers by various conversion technologies as previously described. Nevertheless, various platform chemicals derived require different conversion technologies, and the amounts of cellulose, hemicellulose, and lignin content should be investigated before applying the conversion technologies (Banerjee et al. 2010; Cherubini and Strømman 2011; Isikgor and Becer 2015).

The U.S. Department of Energy has proposed several value-added platform chemicals that can be produced from biomass to maintain the sustainability and techno-economic feasibility of the process as shown in Figure 3.2 (Holladay and White 2004; Holladay et al. 2007). In this chapter, we have focused on the production of important platform chemicals and fuel additives such as 5-hydroxymethylfurfural (HMF), levulinic acid (LA), 5-ethoxymethylfurfural (EMF), and ethyl levulinate (EL). These platform chemicals can be produced directly either from biomass or its derivatives in the presence of a suitable acid catalyst. The catalyst can be homogeneous, heterogeneous, or ionic liquids having Brønsted and Lewis acidity or a combination of both. The function of the catalyst is not only to increase the rates of reactions but also to enhance the product selectivity and yield.

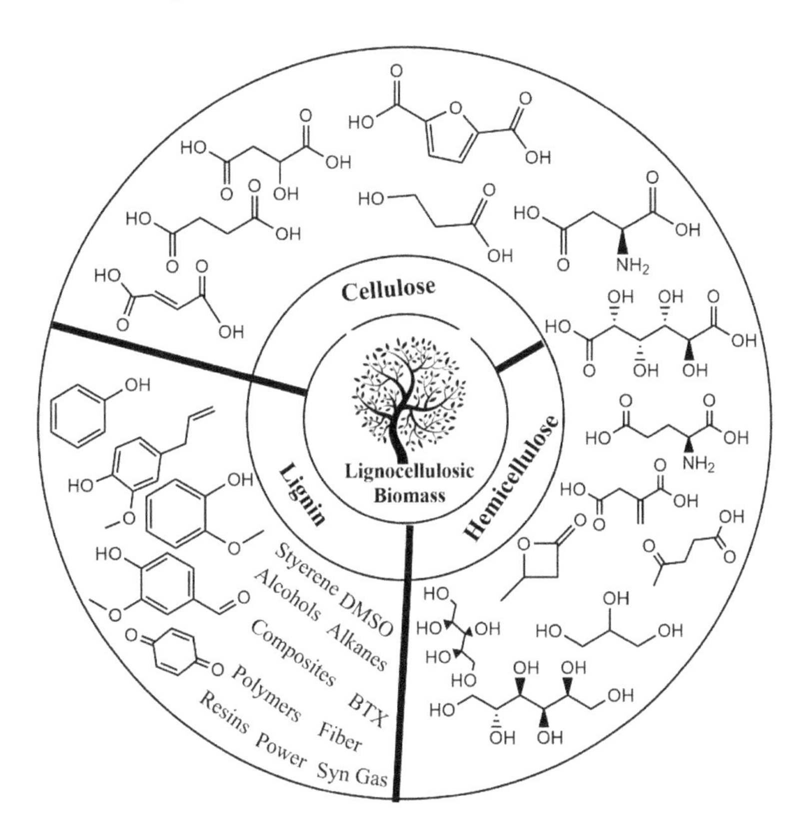

FIGURE 3.2 Value-added chemicals identified by the U.S. Department of Energy that can be produced from biomass.

3.4 CATALYTIC PRODUCTION OF 5-HYDROXYMETHYLFURFURAL FROM BIOMASS

HMF is a potential value-added platform chemical from which various fuels and chemicals can be derived as shown in Figure 3.3. It has wide applications in the production of pharmaceutical products, polymers, biofuels, resins, solvents, and fungicides. It is toxic when consumed in higher concentration; however, small traces of HMF are present in caffeine and dry fruits. In general, HMF is produced directly from lignocellulosic biomass, cellulose, glucose, fructose, or other C_6 sugars in the presence of a suitable acid catalyst. The first step is the depolymerization of lignin, cellulose, and hemicellulose present in the biomass wherein cellulose and some parts of hemicellulose covert to glucose followed by isomerization to fructose, which undergoes catalytic dehydration to yield HMF.

The activity of catalyst, solvent, temperature, and time play a vital role in making the process techno-economically feasible. There are different types of catalysts used for HMF production mainly Brønsted acid, Lewis acid, or a combination of both. Interestingly, isomerization of glucose to fructose requires Lewis acid sites, whereas the conversion of fructose to HMF requires Brønsted acid sites (Ahmad et al. 2016). Therefore, different ratios of Lewis-to-Brønsted acid is required depending on the feed selected for the reaction. In general, inorganic acids, also known as mineral acids, are much stronger than organic acids and are used in HMF production. The Brønsted acids

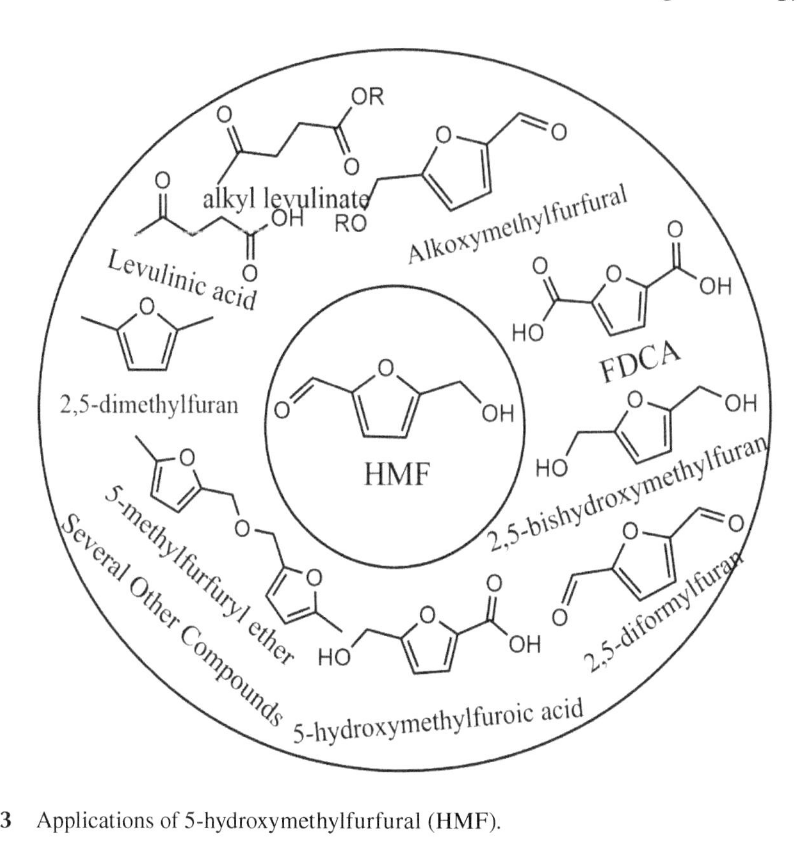

FIGURE 3.3 Applications of 5-hydroxymethylfurfural (HMF).

create an availability of H^+ ions when dissolved in water or other solvents, whereas Lewis acid accepts the lone pair of an electron from another molecule for stabilizing its atom.

On the contrary, application of concentrated mineral acids may lead to corrosion problems in the reactor, catalyst recycling problem, and excessive by-product formation. Therefore, attempts have been made to use mineral acids such as H_2SO_4 in a low quantity. Table 3.1 summarizes different catalysts used for the production of HMF. Recently, Zuo et al. (2017) have reported the production of HMF using a deep eutectic solvent (DES) made from extremely low concentrations of either HCl or H_2SO_4 with choline chloride. Consequently, 88.6% HMF yield was measured in the presence of H_2SO_4 containing DES at 100°C for 240 minutes in an oil bath reactor. Interestingly, when HCl alone was used as the catalyst, a significant HMF yield (57 %) from fructose was found within a short reaction time of 40 minutes at 100°C (Garce et al. 2017).

The catalysts with higher Brønsted acidity enhance the fructose conversion. Nevertheless, other operating factors such as mode of heating and solvents may have contributed towards decreasing the reaction time. However, the major problem with concentrated acids and water is the catalyst separation and recycling, which leads to the assumption that the application of DES may be helpful. On the contrary, a very low HMF yield (3.1%) was measured in the presence of HCl containing DES (Zuo et al. 2017). Interestingly, replacing DES with biphasic solvents (dimethyl sulfoxide or DMSO and water) and the addition of Sn-β-zeolite in the reaction mixture resulted in 63.9% HMF yield from glucose (Jiang et al. 2017). In contrast, when inulin and sucrose were used as the reactants in the presence of HCl containing DES, the HMF yield improved to 61.5% and 38.3%, respectively. This improvement indicates that the structure of the reactant plays a crucial role in the production of HMF.

TABLE 3.1
Catalysts Used for the Production of 5-hydroxymethylfurfural (HMF)

Feed	Catalyst	Temperature (°C)	Time (min)	Solvent	HMF (yield %)	References
Cellulose	[BMIM]Cl-$CrCl_3$	120	300	5% water	80.0	Chiappe et al. (2017)
Cellulose	$InCl_3$/NaCl	200	120	Water/THF	45.0 mol%	Wu et al. (2017)
Fructose	H_2SO_4	100	240	ChCl	88.6	Zuo et al. (2017)
Fructose	HCl	90	40	Water	57.0	Garce et al. (2017)
Fructose	3,3′-(1,2-phenylenebis (methylene) bis(1-benzyl-1H-imidazol-3-ium) hydrogen sulfate (0.2 g)	100	60	DMSO	90.6	Yaman et al. (2017)
Fructose	$AlCl_3$	140	5	DMSO/ NaCl	71.3	De et al. (2011)
Fructose	LaOCl/Nb_2O_5	180	180	Hot water	89.0	Martínez et al. (2017)
Fructose	MDC-SO_3H	120	120	Isopropanol/ DMSO	89.5	Jin et al. (2017)
Fructose	KIT-5-SO_3H	125	45	DMSO	93.9	Najafi and Hamid (2017)
Fructose	KIT-5-Al	125	45	DMSO	88.3	Najafi and Hamid (2017)
Fructose	Amberlyst	160	30	DMSO	90.0	Lu et al. (2017)
Fructose	Ion exchange resin	150	15	Acetone/ water	73.4	Qi et al. (2008)
Glucose	HCl	100	240	ChCl	3.1	Zuo et al. (2017)
Glucose	Sn-CP/HCl	170	240	Water/ DMSO	65.9	Jiang et al. (2017)
Glucose	$CrCl_3$	150	30	ChCl	60.3	Zuo et al. (2017)
Glucose	$AlCl_3$	140	5	DMSO/ NaCl	52.4	De et al. (2011)
Glucose	Al_2B_3	140	120	DMSO	39.9	Zhao et al. (2016)
Glucose	LaOCl/Nb_2O_5	180	180	Hot water	49.0	Martínez et al. (2017)
Glucose	Cr(Salten)-MCM-41-[$(CH_2)_3SO_3$HVIm] HSO_4	140	240	DMSO	50.2	Yuan et al. (2017)
Inulin	HCl	100	240	ChCl	61.5	Zuo et al. (2017)
Inulin	$AlCl_3$	140	5	DMSO/ NaCl	39.2	De et al. (2011)
Microcrystalline cellulose	[PSMIM]HSO_4/ $ZnSO_4.7H_2O$	160	60	Water/THF	58.8	Xuan et al. (2018)
Sucrose	HCl	100	240	ChCl	38.3	Zuo et al. (2017)
Sucrose	$CrCl_3$	150	30	ChCl	69.8	Zuo et al. (2017)
Sucrose	$AlCl_3$	140	5	DMSO/ NaCl	42.5	De et al. (2011)

Similarly, reaction temperature plays an important role in determining HMF yield, thereby creating an opportunity to replace mineral acids with less corrosive acid catalysts. Thus, metal salts containing chloride atoms are found suitable for HMF production. In fact, at the elevated reaction temperature (150°C) both sucrose and glucose yielded more than 60% HMF in the presence of less-corrosive chromium chloride. Interestingly, when these metal chlorides are used with ionic liquids, it is possible to achieve more than 80% HMF yield from cellulose, which otherwise is difficult to convert (Chiappe et al. 2017). Ionic liquids are metal salts which remain liquid at room temperature and commonly referred as green solvents. High thermal stability, low volatility, and low flammability make them useful for applications in biomass conversion reactions.

Ionic liquids have received major attention in research because of their unique property of high thermal stability, low volatility, and low flammability, which help in separation and recycling technologies. Yaman et al. (2017) reported the synthesis of dicationic ionic liquid by substituting in ortho-, meta- and para-positions for production of HMF. It was found that ionic liquids with ortho-substituent resulted in 95.7% fructose conversion, thereby yielding 90.5% HMF at 100°C in 60 minutes. Likewise, the production of 58.8% HMF from microcrystalline cellulose in the presence of acidic ionic liquid and a catalyst is reported in a biphasic system (Xuan et al. 2018).

All the catalysts discussed require a long reaction time, which may lead to less production per shift in any commercial unit. Thus, reducing the reaction time is essential to make the HMF production process more efficient and sustainable. In this context, one possible option is to apply instant or targeted heating of the reaction mixture. Thus, microwave-assisted instant heating methods have been used to produce HMF from various substrates in the presence of $AlCl_3$ and dimethyl sulfoxide (DMSO) as the solvent (De et al. 2011). Consequently, the process yielded 71.3, 52.4, 42.5, and 39.2% HMF from fructose, glucose, sucrose and inulin, respectively, at 140°C in 5 minutes (De et al. 2011). Interestingly, $AlCl_3$ is a strong Lewis acid metal chloride, which also creates Brønsted acidity due to Al^{3+} ions for efficient dehydration of fructose. Therefore, it is hypothesized that the application of another aluminum-containing catalyst may be helpful in the production of HMF due to its significant activity in the dehydration reaction.

Zhao et al. (2016) reported the production of 39.9% HMF from glucose in DMSO media using Al_2B_3 as a catalyst at 140°C in 120 minutes. The $InCl_3$ catalyst possesses Lewis acid sites in the biphasic tetrahydrofuran-water (THF/H_2O) solvent for the production of HMF, whereas NaCl inhibits further rehydration of HMF (Wu et al. 2017). The production of HMF from glucose follows a two-step mechanism such as transformation of a pyranose ring to a furanose ring structure followed by the ring-opening of glucose to isomerize into fructose, which requires 40.1 kilocalories per mole (kcal/mol) of activation energy. On the other hand, the transformation of fructose to HMF occurs through dehydrating three molecules of water, which requires 38.3 kcal/mol of activation energy for retro-aldol reaction (Wang et al. 2017a). The isomerization of glucose to fructose requires more Lewis active sites, whereas the dehydration of fructose to HMF requires Brønsted acidity. Moreover, when glucose or cellulose is the chosen reactant, it is preferable to use a catalyst with both Lewis and Brønsted acidity, which could enhance the HMF yield. The use of modified Lewis acid catalyst (i.e., $LaOCl/Nb_2O_5$ providing both acidity sites) yielded 49–89% of HMF from glucose and fructose, respectively (Martínez et al. 2017).

Several other catalysts containing both Lewis and Brønsted acidities are widely used for the production of HMF from various feedstocks. Wang et al. (2016) found sulfonated carbon catalyst an efficient catalyst for the production of HMF. The metal-organic framework carbons (MDC) prepared from the metal organic framework (MOF) have drawn attention due to their adsorption and separation technologies. The MDC acts as a heterogeneous catalyst, which possesses thermal stability and porous structures. The acidity was enriched by treating with sulfonic group $MDC\text{-}SO_3H$ for creating Brønsted active sites for 89.75% production of HMF from fructose in a biphasic solvent (Jin et al. 2017). Similarly, the sulfonated catalyst prepared using MCM-41 to overcome the disadvantages of a homogeneous catalyst Cr(Salten)-MCM-41-$[(CH_2)_3SO_3HVIm]HSO_4$ yielded 50.2% HMF

(Yuan et al. 2017). Similarly, KIT-5 has ordered mesoporous surfaces, which provides a high surface area, and regular and caged pore sizes (Najafi and Hamid 2017). It consists of interconnected frameworks having active sites for further modification and for enhancing its functionalities. KIT-5-SO_3H prepared by doping with Lewis acidity (Al) and Brønsted acidity (SO_3H) showed 93.9% HMF yield in DMSO from fructose (Najafi and Hamid 2017). Additionally, ion exchange resins also increase the conversion of fructose up to 95.1%, thereby resulting in 73.4% HMF yield in water-acetone media (Qi et al. 2008).

There are several metals and metal oxides that have been widely used for the production of HMF from bio-renewable feedstocks. Table 3.2 summarizes a few metal catalysts used in the production of HMF. For example, niobium (Nb) posses both Brønsted as well as Lewis acidity. Moreover, the Brønsted/Lewis acid ratio is enhanced by doping with other catalysts such as tungsten oxides in different quantities which in turn forms Nb_7W_5. The Nb_7W_5 exhibits higher Brønsted/Lewis acid ratio of 1.83 (Brønsted acidity of 154 μ/mol and Lewis acidity of 84 μ/mol), thereby yielding 52 % HMF from glucose. In contrast, Nb_5W_5 having a 1.26 Brønsted/Lewis acid ratio yielded 47% HMF (Guo et al. 2017). The Brønsted/Lewis acid ratio also can be maintained by niobia/carbon/other supports composition from 0.3 to 4.1 for the conversion of biomass-derived feedstocks to HMF (Li et al. 2018). In this

TABLE 3.2
Production of 5-hydroxymethylfurfural (HMF) in the Presence of Metal Catalysts

Reactant	Catalyst	Temperature (°C)	Time (min)	Solvent	HMF (yield %)	References
Carbohydrate	MeSAPOs-11	170	150	Water/ DMSO	65.1	Sun et al. (2017)
Cellulose	Nb/C-50	170	480	THF/ H_2O-NaCl	53.3	Li et al. (2017)
Cellulose	Fe_3O_4/SBA-15 and 2-ML-SZ/SBA-15	120	360	Isopropanol/ water	43.6	Zhang et al. (2017)
Fructose	Nb-P/SBA-15	160	90	Water/MIBK	92.6	Zhu et al. (2017)
Fructose	Cu-KOMS-2	110	360	DMSO	50.0	Lu et al. (2017)
Fructose	Mesoporous TiO_2 nanoparticles	120	5	DMSO	54.1	Dutta et al. (2011)
Fructose	Ti_7Mo_3	120	60	DMSO	50.3	Qiuyun et al. (2017)
Fructose	Ti_7Mo_3	120	60	Water	21.1	Qiuyun et al. (2017)
Glucose	Nb_7W_5	140	120	2-butanol/ water	52	Guo et al. (2017)
Glucose	NbO/NbP	151.8	120	Water	55.8	Catrinck et al. (2017)
Glucose	Nb/C-50	160	240	THF/ H_2O-NaCl	59.3	Li et al. (2017)
Glucose	BZA-0.20	150	240	DMSO	41.2	Han et al. (2017)
Glucose	PA/TiO_2-ZrO_2	160	240	NaCl/THF	51.3	He et al. (2018)
Glucose	Mesoporous TiO_2 nanoparticles	140	5	DMSO	37.2	Dutta et al. (2011)
Glucose	Ti_7Mo_3	120	180	DMSO	20.3	Qiuyun et al. (2017)
Sucrose	Ti_7Mo_3	140	180	DMSO	37.5	Qiuyun et al. (2017)

regard, the NbO, NbP, and mixture of both NbO/NbP have been investigated for the presence of Lewis and Brønsted active acid sites for the conversion glucose to HMF (Catrinck et al. 2017).

It was found that the mixture of NbO/NbP having a Brønsted/Lewis ratio of 0.758 led to 55.8% HMF (Catrinck et al. 2017). However, an increase in niobia composition leads to an increase in the total active acid sites. For example, Nb/C-50 having a Brønsted/Lewis ratio of about 0.9 showed better catalytic effect than other compositions and yielded 59.3% HMF from glucose at 160°C and 53.3% HMF from cellulose at 170°C, respectively (Li et al. 2018). Similarly, Zhu et al. (2017) reported 92.6% HMF production from fructose at 160°C for 90 minutes in a water/MIBK (methyl isobutyl ketone) media in a volume ratio of 0.5. In Nb-P/SBA-15, the Nb-OH shows Brønsted active sites but Nb^{5+} shows Lewis acid sites due to an unsaturated coordinate. In contrast, Zhang et al. (2017) prepared two mesoporous catalysts from Fe_3O_4 nanoparticles encapsulated in SBA-15 and 2-ML-SZ/SBA-15 that yielded up to 43.6% HMF at 120°C in 360 minutes.

Molecular sieves having metal-containing silicoaluminophosphate such as MeSAPOs-11 show more acid sites with HMF yields approaching 65.1% from carbohydrates (Sun et al. 2017). Similarly, other catalysts such as heteropoly acid having high protonic acidity doped with TiO_2 and ZrO_2 can be useful for enhancing the catalytic activity. For example, the catalyst that formed 5% HPA/TiO_2–ZrO_2 showed 51.3% HMF yield from glucose in the biphasic medium (He et al. 2018). Similarly, B_2O_3/ZrO_2–Al_2O_3 (BZA) solid acid yielded 41.2% HMF from glucose (Han et al. 2017). Similarly, mesoporous TiO_2 nanoparticles used for the dehydration of fructose produced 54.1% HMF by creating Lewis acidity, high surface area, and uniform morphology of TiO_2 (Dutta et al. 2011). Nevertheless, the overall HMF yield decreased drastically when fructose was replaced with glucose as the reactant.

Qiuyun et al. (2017) studied the production of HMF in a batch autoclave from sugars (mainly glucose, fructose and sucrose) with a mesoporous Ti-Mo mixed catalyst in a DMSO medium. Nearly 50.3%–5.2% HMF yields were measured from fructose and glucose, respectively, at 120°C in 60 minutes using 1 gram (g) of DMSO, while 37.5% of HMF from sucrose was obtained at 140°C in 180 minutes. The production of HMF is carried out with different solvents, catalyst concentration, time, temperature, and water amount. It also is found that on increasing the water concentration up to 6 milliliters (mL), the production of HMF reduced to 21.31% at 120°C in 180 minutes (Qiuyun et al. 2017).

3.5 EFFECTS OF POLAR AND APROTIC SOLVENTS ON 5-HYDROXYMETHYLFURFURAL PRODUCTION

Another important factor that plays a vital part in the production of HMF is the effect of solvent. Interestingly, solvents may act as reactant and as a catalyst by providing the dissolved medium for a reaction to take place. Nevertheless, the use of solvents enhances the product quality thermodynamically by eliminating many side reactions. There are two types of polar solvent and non-polar solvent based on their polarity, dipole moment, and the bonding. Polar solvents have a high dielectric constant approximately five times as compared to non-polar solvents. Some non-polar solvents are hexane, benzene, toluene, diethyl ether, and chloroform. Furthermore, polar solvents are classified as protic and aprotic solvents. The protic polar solvents dissociate to provide H^+ ions mainly in subcritical water, whereas the aprotic solvents are solvents having a high dipole moment and their positive species are loosely bonded with a negative dipole. Some of the polar aprotic solvents are ethyl acetate, tetrahydrofuran, dichloromethane, acetone, acetonitrile, dimethylformamide, and dimethylsulfoxide. Acetic acid, n-butanol, isopropanol, n-propanol, ethanol, methanol, formic acid, and water are considered polar protic solvents. Overall, catalyst acidity, solvent, and reactant used are essential for the production of HMF along with the heating system whether it is conventional heating or microwave heating.

The polar protic solvents serve a vital role by providing an aqueous phase for the conversion of lignocellulosic biomass to cellulose followed by the extraction of sugars through the isomerizing of glucose to fructose by providing protons for initiating the reaction (Agmon 1995; Xia et al. 2007; Zhou et al. 2017). In this regard, several reviews and research articles are available that justify the increase in the rate of reaction and selectivity of HMF by the minimal addition of water for the proton transfer through the Grotthuss mechanism (Agmon 1995; Xia et al. 2007; Zhou et al. 2017). In another mechanism, Zhou et al. (2017) proposed that the formation of HMF from glucose follows six steps, such as:

1. Ring opening of glucose
2. Formation of intermediates
3. Closing of fructose ring
4. Protonation step
5. Dehydration of fructose
6. Deprotonation where water provides the pathway for the transfer of proton as well as preventing the side reactions.

Qiuyun et al. (2017) reported 21.1% HMF yield in protic solvent water, although 50.3% HMF yield was measured in an aprotic solvent DMSO at 120°C in 180 minutes and under identical reaction conditions. DMSO increases the selectivity of HMF by eliminating the further rehydration of HMF into other products. Ren et al. (2017) studied the DMSO nature for dehydration of fructose. Fructose exists in a different form when it dissociates in another medium, and accordingly, its stability differs. β-D-fructofuranoses and cis-HMF are reported to have more stable structures in DMSO solvent (Ren et al. 2017). Even when no catalyst is used with DMSO, it shows the catalytic effect on the removal of the first and third water molecule. This catalytic effect is due to the valence unsaturated sulfur and oxygen, and a double bond between sulfur and oxygen. In contrast, when Brønsted acid is used, the H^+ ion also interacts with DMSO to produce $[DMSOH]^+$, which shows high catalytic activity in the removal of all three water molecules (Ren et al. 2017). Nevertheless, it is difficult to separate HMF from DMSO due to its high boiling point. Interestingly, decomposition of sugars present in the lignocellulosic biomass is also possible in the aqueous system. Zhu et al. (2017) used the biphasic solvent system of the aqueous solution and MIBK for the conversion of fructose to HMF. The presence of aqueous solution enhances the solubility, while MIBK increases the selectivity and separation of HMF.

3.6 CATALYTIC PRODUCTION OF 5-ETHOXYMETHYLFURFURAL FROM BIOMASS

5-ethoxymethylfurfural (EMF) is a new and promising building block chemical that either can be used as a fuel or as a feedstock to produce value-added chemicals. It has an energy density (8.7 kilowatt-hour per liter (kWh/L)) equivalent to gasoline (8.8 kWh/L), slightly less than diesel (9.7 kWh/L), and greater than ethanol (6.1 kWh/L) (Alipour et al. 2017). Nevertheless, it can be blended with diesel because of its high boiling point of 274°C, which is higher than gasoline, ethanol, dimethylformamide (DMF), and 2-methylfuran (2-MF) (Bohre et al. 2015). This can improve the blended fuel's overall energy density, stability, and flow properties. Moreover, it reduces the SO_x emissions, particulates, deposition of unburnt carbon in engines, and enhances the smooth running of the engines when blended with diesel. Apart from blending, it also is used in flavoring and aromatic industries, basically in wines and beers (Bohre et al. 2015). Nonetheless, the carbohydrates present in lignocellulosic biomass are the main source of EMF. HMF acts as an intermediate product because the EMF is formed by esterification of HMF in ethanol medium in most of the reaction mechanisms. The conversion of fructose to EMF requires high temperature and longer reaction times compared to the direct esterification of HMF to EMF.

Although EMF is a young member in the family of building-block platform chemicals, a wide range of catalysts has been used in its production from bio-renewable feedstocks. Like other catalytic processes for biomass conversion into value-added chemicals, the initial attempts to produce EMF were made in the presence of H_2SO_4. Table 3.3 gives a list of catalysts used in the production of EMF. Xu (2017) reported the production of both EMF and (EL) from glucose and fructose in the

TABLE 3.3
Catalysts Used in the Production of 5-ethoxymethylfurfural (EMF)

Reactant	Catalyst	Temperature (°C)	Time (min)	Solvent	EMF (yield %)	References
Fructose	H_2SO_4	120	180	Ethanol/hexane	66.29	Xu (2017)
Fructose	Cellulose and H_2SO_4	100	720	Ethanol	72.5	Liu et al. (2013)
Fructose	[Bmim]Cl and $FeCl_3$	100	720	Ethanol	30.1	Zhou et al. (2014)
Fructose	Fe_3O_4/C- SO_3H	100	600	Ethanol/DMSO	64.2	Yao et al. (2016)
Fructose	OMC-SO_3H	100	1440	Ethanol	55.7	Wang et al. (2017a)
Fructose	MIL-101- SO_3H(100)	130	900	Ethanol/THF	67.7	Liu et al. (2016)
Fructose	Lys/PW(2)	120	900	Ethanol/DMSO	76.6	Li et al. (2014)
Fructose	Ag_1H_2PW	100	1440	Ethanol	69.5	Ren et al. (2015)
Fructose	30 wt% K-10 clay –HPW	100	1440	Ethanol	61.5	Liu et al. (2014a)
Fructose	40 wt% MCM-41-HPW	100	1440	Ethanol	42.9	Liu et al. (2014b)
Glucose	Cobalt phthalocyanine and [EMIm]Cl	90	180	Ethanol	80.0	Yadav et al. (2014)
Glucose	OMC-SO_3H	100	1440	Ethanol	26.8	Wang et al. (2017a)
HMF	Cellulose and H_2SO_4	100	600	Ethanol	84.4	Liu et al. (2013)
HMF	Cobalt phthalocyanine and [EMIm]Cl	90	120	Ethanol	92.0	Yadav et al. (2014)
HMF	Fe_3O_4@C- SO_3H	100	600	Ethanol	85.6	Yao et al. (2016)
HMF	PY-PW-1	80	1440	Ethanol	90.0	Wang et al. (2017c)
HMF	Ag_1H_2PW	100	600	Ethanol	88.7	Ren et al. (2015)
HMF	30wt% K-10 clay –HPW	100	600	Ethanol	91.5	Liu et al. (2014a)
HMF	K-10 clay-Al	120	480	Ethanol	89.5	Liu et al. (2015)
HMF	40 wt% MCM-41-HPW	100	720	Ethanol	83.4	Liu et al. (2014b)
Inulin	Fe_3O_4/C- SO_3H	100	600	Ethanol/DMSO	50.1	Yao et al. (2016)
Inulin	OMC-SO_3H	100	1440	Ethanol	53.6	Wang et al. (2017b)
Inulin	MIL-101- SO_3H(100)	130	900	Ethanol/THF	54.2	Liu et al. (2016)
Inulin	Lys/PW(2)	120	900	Ethanol/DMSO	58.5	Li et al. (2014)
Sorbose	Lys/PW(2)	120	900	Ethanol/DMSO	42.4	Li et al. (2014)
Sucrose	Fe_3O_4/C- SO_3H	100	600	Ethanol/DMSO	31.2	Yao et al. (2016)
Sucrose	OMC-SO_3H	100	1440	Ethanol	26.8	Wang et al. (2017c)
Sucrose	Lys/PW(2)	120	900	Ethanol/DMSO	36.5	Li et al. (2014)

presence of a H_2SO_4 catalyst in a media containing both ethanol and hexane at 120°C in a reaction time of 180 minutes. However, considering the corrosive nature of H_2SO_4, many researchers have attempted to use various supports to reduce the overall corrosiveness.

Liu et al. (2013) have studied and explored a combination of cellulose and H_2SO_4 catalysts for the production of EMF and HMF from fructose. Because cellulose itself may react in the presence of H_2SO_4, the application of cellulose-supported H_2SO_4 catalyst needs to be explored further. Interestingly, some reports suggest that ionic liquids with metal salts are effective catalysts for the hydrolysis of cellulose. Thus, cobalt phthalocyanine having Lewis acid character with 1-Ethyl-3-methylimidazolium chloride ([EMIm]Cl) is used for the production of EMF in ethanol media (Yadav et al. 2014). Subsequently, a wide range of ionic liquids has been used for the production of EMF from bio-renewable feedstock. For example, Zhou et al. (2014) reported a combination of [Bmim]Cl and $FeCl_3$ as the catalyst in ethanol medium to produce EMF from fructose. On the contrary, a low yield of EMF (30.1%) was measured, thereby necessitating synthesis and application of more effective and active catalysts.

Interestingly, iron-based catalysts are magnetically separable and very active for hydrolysis and alcoholysis reactions if functionalized properly. For example, SO_3H-functionalized magnetically recovered iron catalysts have been reported to EMF yields of 85.6%, 64.2%, 50.1% and 31.2%percent from HMF, fructose, inulin, and sucrose, respectively, at 100°C in 600 minutes (Yao et al. 2016). The presence of iron helps in the separation of catalysts magnetically, whereas the presence of a functional group (SO_3H) provides the required acidic sites for the reaction to proceed. In addition, the activity of the catalysts also was found to be dependent on the nature of the feedstock. Consequently, several other functionalized catalysts have been reported for the efficient production of EMF from biomass-derived feedstocks. For example, a new hybrid heterogeneous catalyst can be synthesized using an inorganic and organometallic porous structure of polymer having both metal centers attached to the ligand. It may enhance the porous catalyst structure by providing the high surface area and metal sites as well as hydrothermal and chemical stability.

Wang et al. (2017b) studied the sulfonic acid functionalized ordered mesoporous carbon (OMC-SO_3H) effective and stable catalyst for a one-pot synthesis of EMF from low-cost carbohydrates in 1440 minutes at 140°C. The catalyst OMC-SO_3H showed significant production of EMF when fructose and inulin were used as the feedstocks, whereas EMF yield decreased when glucose and sucrose were used as the feedstocks. One possible reason for the reduction in EMF yield can be attributed to the fact that glucose and sucrose essentially require Lewis acidity for isomerization into fructose before their conversion into EMF. Similarly, Liu et al. (2016) studied the effect of MIM-101-SO_3H(x) catalyst by varying the terephthalic and sulfonated terephthalic acid sites from the SO_3H functionalized group. Among all the varying concentration of sulfonic acid groups, MIM-101-SO_3H(100) showed a higher affinity towards the formation of EMF from fructose due to its high Brønsted acidity (1.01 millimoles per gram (mmol/g)) and sulfur content, describing high SO_3H group (2.49 mmol/g) (Liu et al. 2016).

Nevertheless, other catalysts with high Brønsted acidity such as phosphotungstic acid are also very effective for the production of EMF from biomass-derived products (Wang et al. 2017c). Subsequently, the amino acids with phosphotungstic acid creating acid-base bifunctional hybrid catalysts are investigated and found effective for the production of 76.6, 58.5, 42.4, and 36.5% EMF from fructose, inulin, sorbose, and sucrose, respectively, using ethanol and DMSO at 120°C in 900 minutes (Li et al. 2014). Similarly, a silver exchange heterogeneous heteropoly acid is reported to have 88.7% EMF yield from HMF in ethanol medium (Ren et al. 2015). Since silver is a precious and expensive metal, attempts have been made to synthesize heterogenized heteropoly acid catalysts through the addition of supporting material. In this regard, the highest EMF yields of 91.5–61.5% from HMF and fructose, respectively, were measured using 30 wt% K-10 clay-HPW (Liu et al. 2014a).

High Brønsted acidity due to the presence of Keggin heteropoly acid supported over K-10 clay provides high cation exchange and large surface area. The cationic exchange property of K-10 clay was investigated by doping with other metals. It was found that K-10 clay-Al plays a vital role in esterification of HMF to produce 89.5% EMF at 120°C in 480 minutes (Liu et al. 2015). In contrast, when the activity of this catalyst is tested for the production of EMF from fructose, it does not show any significant results in both ethanol and DMSO. Similarly, Liu et al. (2014b) reported significant catalytic activity of 40 wt% MCM-41-HPW in EMF yields of 83.4% and 42.9% from HMF and fructose, respectively.

3.7 EFFECTS OF SOLVENT FOR THE PRODUCTION OF 5-ETHOXYMETHYLFURFURAL

It has been observed that using hexane as a solvent increases the production of EMF, which necessitates further studies on the solvent effect. It is observed that the effect of solvents followed the order: hexane < cyclohexane < toluene < benzene < MIBK < THF < acetone (Xu 2017). This trend was dependent on the polarity of the solvent, which increases with a decrease in production of EMF. However, introducing water as a solvent accelerates the production of LA by inhibiting the esterification of HMF to EMF and EL (Xu 2017). THF as a solvent shows higher affinity towards the formation of EMF in the presence of functionalized catalysts such as MIL-101-SO_3H(100) because it inhibits the interactions of the SO_3H functionalized group and by-product formation (Liu et al. 2016). The β-D-fructofuranoses, trans-HMF, and trans-EMF are in stable forms in ethanol medium. When Brønsted acidity is used, the fructose gets first dehydrated to HMF which is further esterified to EMF in ethanol medium. In the case when the acid catalyst is not used, $[H]^+$ and $[C_2H_5OH_2]^+$ exhibit the catalytic activity for the dehydration of fructose to HMF and esterification of HMF to EMF (Xiang et al. 2017).

3.8 CATALYTIC PRODUCTION OF LEVULINIC ACID FROM BIOMASS

Levulinic acid (LA) is the versatile building block of value-added chemicals and fuel additives and is an intermediate product in the production of EL. The production of LA from lignocellulosic biomass follows (also shown in Figure 3.4) (Morone et al. 2015):

1. Hydrolysis and depolymerization of cellulose to glucose
2. Isomerisation of glucose to fructose due to Lewis acidity
3. Dehydration ($-3H_2O$) of fructose to produce HMF due to Brønsted acidity
4. Rehydration ($+2H_2O$) of HMF to yield LA

LA is used for the production of γ-valerolactone, flavoring agents, EL, and pharmaceuticals agents (Rackemann and Doherty, 2011). Similar to HMF and EMF, the production of LA depends on the catalyst, reaction temperature, solvent, substrate, and solvent media. The various catalysts reported in this regard are metal chlorides, inorganic acids, heteropoly acids, ionic liquids, and various other heterogeneous solid acid catalysts (Fu et al. 2016).

Table 3.4 summarizes a few catalysts used in the production of LA. Various peer reviews are available which show the application of a variety of catalysts such as HCl, H_2SO_4, and ionic liquids that have been used for the production of LA from bio-based sources (Morone et al. 2015). Fu et al. (2016) reported the LA production from 18 different types of ionic liquids consisting of haloids, hydrogen sulfates, methanesulfonates, tosylates, trifluoromethane sulfonates, acetates, and formates. The catalytic activity was in the order of $Cl^- > HSO_4^- > CH_3SO_3^-$ indicating that haloids and hydrogensulfate ionic liquids show a high affinity towards the production of LA using fructose, glucose, and cellulose. Moreover, the ionic liquid [BSO_3HMIm]HSO_4 yielded 60.8% and 54.5% LA from glucose and cellulose, respectively, but is limited by recycling and separation issues. On the contrary, [$PrSO_3$HMIm]Cl showed 78.6%, 70.5%, and 65.1% yields from fructose, glucose, and cellulose, respectively, at 180°C in 180 minutes.

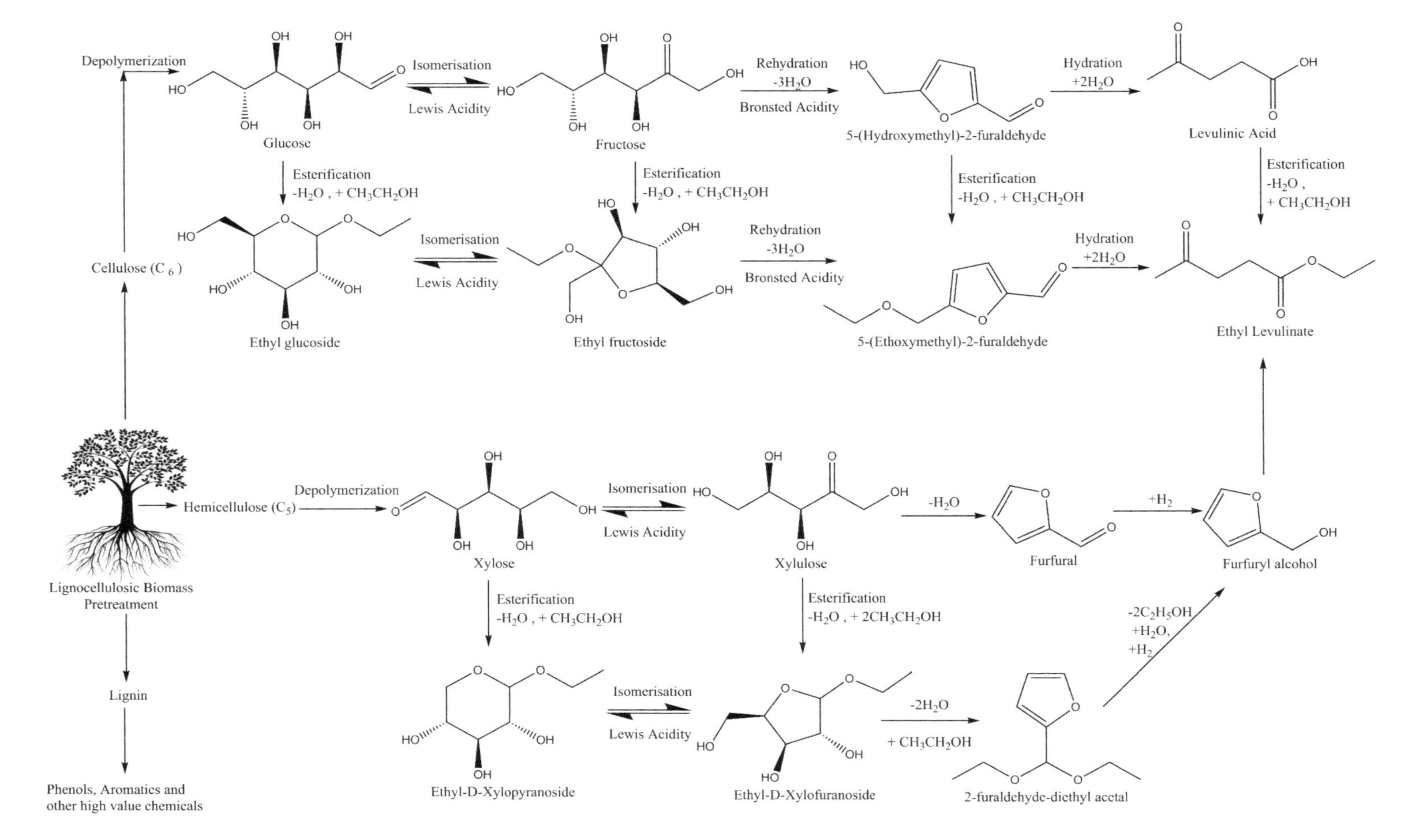

FIGURE 3.4 Reaction mechanism and type of acidity required to produce 5-hydroxymethylfurfural (HMF), 5-ethoxymethylfurfural (EMF), levulinic acid (LA), and ethyl levulinate (EL) from various substrates.

TABLE 3.4
Catalysts Used in the Production of Levulinic Acid (LA)

Reactant	Catalyst	Temperature (°C)	Time (min)	Solvent	Levulinic acid (yield %)	References
Cellobiose	SA- SO_3H	180	720	Water	58.9	Shen et al. (2017)
Cellulose	$[PrSO_3HMIm]Cl$	180	180	Water	65.1	Fu et al. (2016)
Cellulose	$CrCl_3$-HY zeolite	100	120	[EMIM]Cl	46	Aishah and Amin (2013)
Cellulose	SA- SO_3H	180	720	Water	46	Shen et al. (2017)
Empty fruit bunch	$CrCl_3$-HY zeolite	100	120	[EMIM]Cl	20	Aishah and Amin (2013)
Fructose	$[PrSO_3HMIm]Cl$	180	180	Water	78.6	Fu et al. (2016)
Fructose	HCl	120	180	Water	83	Garce et al. (2017)
Fructose	HCl	95	90	Water	<60	Alipoura and Omidvarborna (2016)
Fructose	Amberlyst-15	140	480	Water	52.9	Acharjee and Lee (2018)
Furfuryl alcohol	H-ZSM-5	120	30	THF/water	>70	Mellmer et al. (2015)
Glucose	$[PrSO_3HMIm]Cl$	180	180	Water	70.5	Fu et al. (2016)
Glucose	Amberlyst-15	140	1440	Water	37	Acharjee and Lee (2018)
Glucose	Amberlyst-15/Sn-β	140	1440	Water	44	Acharjee and Lee (2018)
Glucose	SA-SO_3H	180	720	Water	61.3	Shen et al. (2017)
Glucose	Fe-NbP	180	180	Deionized water	64.2	Liu et al. (2017)
Kenaf	$CrCl_3$-HY zeolite	100	120	[EMIM]Cl	17	Aishah and Amin (2013)
Microcrystalline cellulose	$[P_4PS]_3PW_{12}O_{40}$	150	300	Water	18.1	Song et al. (2016)
Microcrystalline cellulose	SA-SO_3H	180	720	Water	51.5	Shen et al. (2017)
Starch	SA-SO_3H	180	720	Water	54.3	Shen et al. (2017)
Vegetable waste	Amberlyst-36	150	5	DMSO/water	17	Chen et al. (2017)
Xylose	AZY0.25	170	180	Hot compressed water	30	Chamnankid et al. (2014)

Several other heteropolyanions-based ionic liquids are used in the production of LA, althought $[P_4PS]_3PW_{12}O_{40}$ was found to be more effective in water (Song et al. 2016). Therefore, attempts have been made to use a combination of ionic liquids and heterogeneous catalysts for the efficient production of LA from renewable feedstocks. However, no significant LA yield was obtained, thereby necessitating the need for new catalysts (Aishah and Amin 2013). Nevertheless, a combination of metal salts, zeolite, and ionic liquids yielded more LA compared to that from $[P_4PS]_3PW_{12}O_{40}$ alone. Interestingly, the homogeneous catalysts are found to be more effective for LA production from biomass. For example, a multi-step production of more than 60% LA was obtained from corn stover using HCl as the catalyst in deionized water within 30 minutes (Alipoura and Omidvarborna 2016). A further enhancement of LA yields up to 80% was obtained with an increase in the reaction temperature up to 120°C. Similarly, when biomass was pretreated with phosphoric acid, hydrochloric acid, and ionic liquid to break the lignin seal and extract cellulose, maximum productions of 40.1%,

49.2% and 60.7% LA were obtained (Muranaka et al. 2014). According to Schmidt et al. (2017), there are two pretreatment steps when lignocellulosic biomass is used directly for the production of LA, as follows:

1. Thermal pretreatment or hydrolysis for separating cellulose, hemicellulose and lignin
2. Conversion of cellulose and hemicellulose (C_6 sugars) by depolymerizing to glucose

Nevertheless, the separation of products and homogeneous catalysts is a challenging task. Thus, recent efforts have been directed towards synthesis and application of heterogeneous catalysts for LA production from various bio-renewable feedstocks. The various types of solid acid catalysts such as ZSM-5, Nafion SAC-13, and Amberlyst-15 have been used for the production of LA. Furthermore, their activity also can be enhanced by doping with other precursors and with NaI or NaCl. The increment in LA yield (44%) was observed while using Amberlyst-15 with Sn-BETA, whereas pure Amberlyst-15 yielded only 37% LA from glucose (Acharjee and Lee 2018). Interestingly, Amberlyst-36 having Brønsted acidity was found to produce up to 17% of LA from food waste at 150°C in 5 minutes (Chen et al. 2017). Similarly, furfuryl alcohol was converted to 70% LA by using HZSM-5 in monophasic THF and water solvent. The activity for producing 70% yield of LA from furfuryl alcohol was due to the enhanced structural property of HZSM-5 catalyst, proper ratio of THF and water (4:1), and a Brønsted acid site (Mellmer et al. 2015). However, a decrease in LA yield was observed by using only water and only THF. The alkaline treatment of zeolite in hot compressed water increased LA yields up to 30% at 170°C in 180 minutes, a production rate nearly five times more than pure zeolite under same optimum conditions (Chamnankid et al. 2014).

Recently, a new carbon composite catalyst was synthesized by sucralose (SA) or sucrose (SO) consisting of chlorine and SO_3H groups, which provide suitable acidity. Accordingly, SA-SO_3H and SO-SO_3H are used for LA production. SA-SO_3H showed up to 51.5% LA yield from microcrystalline cellulose due to the chlorine group binding domain, which is present in SA-SO_3H (Shen et al. 2017). Similarly, the incorporated iron in niobium phosphate also creates the higher acid density (3.59 mmol/g), active sites (11.77 μmol/m^2), and optimum Brønsted-to-Lewis acid ratio (1.32) (Liu et al. 2017). The niobium pentoxide is considered to be a stable acid catalyst, and it is doped with other oxides to increase its effect on acidity and surface area. Fe-NbP shows 64.2% production due to its hydrothermally synthesized nature, acidic effects, and surface area which increases by treating with phosphoric acid.

3.9 CATALYTIC PRODUCTION OF ETHYL LEVULINATE

EL is another interesting fuel additive and building block chemical. Furthermore, it can be used as a potential candidate to replace carcinogenic aromatics from gasoline. In general, EL is a better additive than other alkyl levulinates such as methyl levulinate and butyl levulinate. The methyl levulinate is soluble in water and it separates out from gasoline when blended with other fuels in the cold weather conditions. Similarly, the use of butyl levulinate requires a modification in running engines. On the contrary, EL is one of the most promising fuel additives because of its reduction in cloud point, pour point, and cold filter plugging point when blended with cottonseed oil methyl ester (CSME) and poultry fat methyl ester (PFME). It also improves the induction period with a decrease in kinematic viscosity and flash point (Joshi et al. 2011).

Besides its application as a fuel additive, it is primarily used in fragrance and flavoring industries and in the production of diphenolic esters, plasticizers, resins, solvents, and other fuels and additives. Interestingly, the production of EL is analogous to the production of LA, which requires similar catalysts and conditions except that the water is replaced by ethanol as shown in Figure 3.4. A critical review by Ahmad et al. (2016) can be referred to for a detailed understanding of EL production from various substrates and in the presence of various catalysts. There are various other process parameters like temperature, time, and solvent that affect the route for EL production (Quereshi et al. 2016). The reaction mechanism changes according to various reactant and catalyst chosen for the reaction.

3.10 CONCLUSIONS

Biomass-derived chemicals and fuel have become a necessity of the current society to counter environmental damages as well as the growing demand for chemicals. Interestingly, biomass-derived HMF, EMF, LA, and EL have emerged as the most promising and highly reported building-block platform chemicals. Overall, the production of HMF, EMF, LA, and EL requires acid catalysts, which may be further categorized into Brønsted and Lewis acid catalysts. Furthermore, the requirement of an optimal Brønsted-to-Lewis acid ratio depends upon the starting feedstock material. A reactant may undergo dehydration reaction and not isomerization such as fructose. Therefore, catalysts with high Brønsted acidity will be more effective. On the contrary, the reactants which undergo isomerization reaction such as glucose require both Brønsted and Lewis acidities for efficient production of HMF, EMF, LA, and EL. Furthermore, the operating conditions such as reaction temperature, reaction time, feed concentration, catalysts concentration, solvent volume, and the type of reactor used also affect the overall EL yield. Nevertheless, the type of catalysts and their acidity remain a dominating factor in determining the overall efficiency of the biomass conversion process.

REFERENCES

Abdelaziz, E. A., R. Saidur, and S. Mekhilef. 2011. A review on energy saving strategies in industrial sector. *Renewable and Sustainable Energy Reviews* 15:150–168.

Acharjee, T. C., and Y. Y. Lee. 2018. Production of levulinic acid from glucose by dual solid-acid catalysts. *Environmental Progress & Sustainable Energy* 37:471–480.

Agmon, N. 1995. The Grotthuss mechanism. *Chemical Physics Letters* 244:456–462.

Ahmad, E., M. I. Alam, K. K. Pant, and M. A. Haider. 2016. Catalytic and mechanistic insights into the production of ethyl levulinate from biorenewable feedstocks. *Green Chemistry* 18:4804–4823.

Aishah, N., and S. Amin. 2013. Catalytic conversion of lignocellulosic biomass to levulinic acid in ionic liquid. *Bioresources* 8:5761–5772.

Alipour, S., and H. Omidvarborna. 2016. High concentration levulinic acid production from corn stover. *RSC Advances* 6:111616–111621.

Alipour, S., H. Omidvarborna, and D. Kim, D. 2017. A review on synthesis of alkoxymethyl furfural, a biofuel candidate. *Renewable and Sustainable Energy Reviews* 71:908–926.

Alonso, D. M., J. Q. Bond, and J. Dumesic 2010. Catalytic conversion of biomass to biofuels. *Green Chemistry* 12:1493–1513.

Asif, M., and T. Ã. Muneer. 2007. Energy supply, its demand and security issues for developed and emerging economies. *Renewable and Sustainable Energy Reviews* 11:1388–1413.

Banerjee, S., S. Mudliar, R. Sen, B. Giri, D. Satpute, T. Chakrabarti, and R. A. Pandey. 2010. Commercializing lignocellulosic bioethanol: Technology bottlenecks. *Biofuels, Bioproducts and Biorefining* 4:77–93.

Bohre, A., S. Dutta, B. Saha, and M. M. Abu-omar. 2015. Upgrading furfurals to drop-in biofuels: an overview. *ACS Sustainable Chemistry & Engineering* 3:1263–1277.

Catrinck, M. N., E. S. Ribeiro, R. S. Monteiro, R. M. Ribas, M. H. P. Barbosa, and R. F. 2017. Direct conversion of glucose to 5-hydroxymethylfurfural using a mixture of niobic acid and niobium phosphate as a solid acid catalyst. *Fuel* 210:67–74.

Chamnankid, B., C. Ratanatawanate, and K. Faungnawakij. 2014. Conversion of xylose to levulinic acid over modified acid functions of alkaline-treated zeolite Y in hot-compressed water. *Chemical Engineering Journal* 258:341–347.

Chen, S. S., I. K. M. Yu, D. C. W. Tsang, A. C. K. Yip, E. Khan, L. Wang, Y. S. Ok, and C. S. Poon. 2017. Valorization of cellulosic food waste into levulinic acid catalyzed by heterogeneous Brønsted acids: Temperature and solvent effects. *Chemical Engineering Journal* 327:328–335.

Cherubini, F., and A. H. Strømman. 2011. Chemicals from lignocellulosic biomass: Opportunities, perspectives, and potential of biorefinery systems. *Biofuels, Bioproducts and Biorefining* 5:548–561.

Chiappe, C., M. Jesus, R. Douton, A. Mezzetta, C. S. Pomelli, G. Assanelli, and A. R. de Angelis. 2017. Recycle and extraction: Cornerstones for an efficient conversion of cellulose into 5-hydroxymethylfurfural in ionic liquids. *ACS Sustainable Chemistry & Engineering* 5:5529–5536.

Rackemann, D. W., and W. O. Doherty. 2011. The conversion of lignocellulosics to levulinic acid. *Biofuels, Bioproducts and Biorefining* 5:198–214.

De, S., S. Dutta, and B. Saha. 2011. Green chemistry microwave assisted conversion of carbohydrates and biopolymers to 5-hydroxymethylfurfural with aluminium chloride catalyst in water. *Green Chemistry* 13:2859–2868.

Dutta, S., S. De, A. K. Patra, M. Sasidharan, A. Bhaumik, and B. Saha. 2011. General microwave assisted rapid conversion of carbohydrates into 5-hydroxymethylfurfural catalyzed by mesoporous TiO_2 nanoparticles. *Applied Catalysis A: General* 409–410:133–139.

Fu, J., X. Xu, X. Lu, and X. Lu. 2016. Hydrothermal decomposition of carbohydrates to levulinic acid with catalysis by ionic liquids. *Industrial and Engineering Chemistry Research* 55:11044–11051.

Garce, D., D. Eva, and S. Ordo. 2017. Aqueous phase conversion of hexoses into 5-hydroxymethylfurfural and levulinic acid in the presence of hydrochloric acid: Mechanism and kinetics. *Industrial and Engineering Chemistry Research* 56:5221–5230.

Guo, B., L. Ye, G. Tang, L. Zhang, B. Yue, and C. Edman. 2017. Effect of Brønsted/Lewis acid ratio on conversion of sugars to 5-hydroxymethylfurfural over mesoporous Nb and Nb-W oxides. *Chinese Journal of Chemistry* 35:1529–1539.

Hadar, Y. 2013. Sources for lignocellulosic raw materials for the production of ethanol. In *Lignocellulose Conversion: Enzymatic and Microbial Tools for Bioethanol Production*, ed. V. Faraco, 21–39. Heidelberg, Germany: Springer.

Han, B., P. Zhao, R. He, T. Wu, and Y. Wu. 2017. Catalytic conversion of glucose to 5-hydroxymethyfurfural over B_2O_3 supported solid acids catalysts. *Waste and Biomass Valorization*. doi:10.1007/s12649-017-9971-4.

He, R., X. Huang, P. Zhao, B. Han, T. Wu, and Y. Wu. 2018. The synthesis of 5-hydroxymethylfurfural from glucose in biphasic system by phosphotungstic acidified titanium—Zirconium dioxide. *Waste and Biomass Valorization* 9:657–668.

Hendriks, A. T. W. M., and G. Zeeman. 2009. Pretreatments to enhance the digestibility of lignocellulosic biomass. *Bioresource Technology* 100:10–18.

Holladay J., and J. White (PNNL); Amy Manheim (DOE-HQ). 2004. Top value added chemicals from biomass Volume I —Results of screening for potential candidates from sugars and synthesis gas. U.S. Department of Energy. Pacific Northwest National Laboratory, Oak Ridge, TN.

Holladay, J. E., J. F. White, J. J. Bozell, and D. Johnson. 2007. Top value-added chemicals from biomass Volume II—Results of screening for potential candidates from biorefinery lignin. U.S. Department of Energy. Pacific Northwest National Laboratory, Oak Ridge, TN.

Isikgor, F. H., and C. R. Becer. 2015. Lignocellulosic biomass: A sustainable platform for production of bio-based chemicals and polymers. *Polymer Chemistry* 6:4497–4559.

Jiang, N., W. Qi, Z. Wu, R. Su, and Z. He. 2017. "One-pot" conversions of carbohydrates to 5-hydroxymethylfurfural using Sn-ceramic powder and hydrochloric acid. *Catalysis Today* 302:94–99.

Jin, P., Y. Zhang, Y. Chen, J. Pan, X. Dai, and M. Liu. 2017. Facile synthesis of hierarchical porous catalysts for enhanced conversion of fructose to 5-hydroxymethylfurfural. *Journal of the Taiwan Institute of Chemical Engineers* 75:59–69.

Joshi, H., B. R. Moser, J. Toler, W. F. Smith, and T. Walker. 2011. Ethyl levulinate: A potential bio-based diluent for biodiesel which improves cold flow properties. *Biomass and Bioenergy* 35:3262–3266.

Kumar, P., D. M. Barrett, M. J. Delwiche, and P. Stroeve. 2009. Methods for pretreatment of lignocellulosic biomass for efficient hydrolysis and biofuel production. *Industrial and Engineering Chemistry Research* 48:3713–3729.

Li, H., K. Santosh, R. Kotni, and S. Shunmugavel. 2014. Direct catalytic transformation of carbohydrates into 5-ethoxymethylfurfural with acid—Base bifunctional hybrid nanospheres. *Energy Conversion and Management* 88:1245–1251.

Li, X., K. Peng, Q. Xia, X. Liu, and Y. Wang. 2018. Efficient conversion of cellulose into 5-hydroxymethylfurfural over niobia/carbon composites. *Chemical Engineering Journal* 332:528–536.

Liu, A., B. Liu, Y. Wang, R. Ren, and Z. Zhang. 2014a. Efficient one-pot synthesis of 5-ethoxymethylfurfural from fructose catalyzed by heteropolyacid supported on K-10 clay. *Fuel* 117:68–73.

Liu, A., Z. Zhang, Z. Fang, B. Liu, and Z. Huang. 2014b. Synthesis of 5-ethoxymethylfurfural from 5-hydroxymethylfurfural and fructose in ethanol catalyzed by MCM-41 supported phosphotungstic acid. *Journal of Industrial and Engineering Chemistry* 20:1977–1984.

Liu, B., Z. Gou, A. Liu, and Z. Zhang. 2015. Synthesis of furan compounds from HMF and fructose catalyzed by aluminum-exchanged K-10 clay. *Journal of Industrial and Engineering Chemistry* 21:338–339.

Liu, B., Z. Zhang, and K. Huang. 2013. Cellulose sulfuric acid as a bio-supported and recyclable solid acid catalyst for the synthesis of 5-hydroxymethylfurfural and 5-ethoxymethylfurfural from fructose. *Cellulose* 20:2081–2089.

Liu, X., H. Li, H. Pan, H. Zhang, S. Huang, K. Yang, and W. Xue. 2016. Efficient catalytic conversion of carbohydrates into 5-ethoxymethylfurfural over MIL-101-based sulfated porous coordination polymers. *Journal of Energy Chemistry* 25:523–530.

Liu, Y., H. Li, J. He, W. Zhao, T. Yang, and S. Yang. 2017. Catalytic conversion of carbohydrates to levulinic acid with mesoporous niobium-containing oxides. *Catalysis Communications* 93:20–24.

Lu, X., H. Zhao, W. Feng, and P. Ji. 2017. A non-precious metal promoting the synthesis of 5-hydroxymethylfurfural. *Catalysts* 7:1–10.

Martínez, J. J., D. F. Silva, E. X. Aguilera, H. A. Rojas, M. H. Brijaldo, F. B. Passos, and G. P. Romanelli. 2017. Dehydration of glucose to 5-hydroxymethylfurfural using $LaOCl/Nb_2O_5$ catalysts in hot compressed water conditions. *Catalysis Letters* 147:1765–1744.

Mckendry, P. 2002. Energy production from biomass (Part 1): Overview of biomass. *Bioresource Technology* 83:37–46.

Mellmer, M. A., J. M. R. Gallo, D. M. Alonso, and J. A. Dumesic. 2015. Selective production of levulinic acid from furfuryl alcohol in thf solvent systems over H-ZSM-5. *ACS Catalysis* 5:3354–3359.

Morone, A., M. Apte, and R. A. Pandey. 2015. Levulinic acid production from renewable waste resources: Bottlenecks, potential remedies, advancements and applications. *Renewable and Sustainable Energy Reviews* 51:548–565.

Mosier, N., C. Wyman, B. Dale, R. Elander, Y. Y. Lee, M. Holtzapple, and M. Ladisch. 2005. Features of promising technologies for pretreatment of lignocellulosic biomass. *Bioresource Technology* 96:673–686.

Muranaka, Y., T. Suzuki, H. Sawanishi, I. Hasegawa, and K. Mae. 2014. Effective production of levulinic acid from biomass through pretreatment using phosphoric acid, hydrochloric acid, or ionic liquid. *Industrial and Engineering Chemistry Research* 53:11611–11621.

Najafi, A., and C. Hamid. 2017. The catalytic effect of Al-KIT-5 and KIT-5-SO_3H on the conversion of fructose to 5-hydroxymethylfurfural. *Research on Chemical Intermediates* 43:5507–5521.

Pasangulapati, V., K. D. Ramachandriya, A. Kumar, M. R. Wilkins, C. L. Jones, and R. L. Huhnke. 2012. Effects of cellulose, hemicellulose and lignin on thermochemical conversion characteristics of the selected biomass. *Bioresource Technology* 114:663–669.

Qi, X., M. Watanabe, M. Aida, and R. L. Smith. 2008. Catalytic dehydration of fructose into 5-hydroxymethylfurfural by ion-exchange resin in mixed-aqueous system by microwave heating. *Green Chemistry* 10:799–805.

Qiuyun, Z., W. Fangfang, L. Dan, M. Peihua, and Z. Yutao. 2017. Mesoporous Ti-Mo mixed oxides catalyzed transformation of carbohydrates into 5-hydroxymethylfurfural. *China Petroleum Processing and Petrochemical Technology* 19:26–32.

Quereshi S., E. Ahmad, K. K. Pant, and S. Dutta. 2016. Insights into the metal salt catalyzed ethyl levulinate synthesis from biorenewable feedstocks. *Catalysis Today* 291:187–194.

Ren, L. K., L. F. Zhu, T. Qi, J. Q. Tang, H. Q. Yang, C. W. Hu. 2017. Performance of dimethylsulfoxide and brønsted acid catalysts in fructose conversion to 5-hydroxymethylfurfural. *ACS Catalysis* 7:2199–2212.

Ren, Y., B. Liu, Z. Zhang, and J. Lin. 2015. Silver-exchanged heteropolyacid catalyst (Ag_1H_2PW): An efficient heterogeneous catalyst for the synthesis of 5-ethoxymethylfurfural from 5-hydroxymethylfurfural and fructose. *Journal of Industrial and Engineering Chemistry* 21:1127–1131.

Schmidt, L. M., L. D. Mthembu, P. Reddy, N. Deenadayalu, M. Kaltschmitt, and I. Smirnova. 2017. Levulinic acid production integrated into a sugarcane bagasse based biorefinery using thermal-enzymatic pretreatment. *Industrial Crops and Products* 99:172–178.

Shen, F., R. L. Smith, L. Li, L. Yan, and X. Qi. 2017. Eco-friendly method for efficient conversion of cellulose into levulinic acid in pure water with cellulase-mimetic solid acid catalyst. *ACS Sustainable Chemistry & Engineering* 5:2421–2427.

Song, C., S. Liu, X. Peng, J. Long, W. Lou, and X. Li. 2016. Catalytic conversion of carbohydrates to levulinate ester over heteropolyanion-based ionic liquids. *ChemSusChem* 8:3307–3316.

Sun, X., J. Wang, J. Chen, J. Zheng, H. Shao, and C. Huang. 2017. Dehydration of fructose to 5-hydroxymethylfurfural over MeSAPOs synthesized from bauxite. *Microporous and Mesoporous Materials* 259:238–243.

Wang, J., J. Xi, Q. Xia, X. Liu, and Y. Wang. 2017a. Recent advances in heterogeneous catalytic conversion of glucose to 5-hydroxymethylfurfural via green routes. *Green Chemistry* 60:870–886.

Wang, J., L. Zhu, Y. Wang, H. Cui, Y. Zhang, and Y. Zhang. 2016. Fructose dehydration to 5-hmf over three sulfonated carbons: effect of different pore structures. *Chemical Technologyand Biotechnology* 92:1454–1463.

Wang, J., Z. Zhang, S. Jin, and X. Shen. 2017b. Efficient conversion of carbohydrates into 5-hydroxylmethylfurfan and 5-ethoxymethylfurfural over sulfonic acid-functionalized mesoporous carbon catalyst. *Fuel* 192:102–107.

Wang, Z., H. Li, C. Fang, W. Zhao, T. Yang, and S. Yang. 2017c. Simply assembled acidic nanospheres for efficient production of 5-ethoxymethylfurfural from 5-hydromethylfurfural and fructose. *Energy Technology* 5:2046–2054.

Wu, W. Q., and S. B. Wu. 2017. Conversion of Eucalyptus cellulose into 5-hydroxymethylfurfural using Lewis acid catalyst in biphasic solvent system. *Waste and Biomass Valorization* 8:1303–1311.

Xia, Y., Y. Liang, Y. Chen, M. Wang, L. Jiao, F. Huang, and S. Liu. 2007. An unexpected role of a trace amount of water in catalyzing proton transfer in phosphine-catalyzed (3 + 2) cycloaddition of allenoates and alkenes. *Journal of American Chemical Society* 8:3470–3471.

Xiang, B., Y. Wang, T. Qi, H. Yang, and C. Hu. 2017. Promotion catalytic role of ethanol on Brønsted acid for the sequential dehydration-etherification of fructose to 5-ethoxymethylfurfural. *Journal of Catalysis* 352:586–598.

Xu, G. 2017. One-pot ethanolysis of carbohydrates to promising biofuels: 5-ethoxymethylfurfural and ethyl levulinate. *Asia Pacific Journal of Chemical Engineering* 12:527–535.

Xuan, Y., R. He, B. Han, T. Wu, and Y. Wu. 2018. Catalytic conversion of cellulose into 5-hydroxymethylfurfural using [PSMIM]HSO_4 and $ZnSO_4.7H_2O$ co-catalyst in biphasic system. *Waste and Biomass Valorization* 9:401–408.

Yadav, K. K., S. Ahmad, and S. M. S. Chauhan. 2014. Chemical elucidating the role of cobalt phthalocyanine in the dehydration of carbohydrates in ionic liquids. *Journal of Molecular Catalysis. A, Chemical* 394:170–176.

Yaman, S. M., S. Mohamad, and N. S. A. Manan. 2017. How do isomeric ortho, meta and para dicationic ionic liquids give impact to the production of 5-hydroxymethylfurfural. *Journal of Molecular Liquids* 238:574–581.

Yao, Y., Z. Gu, Y. Wang, H. Wang, and W. Li. 2016. Magnetically-recoverable carbonaceous material: An efficient catalyst for the synthesis of 5-hydroxymethylfurfural and 5-ethoxymethylfurfural from carbohydrates 1. *Russian Journal of General Chemistry* 86:1698–1704.

Yokoyama, S. and Matsumura, Y. 2008. *Asian Biomass Handbook: A Guide for Biomass Production and Utilization* Japan: The Japan Institute of Technology.

Yuan, W., Y. Huang, C. Wu, X. Liu, Y. Xia, and H. Wang. 2017. MCM-41 immobilized acidic functional ionic liquid and chromium (III) complexes catalyzed conversion of hexose into 5-hydroxymethylfurfural. *Chinese Journal of Chemical Engineering* 35:1739–1748.

Zhang, L., C. Charles, and P. Champagne. 2010. Overview of recent advances in thermo-chemical conversion of biomass. *Energy Conversion and Management* 51:969–982.

Zhang, Y., Y. Chen, P. Jin, M. Liu, J. Pan, and Y. Yan. 2017. A novel route for green conversion of cellulose to HMF by cascading enzymatic and chemical reactions. *American Institute of Chemical Engineers* 63:4920–4932.

Zhao, W., T. Yang, H. Li, and Y. Lu. 2016. Efficient production of 5-hydroxymethylfurfural from carbohydrates catalyzed by mesoporous Al-B hybrids. *Waste and Biomass Valorization* 8:1371–1378.

Zhou, F., X. Sun, D. Wu, Y. Zhang, and H. Su. 2017. Role of water in catalyzing proton transfer in glucose dehydration to 5-hydroxymethylfurfural. *ChemCatChem* 9:2784–2789.

Zhou, X., Z. Zhang, B. Liu, Q. Zhou, S. Wang, and K. Deng. 2014. Catalytic conversion of fructose into furans using $FeCl_3$ as catalyst. *Journal of Industrial and Engineering Chemistry* 20:644–649.

Zhu, L. W., J. G. Wang, P. P. Zhao, F. Song, X. Y. Sun, L. H. Wang, H. Y. Cui, and W. M. Yi. 2017. Preparation of the Nb-P/SBA-15 catalyst and its performance in the dehydration of fructose to 5-hydroxymethylfurfural. *Journal of Fuel Chemistry and Technology* 45:651–659.

Zuo, M., K. Le, Z. Li, Y. Jiang, X. Zeng, X. Tang, Y. Sun, and L. Lin. 2017. Green process for production of 5-hydroxymethylfurfural from carbohydrates with high purity in deep eutectic solvents. *Industrial Crops and Products* 99:1–6.

4 Production and Characterization of Biodiesel through Catalytic Routes

Yun Hin Taufiq-Yap and Nasar Mansir

CONTENTS

4.1 Introduction 53
4.2 Biodiesel as a Reliable Fuel 55
4.3 Feedstocks for Biodiesel Production 55
4.3.1 Edible Oils 56
4.3.2 Non-edible Oils 56
4.3.3 Microalgal Oil 57
4.3.4 Waste Cooking Oil 58
4.4 The Composition of Vegetable-Based Oils 59
4.5 Biodiesel Production Processes 60
4.6 Catalytic System for Biodiesel Synthesis 61
4.6.1 Biodiesel Production Using Homogeneous Catalysts 62
4.6.2 Biodiesel Production Using Heterogeneous Catalysts 63
4.6.3 Biodiesel Production Using Heterogeneous Acid Catalysts 64
4.6.4 Biodiesel Production from Heterogeneous Solid Base Catalysts 65
4.7 Biodiesel Characterization and Yield Evaluation Techniques 65
4.7.1 Gas Chromatography 66
4.7.2 Fourier Transform Infrared Spectroscopy 67
4.8 Conclusions 67
Acknowledgments 67
References 67

4.1 INTRODUCTION

Current global energy crises are the result of human population growth as well as industrial and technological advancements (Mansir et al. 2017c). Environmental policies concerning global warming and the release of toxic and harmful gases into the environment caused by the use of conventional fossil fuels in internal combustion engines slowly encourage reduction in the continual use of petroleum-based fuels. Moreover, the possibility of the global depletion of fossil fuel reserves in the future has contributed to the shortage and the high cost of conventional fossil fuels. The aforementioned reasons and many others entailed searching for an alternative to fossil fuels that would be renewable, biodegradable, and less toxic (Nizah et al. 2014). The option of renewable energy sources rather than fossil fuels includes biomass-based fuels, hydropower, solar power, and wind power, among others.

Biomass is one of the major sources of renewable energy receiving the attention of researchers from academia and industries. Biomass generally refers to plant materials (e.g., crops, woody material, organic waste, and vegetable oil) that can either be converted to biofuels or directly be combusted for energy through heating (Cheng 2009). Biofuel is normally considered as fuel either in liquid or gaseous form generated from biomass and utilized as fuel for internal combustion engines (i.e., airplanes, cars, generators, and ships). Biofuels have proved to be economically viable considering the availability of feedstocks and eco-friendly as a result of biodegradability and less toxicity (Nigam and Singh 2011).

Conventional techniques used to produce biofuels from biomass includes liquefaction and pyrolysis for bio-oils, gasification for syngas, hydrolysis of sugars to bioethanol, direct combustion of wood to generate heat, and transesterification and esterification of vegetable oil or lipids to biodiesel (Lee et al. 2015). Biodiesel was found to be one of the most reliable of renewable biofuel alternatives to petroleum-based fuel for automobiles and other internal combustion engines (Mansir et al. 2017b). The reason for considering biodiesel as sustainable biofuel is because of feedstock availability, biodegradable nature, less toxicity, and low sulfur composition (Galadima and Muraza 2014).

The physical and chemical properties of biodiesel that are like those of conventional petroleum fuels is another reason for its wide acceptance (Nigam and Singh 2011). Despite the promising sustainability demonstrated by biodiesel, its production on the commercial scale remains the major obstacle for its global commercialization. The current biodiesel production adopts the utilization of food grades oils, which are very expensive and eventually makes biodiesel production very expensive. The cost of feedstock alone covers about 80% of the overall biodiesel production cost (Galadima and Muraza 2014). Therefore, adopting the utilization of food grade oil for biodiesel production turns out to be more expensive than the conventional petroleum-based fuels.

The conventional homogeneous catalytic system is very sensitive to free fatty acid (FFA) and other impurities such as water, which eventually form soap through hydrolysis when high FFA feedstock is used, hence only food grade oil applies (Shu et al. 2010). Moreover, recently, global environmental policies have criticized the use of a homogeneous catalyst for biodiesel production. This criticism is due to the generation of wastewater produced during the washing process of biodiesel production, which may eventually pollute the immediate environment, this affecting the biotic and abiotic ecosystems. Biodiesel is popularly manufactured by the catalytic transesterification reaction of triglycerides (a major component of vegetable oil) or esterification of animal fat using short chain alcohol (Wong et al. 2015). However, related research has completely confirmed several issues concerning biodiesel that are still not resolved persuasively. Reaction of vegetable oil components (triglycerides) to biodiesel involves a reaction of the triglycerides with methanol. The short chain monohydric alcohol, usually methanol, are recommended by most of the researchers without distinct reason, which offers the finest viscosity requirements in line with the provision of American Society of Testing and Materials (ASTM) or related international agencies (Galadima and Muraza 2014).

Biodiesel production on commercial scale at low cost and in an environmentally benign condition is the only way for it to be globally accepted as fuel for industrial and transportation services. This acceptance can only be achieved through diversifying and using different low-cost starting materials for biodiesel production. The low-cost starting materials for biodiesel include palm fatty acid distillate (PFAD), waste cooking oil, *Jatropha curcas* oil, animal fat, and grease (Bhuiya et al. 2014). Considering the FFA composition and other contaminants of these feedstocks, separation difficulty, and other related environmental issues of the homogeneous catalyst after a reaction, the heterogeneous catalyst appears to be the best option for the sustainable commercial production of biodiesel with lesser difficulties.

This chapter will specifically focus on the production and characterization of biodiesel through the catalytic routes. This focus will involve the biodiesel as fuel, as well as the information about the feedstock and the catalyst system that will be sustainable in the near future. The biodiesel characterization techniques and fuel standard properties will also be discussed.

4.2 BIODIESEL AS A RELIABLE FUEL

Biodiesel or fatty acid methyl ester (FAME) is conventionally produced from a renewable-based starting material, usually vegetable oil or animal fat. Biodiesel is technically manufactured using chemical reactions aided by catalyst, methanol, and temperature. The adopted chemical reaction processes used for the synthesis is esterification of fatty acid to produce biodiesel and water or transesterification of triglycerides to produce biodiesel and glycerol (Shin et al. 2012). The recent attention received by renewable fuels particularly biodiesel was because of its environmental benign nature, availability of substrate, naturally biodegradable, and less toxic components such as sulfur and nitrous oxides. Besides its environmentally benign nature and low-cost starting material, biodiesel operates conventionally in the same way that petroleum-based fuels work in the engines without any identified obstacles (Taufiq-Yap et al. 2014).

Regardless of little variation observed in biodiesel fuel properties when compared to that of petroleum-based fuel, some properties such as cetane number and kinematic viscosity are the same (Lokman et al. 2014). Accordingly, the excellent lubricating property of such fuel leads to better and long engine life. The overall properties of the standard biodiesel were evaluated and compared to those of petroleum-based fuel and were within the accepted range based on ASTM as presented in Table 4.1. Based on this comparison, the biodiesel fuel could be accepted as a reliable fuel of the future.

4.3 FEEDSTOCKS FOR BIODIESEL PRODUCTION

There are about three different categories of feedstocks used for biodiesel synthesis. These include edible oils such as soybean, sunflower oil, and so on, which are termed as first-generation feedstock. The second-generation feedstocks are mostly non-edible oils, which include *Jatropha* oil, *Moringa oleifera* oil, and so on. The microalgae oil, although not yet fully used for biodiesel production, is considered a third-generation feedstock. Microalgae oil is now regarded as one of the most sustainable feedstocks because of its easy cultivation and high oil content. Other non-edible feedstocks used for biodiesel production include PFAD, waste cooking oil, and animal fat.

TABLE 4.1
ASTM Standard Fuel Properties Comparison between Biodiesel and Petro-Diesel

Fuel Property	Petro-Diesel	Biodiesel
Standard Method	ASTM D975	ASTM D6751
Fuel composition	Hydrocarbon (C10–C21)	FAME (C12–C22)
Cetane number	40–55	48–60
Density (g/cm^3)	0.85	0.88
Cloud point (°C)	−15 to 5	−3 to 12
Flash point (°C)	60 to 80	100 to 170
Pour point (°C)	−30 to −15	−15 to 5
Carbon content (wt%)	87	77
Hydrogen content (wt%)	13	12
Oxygen content (wt%)	0	11
Sulfur content (wt%)	0.05	0.05
Water content (vol%)	0.05	0.05

Source: Lotero, E. et al., *Ind. Eng. Chem. Res.*, 44, 5353–5363, 2005.

4.3.1 Edible Oils

Edible oils used for biodiesel production include soybean oil, palm oil, sunflower oil, and cottonseed oil (Rathore and Madras 2007). These feedstocks are considered first-generation feedstocks for biodiesel production. Edible feedstocks are the early vegetable oils used for biodiesel production considering their high oil content (see Table 4.2). Moreover, they produce excellent biodiesel with less sulfur and CO_2 content. Because of their purity and low FFA content, a conventional homogeneous catalyst system usually was used for FAME production from such feedstocks.

The reaction rate of the conventional catalyst system is faster than the heterogeneous catalyst system. However, continuous use of edible feedstocks for fuel production is now regarded as non-sustainable due to the possibility of creating food versus fuel competition, which may eventually lead to hunger and famine. Besides, edible oils are expensive and using them as feedstocks for the commercialization of biodiesel would make biodiesel very expensive compared to conventional fossil fuel. Figure 4.1 presents the vegetable oil export for biodiesel from different countries.

4.3.2 Non-edible Oils

Non-edible oils are second-generation feedstocks for biodiesel production. The categories of these feedstocks possess various oil contents and are presented in Table 4.3. Other feedstocks that

TABLE 4.2
Oil Content of Selected Edible Feedstocks for Biodiesel

Oil Seed Crop	Oil Yield (%)
Palm oil (mesocarp) *(Elaeis guineensis)*	20.1
Palm kernel *(Elaeis guineensis)*	45.4
Cottonseed *(Gossypium hirsutum)*	14.7
Groundnut *(Arachis hypogaea)*	43.2
Sunflower *(Helianthus annuus)*	41.2
Rapeseed *(Brassica napus)*	39.7
Coconut *(Cocos nucifera)*	66.1

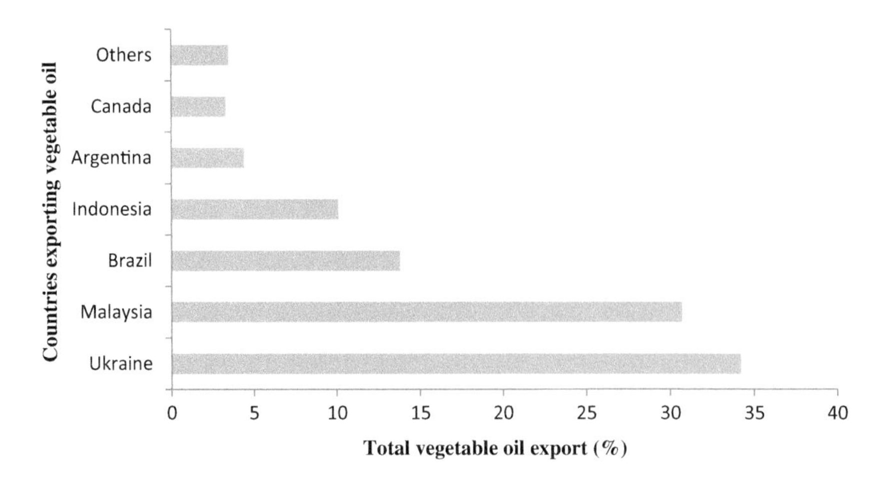

FIGURE 4.1 Expected distributions of vegetable oil exports by 2022. (Data adapted from FAO, Food and Agriculture Organization, Oil Seeds Market Summary Trade and Markets Division, Food Outlook, Food and Agriculture Organization, http://www.fao.org/fileadmin/templates/est/COMM_M, 2014.)

TABLE 4.3
Oil Content of Selected Non-edible Feedstocks for Biodiesel

Oil Seed Crop	Oil Yield (%)
Babassu oil *(Attalea speciosa)*	60–70
Borage oil *(Borago officinalis)*	20
Camelina oil *(Camelina sativa)*	38–40
Castor oil *(Ricinus communis)*	45–50
Cuphea oil *(Cuphea viscosissima)*	25–43
Hemp oil *(Cannabis sativa)*	33
Jatropha oil *(Jatropha curcas)*	45
Jojoba oil *(Simmondsia chinensis)*	44
Karanja oil *(Pongamia pinnata)*	27–39
Linseed oil *(Linum usitatissimum)*	37–42
Neem oil *(Azadirachta indica)*	40–50

Source: Ramachandran, K. et al., *Renew. Sust. Energ. Rev.*, 22, 410–418, 2013.

are categorized as non-edible oil crops include PFAD, waste cooking oil, grease, and animal fat (Wilson and Clark 2000). Non-edible oils are sustainable biodiesel feedstocks due to their low-cost and availability worldwide.

Different research conducted on these feedstocks has recorded excellent biodiesel yield (Amani et al. 2014). However, non-edible oils are not pure feedstocks because they possess high FFA and some water contents. Hence, they cannot be converted to biodiesel using the homogeneous catalyst system (e.g., NaOH, KOH, and so on) because they produce soap through hydrolysis instead of biodiesel (Muhammad et al. 2014). Heterogeneous solid acid catalysts have been used for the transesterification of low-cost non-edible feedstock such as waste cooking oil, such as *Jatropha curcas* oil, to biodiesel. However, heterogeneous acid catalysts have a slow reaction rate and require a reaction temperature and are thus not economical. Base catalysts are known for their excellent performance for the transesterification reaction considering their high rate of reaction, which is approximately 4,000 times faster than heterogeneous acid catalysts (Avhad and Marchetti 2016). Nevertheless, the solid base catalyst can only transesterify feedstocks with less than 3% FFA, otherwise, a two-step process is necessary for most of the non-edible feedstocks. Non-edible feedstocks are made up of a mixture of FFA, triglycerides, and, in some cases, water content. Non-edible feedstocks with a modified heterogeneous solid catalyst having both acid and base active sites can successfully make sustainable biodiesel production. The oil content of different non-edible oils is presented in Table 4.3.

4.3.3 Microalgal Oil

Microalgal oil is a third-generation feedstock for biodiesel production. It is one of the most promising and inexhaustible feedstock for the economical production of commercial biodiesel when compared to edible and non-edible oils (first- and second-generation) (Mata et al. 2010). Oil content-rich microalgae could potentially produce an oil yield 25 times higher than the oil yield produced from conventional biodiesel crops such as palm oil or *Jatropha curcas* oil. Microalgae with an average oil production of at least 70% by weight of dry biomass needed just 0.1 square meter per year per kilogram (m^2 yr/km) of biodiesel to generate 121,104 kg of biodiesel per year. The huge production value of this plant makes it an excellent and sustainable source of feedstock for biodiesel (Cabanelas et al. 2013).

Microalgae reproduce by completing an entire lifecycle within a few days in any environment. Hence, there is no need for fresh water because they can survive in salt or contaminated water. They do not require herbicides and pesticides for cultivation. Despite the high oil content of microalgae plant, another advantage is bio-compensation of CO_2 emissions. About 1.83 kg of CO_2 is absorbed by only 1 kg of dry algal biomass, which therefore improves the air quality (Cabanelas et al. 2013). However, serious technological problems related to the microalgae production should be addressed before they become plausible candidates for biodiesel production on the commercial level (Galadima and Muraza 2014).

4.3.4 Waste Cooking Oil

Although biodiesel is widely considered as an acceptable optional renewable source of energy to petroleum-based fuel, there is concern for the sustainability of using conventional edible oils as feedstock. The present production of biodiesel on a commercial scale depends completely on food grade oil such as rapeseed (6.01 million tons), palm oil (6.34 million tons), and soybean oil (7.08 million tons). Recently, some non-edible oils such as Castor oil and *Jatropha curcas oil* have been used as feedstocks for production globally (Lee et al. 2014).

Besides the food versus fuel rivalry in the foreseeable future that might be caused because of continual use of edible vegetable oils as biodiesel feedstocks, the primary cost of biodiesel depends entirely on the cost of starting material. Hence, frequent use of food grade oils will make the biodiesel more expensive compared to petroleum-based fuel, which breaches one of the fundamental purposes of biodiesel production (Sarin et al. 2007). *Jatropha curcas* oil is predominantly one of the outstanding feedstocks for biodiesel production because it has around 45% oil content and can be cultivated on non-fertile land, thus avoiding rivalry with food crops and cultivation of farmlands. Nevertheless, *Jatropha curcas* oil seeds are highly toxic, making the harvest extremely difficult (Escobar et al. 2009).

Recent studies have suggested that the biodiesel production cost could probably be reduced to at least half or more by adopting low-cost starting materials such as waste cooking oil rather than the food grade edible oils (Escobar et al. 2009). It is estimated that large amounts of waste oil totaling millions of tons are annually generated through various food processing activities and disposed on land and various water bodies worldwide. A low-cost starting material of that magnitude for biodiesel production would help significantly in subsidizing the production on a commercial scale (Lee et al. 2014). Moreover, converting waste cooking oil into biodiesel could significantly aid in mitigating the issues of pollution caused by its disposal into water bodies and on land worldwide. However, waste cooking oil undergoes some chemical changes during the frying process that results in high FFA levels and some impurities in the oil. However, fresh vegetable oil is predominantly composed of triglyceride and some traces of diglyceride, monoglyceride, and FFA (typically <1%). Feedstock with high FFA and water composition hinders the reaction process of biodiesel production because of the emulsion and soap formation through the hydrolysis process when homogeneous basic catalysts are used (Lokman et al. 2014).

Before using such materials as feedstocks for biodiesel, the aforementioned issues must be considered and addressed. Fatty acid composition, which influences the biodiesel composition, also differs from one feedstock to another. Such properties include cold filter and plugging point, cetane number, cloud point, flash point, and oxidation stability (Nakajima and Hara 2012). Typically, vegetable oils are composed of five major fatty acids with compositions depicted in Table 4.4 (i.e., palmitic (16:0), stearic (18:0), oleic (18:1), linoleic (18:2), and linolenic (18:3) acids).

Waste cooking oil is now regarded as one of the most inexpensive feedstocks to use to produce biodiesel, with dual benefit. First, the waste cooking oil can be used as a feedstock for biodiesel. Second, its use as a biodiesel feedstock can prevent it from being disposed into water bodies or on land and therefore preventing water pollution and keeping the environment cleaner. Hence, waste cooking oil is now regarded as a viable and sustainable starting material for commercial quantity production of biodiesel with a dual advantage.

TABLE 4.4
Common Fatty Acid Components of Vegetable-Based Oils

Fatty Acid	Chemical Formula	Structure (Carbon number: Degree of unsaturation)	Weight %
Palmitic acid	$C_{16}H_{32}O_2$	16:0	7.3
Stearic acid	$C_{18}H_{36}O_2$	18:0	4.0
Oleic acid	$C_{18}H_{34}O_2$	18:1	26.9
Linoleic acid	$C_{18}H_{32}O_2$	18:2	60.0
Linolenic acid	$C_{18}H_{30}O_2$	18:3	1.8

Source: Muppaneni, T. et al., *Fuel*, 107, 633–640, 2013.

4.4 THE COMPOSITION OF VEGETABLE-BASED OILS

Triglycerides are known to be the major components of vegetable oil, which are composed of a mole of glycerol and three moles of fatty acids. The fatty acids vary in respect to the length of the carbon chain and the number of unsaturated bonds (double bonds). The vegetable-based oil is characterized by hydrophobic properties consisting of about 98% triglycerides and trace amounts of monoglycerides and diglycerides (Demirbas and Kara 2006). The regular fatty acid compositions in different vegetable oils are depicted in Table 4.5. Vegetable oils generally differ in the type of fatty acid joined to the triglyceride molecule. The chemical traits of fatty acids are jointly explained by their carbon number and the degree of unsaturation. The fatty acid composition of a specific vegetable oil determines the fuel properties of the biodiesel produced from that vegetable oil.

TABLE 4.5
Fatty Acid Composition of Different Vegetable Oils

	Fatty Acid Composition (wt%) (Carbon Number: Degree of Unsaturation)										
Vegetable Oils	12:0	14:0	16:0	16:1	18:0	18:1	18:2	18:3	20:0	22:0	22:1
Palm	0.3	1.2	44.3	0	4.3	39.3	10.0	0	0	0	0
Palm kernel	50.1	15.4	7.3	0	18	14.5	2.4	0	0	0	0
Sunflower seed	0	0	6.4	0.1	2.9	17.7	72.9	0	0	0	0
Groundnut	0	0	11.2	0	3.6	41.1	35.5	0.1	0	0	0
Jatropha	0	0	18.5	0	2.3	49.0	29.7	0	0	0	0
Cottonseed	0	0	23.0	0	2.3	15.6	55.6	0.3	0	0	0
Coconut	50.9	21.1	9.5	0	4.9	8.4	0.6	0	0	0	0
Black mustard	0	1.5	5.3	0.2	1.3	11.7	16.9	2.5	9.2	0.4	41.0
Karanja	0	0	5.8	0	5.7	57.9	10.1	0	3.5	0	0
Neem	0	0	17.8	0	16.5	51.2	11.7	0	2.4	0	0
Olive	0	0	13.8	1.4	2.8	71.6	9.0	1.0	0	0	0
Sesame	0	0	9.6	0.2	6.7	41.1	41.2	0.7	0	0	0
Rapeseed	0	0	3.5	0	0.9	64.1	22.3	8.2	0	0	0

Source: Issariyakul, T. and Dalai, A.K., *Renew. Sust. Energ. Rev.*, 31, 446–471, 2014.

4.5 BIODIESEL PRODUCTION PROCESSES

Biodiesel usually is synthesized by the transesterification reaction process of triglycerides or esterification of FFA (Mansir et al. 2017c). During the transesterification reaction procedure, 1 mole of triglycerides reacts with 3 moles of methanol in the presence of a catalyst and at a temperature to generate 3 moles of biodiesel or fatty acid methyl ester and glycerol as the major by-products. The general equation for the reaction is shown in Figure 4.2. According to Avhad and Marchetti (2016), the transesterification process is preceded in three different stages, converting triglycerides to diglycerides and diglycerides to monoglycerides as shown in Figure 4.3. The reaction process usually moves in both forward and backward directions (reversible). Therefore, to move the reaction in the forward direction, there is a need for a large amount of methanol (Avhad and Marchetti 2016).

After the compilation of the reaction process, the reaction mixture usually is separated into individual components: excess methanol using a simple heating on hot plate at 70°C; the used solid catalyst is separated using a centrifuging technique; and finally the glycerol is removed using a separating funnel by allowing the biodiesel and glycerol mixture to settle down – since

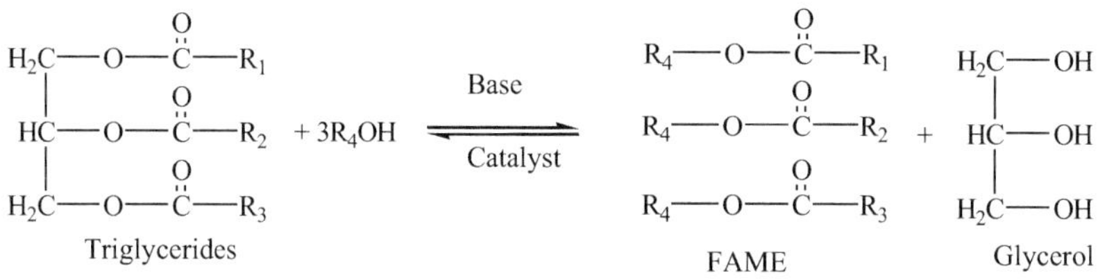

FIGURE 4.2 Biodiesel production reaction process from triglycerides. (From Avhad, M.R. and Marchetti, J.M., *Cataly. Rev.*, 58, 157–208, 2016.)

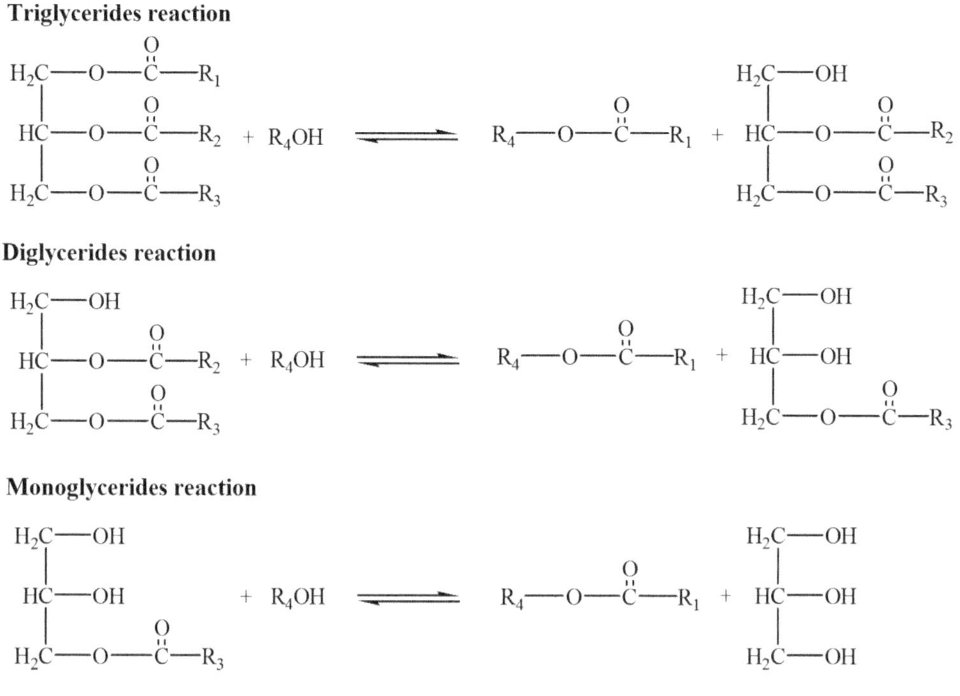

FIGURE 4.3 Biodiesel production reaction process in three steps from triglycerides. (From Avhad, M.R. and Marchetti, J.M., *Cataly. Rev.*, 58, 157–208, 2016.)

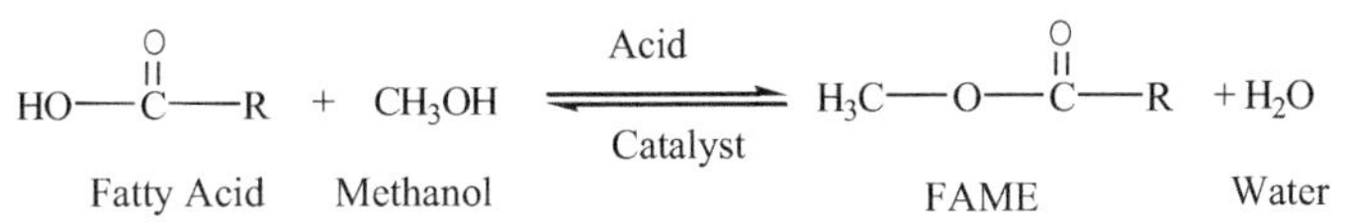

$$\underset{\text{Fatty Acid}}{HO-\overset{O}{\overset{\|}{C}}-R} + \underset{\text{Methanol}}{CH_3OH} \xrightleftharpoons[\text{Catalyst}]{\text{Acid}} \underset{\text{FAME}}{H_3C-O-\overset{O}{\overset{\|}{C}}-R} + \underset{\text{Water}}{H_2O}$$

FIGURE 4.4 Biodiesel production reaction process from a fatty acid. (From Avhad, M.R. and Marchetti, J.M., *Cataly. Rev.*, 58, 157–208, 2016.)

the glycerol is heavier and will settle at the bottom layer while the biodiesel is lighter and will form a top layer (Islam et al. 2015).

The homogeneous catalyst system moves the transesterification reaction about 4,000 times faster than the heterogeneous solid acid catalyst system. The faster reaction rates of the homogeneous base catalyst compared to the solid acid catalyst result from a high catalyst surface contact of the homogeneous base catalyst. The solid acid catalytic reaction rate is hindered by partial contact of solid catalyst's surface with the reaction mixture and therefore leads to a lower catalytic activity compared to the homogeneous base catalyst system (Chen and Fang 2011). However, leaching and separation problems are the major drawbacks associated with such catalysts.

The esterification reaction process is another way in which biodiesel or methyl ester is produced. During the process, a mole of free fatty acid is used to react with a mole of methanol with the aid of a solid acid catalyst to form 1 mole of methyl ester and 1 mole of water. The overall equation of the esterification reaction is shown in Figure 4.4. Chen and Fang (2011) reported that the esterification reaction process is slower and lengthy in catalytic activity when compared to a base-catalyzed transesterification reaction process.

4.6 CATALYTIC SYSTEM FOR BIODIESEL SYNTHESIS

The catalytic transesterification of vegetable-based oil using methanol as the reaction medium to produce biodiesel is an essential industrial process. During the biodiesel production process, the catalyst is needed to enhance the reaction process rate and the overall reaction yield. The transesterification reaction of triglyceride could be either by the homogeneous liquid catalyst or by the heterogeneous solid catalyst. The two major classes of catalysts (liquid and solid) for biodiesel reaction process are compared based on advantage and disadvantages in Table 4.6.

Previously, the conventional homogeneous catalytic system was adopted popularly for the commercial production of biodiesel using food grade oils. This adoption is due to its kinetically faster reaction rate compared to the heterogeneous catalyst system. However, the major drawback to this catalyst system is separation after use, which requires a large amount of water washing and a considerable amount of energy. The extra activity makes the system costly and difficult. Even though a homogeneous acid

TABLE 4.6
Comparison between Homogeneous and Heterogeneous Catalysts Systems to Produce Biodiesel

Features	Heterogeneous Catalyst	Homogeneous Catalyst
Methodology	Continuous fix bed operation is possible	Limitation in continuous fix bed operation
Reaction rate	Slow rate and moderate conversion	Faster reaction rate with high conversion
After process	Easy separation of the catalyst after use	Difficult to recover after use
Effect of high FFA and moisture	Not sensitive to high FFA and water	Very sensitive to high FFA and water
Reusability	Could be reused more than one time	Impossible to be reused
Production cost	Low cost	Comparatively costly

catalyst could be applicable to low quality feedstock such as animal fat, PFAD, and waste cooking oil due to high FFA and moisture content, the catalyst could only be used once, thus making the production costly again. Despite the low reaction rate, recent studies suggested the development of new heterogeneous catalytic systems, which include solid bases such as metal oxides, alkaline earth metal oxides, hydroxides, and simple carbonate salts that demonstrate early success in the biodiesel reaction process (Singh and Fernando 2008).

4.6.1 Biodiesel Production Using Homogeneous Catalysts

Commercialization of biodiesel conventionally relies on food grade oils and a homogeneous liquid catalyst system. The catalyst system is known for its remarkable advantage of high reaction rate because of less mass transfer effect and high catalytic activity compared to a heterogeneous solid acid catalyst system. The homogeneous catalyst system is classified into acidic and basic catalysts. Homogeneous acid catalysts are used for the esterification reaction process to produce methyl esters and water from fatty acids. Such catalysts include hydrochloric acid (HCl), sulfuric acid (H_2SO_4), nitric acid (HNO_3), and so on. Similarly, the homogeneous base catalysts are used for the synthesis of biodiesel and glycerol from triglycerides. Homogeneous base catalysts, particularly potassium hydroxide and sodium hydroxide, are reported to be the best conventional catalysts adopted for the synthesis of biodiesel at a commercial quantity. The catalyst's efficiency in transesterification of triglycerides to biodiesel is a result of the high rate of reaction, which is much faster compared to heterogeneous solid catalysts. Homogeneous-based catalysts production of biodiesel recorded excellent yield at low reaction conditions (Avhad and Marchetti 2016).

Despite the faster reaction rate of such catalysts, moisture content and level of FFA of the feedstock are an essential consideration before selecting a catalyst for a specific type of biodiesel production reaction process. Other homogeneous base catalysts using methyl ester synthesis include sodium methoxide (CH_3ONa) and potassium methoxide (CH_3OK) (Uzun et al. 2012). Besides the low reaction conditions of such catalysts during transesterification of triglycerides, excellent methyl ester yield have been recorded (Avhad and Marchetti 2016). Homogeneous basic and acidic catalysts have been reported for use in biodiesel synthesis from various feedstock recording excellent yields.

Uzun et al. (2012) achieved the methyl ester yield of 96% at mild reaction conditions at 50°C, and reaction time of 30 minutes using homogeneous NaOH catalyst for soybean vegetable oil. However, generally, homogeneous catalysts are associated with so many obstacles that have made biodiesel synthesis hard and costly. Such obstacles peculiar to these catalysts include separation of catalysts after the reaction, the formation of emulsions and soapy materials due to high FFA, and moisture content in the low-quality feedstocks (Mansir et al. 2017a). Therefore, such catalysts are not considered as environmentally and economically feasible.

The catalytic performance of transesterification reaction of triglycerides to biodiesel using homogeneous alkaline catalyst is predominantly affected by two major factors such as the high level of FFA and moisture composition in the feedstock (vegetable oil). FFA is defined as carboxylic acid with a long carbon chain derived from glycerol. Despite the faster reaction rate of homogeneous base catalysts, which run faster than heterogeneous catalysts, liquid base catalyst activity is drastically hampered by the high level of FFA content and eventually leading to poor biodiesel conversion. Moreover, high FFA oils when reacting with homogeneous alkaline catalyst result in the formation of soap (Canakci and Gerpen 2004).

Waste cooking oil generally is believed to contain various impure components including moisture, sterols, and high FFA as a result of frying at high temperature (Mahesh et al. 2015). Specifically, the FFA content in waste cooking oil will have an adverse effect on the transesterification reaction over homogeneous base catalyst. The common homogeneous catalysts that easily form soap when reacting with FFA are NaOH and KOH during transesterification of low-quality feedstock like waste cooking oil, PFAD, and animal fat. The typical saponification reaction process is shown in Figure 4.5.

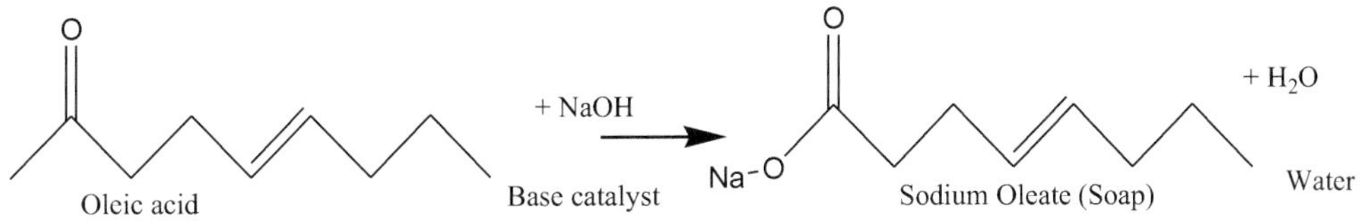

Oleic acid + NaOH Base catalyst → Sodium Oleate (Soap) + H_2O Water

FIGURE 4.5 Alkali catalyst and free fatty acid saponification reaction.

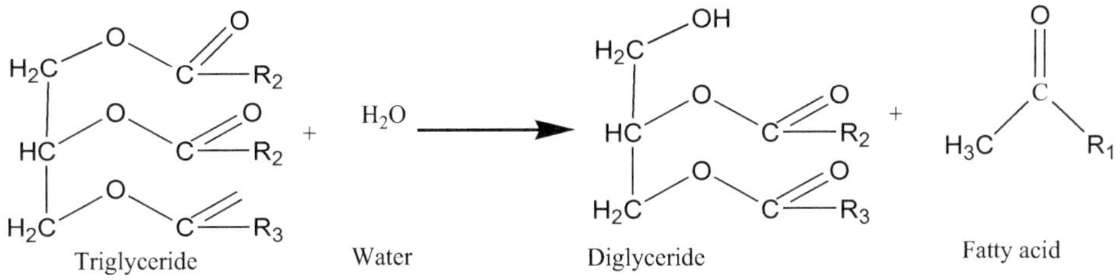

Triglyceride + H_2O Water → Diglyceride + Fatty acid

FIGURE 4.6 Hydrolysis reaction of triglycerides in the presence of water. (Felizardo, P. et al., *Waste Manage.*, 26, 487–494, 2006.)

Water or moisture content is another impurity characterized by low-quality biodiesel feedstocks such as waste cooking oil. The water content in such feedstock could greatly affect the fatty acid methyl ester yield when the homogeneous base catalyst is used for the reaction. The moisture can easily hydrolyze the triglycerides to diglyceride components to form FFA at extreme temperature thus leading to catalyst deactivation (Felizardo et al. 2006). The typical hydrolysis reaction is shown in Figure 4.6. Generally, a homogeneous base-catalyzed transesterification reaction lacks flexible use of various feedstocks because it is limited only to high-grade vegetable oils such as rapeseed oil and soybean oil with little or no impurities (Demirbas 2003).

4.6.2 Biodiesel Production Using Heterogeneous Catalysts

Recently, biodiesel production from vegetable oil-based feedstock using heterogeneous catalysts has received much attention from academia and industry (Tan et al. 2015; Mansir et al. 2017a). Conventionally, biodiesel is produced from vegetable oil using homogeneous catalysts. However, such catalysts are only fit for food-grade oil; hence, they are no longer sustainable. Unlike homogeneous catalyst system, heterogeneous catalysts are not in the same phase with the reaction mixture. Therefore, they can easily be separated after the reaction. Another interesting fact about a heterogeneous catalyst is that it can be used for the reaction more than one time and is easy to separate and dispose of with little environmental effect.

All the identified issues related to homogeneous catalysts are significantly addressed by the heterogeneous catalyst system. Nevertheless, the heterogeneous catalyst system is associated with a slow reaction rate compared to a conventional homogeneous catalyst system. This rate is because of the mass transfer resistance effect of the solid catalysts, which reduces the surface contact during the reaction process. Besides, when the heterogeneous catalysts are properly synthesized, they could resolve the post-reaction issues related to homogeneous catalysts and reduce leaching of catalytic active phases, therefore improving the sustainable biodiesel yields at low-cost (Avhad and Marchetti 2016). The promotion of commercialization of biodiesel is likely by appropriate use of heterogeneous catalysts. This use is due to the reduced leaching of the catalyst active sites in the reaction medium, which helps in elongating the catalyst life. These catalysts could make economical and sustainable biodiesel production in commercial quantities from low-quality vegetable-based feedstocks (Avhad and Marchetti 2016).

4.6.3 Biodiesel Production Using Heterogeneous Acid Catalysts

Besides the conventional way of biodiesel production using homogeneous catalysts, use of heterogeneous acid catalysts was reported for biodiesel production from various low-quality feedstocks. Such catalysts offered remarkable advantages over their homogeneous catalysts counterpart. Apart from easy separation after the reaction, solid acid catalysts could be used more than once. Conventional production of biodiesel used fresh food grade oil rather than liquid catalyst. However, continual use of such feedstock is no longer sustainable because of the possibility of food versus fuel rivalry. Low-grade feedstocks such as PFAD, animal fat, and waste cooking oil can be used for commercialization of biodiesel. However, such feedstock cannot be converted to biodiesel using a conventional catalyst system. This inability is due to the high composition of FFA and other impurities such as moisture. Solid acid catalysts can conveniently convert low-grade feedstock to biodiesel at mild reaction conditions. This possibility is a result of the simultaneous esterification of FFA and transesterification of triglycerides by solid acid catalysts (Figure 4.7) catalyzed by Brønsted acid species and Lewis acid species, respectively (Muhammad et al. 2014).

Remarkable biodiesel yields were reported from low-grade feedstock using heterogeneous solid catalysts. Moreover, the problems related to corrosion of biodiesel reactors using homogeneous acid catalysts can be addressed by using heterogeneous solid acid catalysts (Dehkordi and Ghasemi 2012). The mechanism of esterification and transesterification of the solid heterogeneous acid catalyst is believed to be the same as that of the homogeneous acid and base catalysts. The only difference between the two forms of catalysts is that the reaction of solid catalysts generally occurs at the

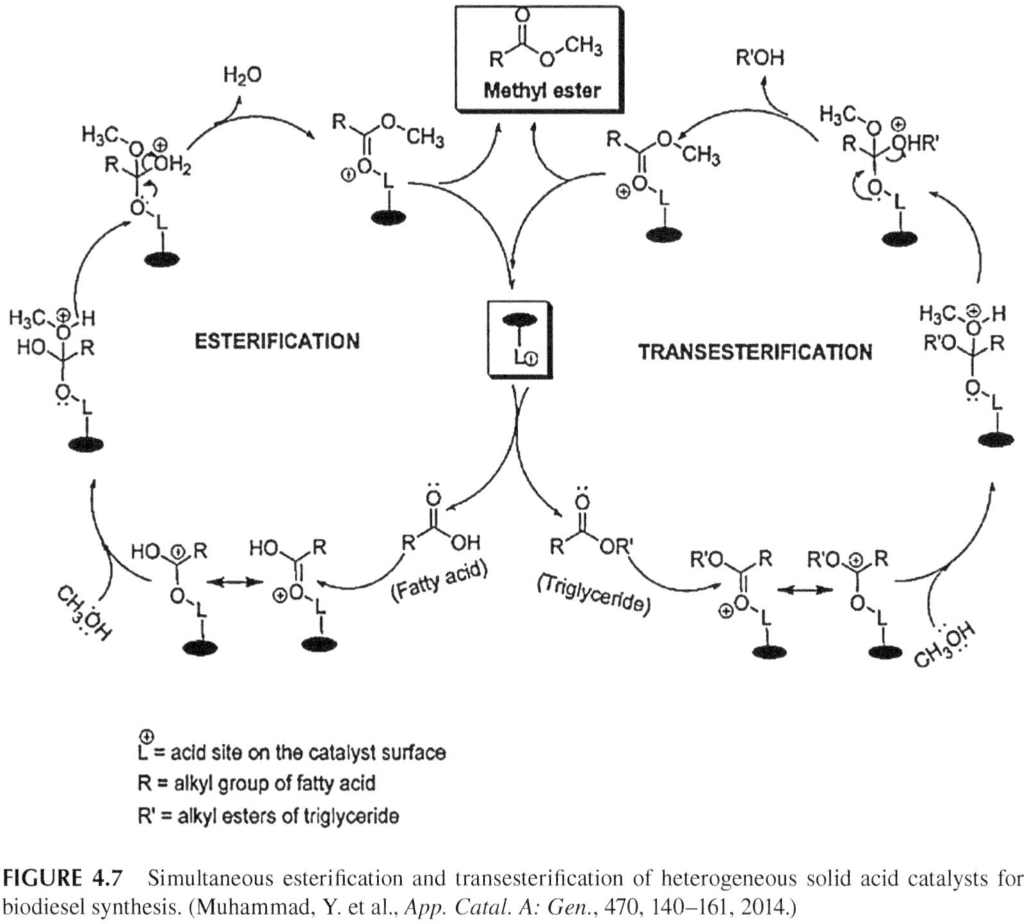

$L^{\oplus}$ = acid site on the catalyst surface
R = alkyl group of fatty acid
R' = alkyl esters of triglyceride

FIGURE 4.7 Simultaneous esterification and transesterification of heterogeneous solid acid catalysts for biodiesel synthesis. (Muhammad, Y. et al., *App. Catal. A: Gen.*, 470, 140–161, 2014.)

catalyst surface, which depends on the surface acid sites and interconnected pore system (Avhad and Marchetti 2016).

Apart from transesterification of triglycerides to biodiesel, solid acid catalysts can conveniently esterify PFAD, typically more than 85% fatty acid, at low reaction conditions (Lokman et al. 2014). However, the conversion of triglycerides to biodiesel by a solid acid catalyst is performed at longer reaction times and high reaction temperatures. This process is one of the down sides of these catalysts and, therefore, recommended for the new catalytic route that can perform transesterification at low reaction conditions.

4.6.4 Biodiesel Production from Heterogeneous Solid Base Catalysts

Although homogeneous basic catalysts are reported to be excellent candidates for biodiesel production from triglycerides on the commercial scale, the environmental issues associated with these catalysts from the post-reaction catalyst removal activities remain the major concern for their use today. Heterogeneous basic catalysts have recently received attention for their role in the transesterification of vegetable-based feedstocks to biodiesel. Like homogeneous catalysts, this category of catalysts can successfully lead to high biodiesel yields under mild reaction conditions.

A basic heterogeneous catalyst system is made up of metal oxides from Group I and II of the periodic table. The oxides of these catalysts could be single or binary mixed metal oxides. Other heterogeneous basic catalysts for biodiesel synthesis include anionic ion exchange resins and hydrotalcites. Moreover, heterogeneous basic catalysts can convert low-grade feedstocks such as waste cooking oil with FFA composition of up to 3 weight percent (wt%) in a single run process as a result of their high strength (Weckhuysen et al. 1998). The presence of M^{2+} cations and O^{2-} anions dispersed in alkaline earth metal oxides and the surface hydroxyl group in various coordination surroundings is the probable reason for the widely distributed basic centers in the basic metal oxides. The calcination temperature and the catalyst's surface composition could temper the surface oxygen of these oxides. Although heterogeneous solid base catalysts are best for transesterification reaction, certain properties such as selectivity specific surface area, basicity, and level of leaching during reaction process are essential in selecting the appropriate catalyst for such reactions (Avhad and Marchetti 2016).

Calcium oxide (CaO) and other alkaline earth metal oxides are the most commonly used heterogeneous solid base catalysts to produce biodiesel from vegetable-based oils. This use is perhaps the result of its basicity strength and abundance in different natural resources in the form of carbonates (Navajas et al. 2013), crab shells (Correia et al. 2014), and so on. CaO-based heterogeneous catalysts have been used for biodiesel production from various vegetable-based oils demonstrating remarkable catalytic activity with excellent yield. However, despite such advantages, CaO solid catalysts are far from perfect because they are affected by leaching of catalytic active sites into the product after the first reaction cycle and therefore cannot be reused many times. The leaching of the catalytic active sites of CaO during the transesterification reaction can be reduced by doping the transition metal oxides over the surface of CaO (Mansir et al. 2018).

4.7 BIODIESEL CHARACTERIZATION AND YIELD EVALUATION TECHNIQUES

Biodiesel is synthesized by the transesterification or esterification reaction process. In each case, the produced biodiesel mixture can be cooled before the separation process begins. A centrifuging technique is used to remove the catalyst from the biodiesel mixture in the case of a heterogeneous catalyst system while water washing is used to remove the homogeneous catalyst. The methanol separation is usually carried out by heating the FAME and methanol mixture at 70°C. The final FAME and glycerol mixture is allowed to settle in a separation funnel where the glycerol settles and forms the bottom layer and the FAME forms the upper layer. The characterization of the synthesized biodiesel is carried out using different techniques to confirm the presence of methyl ester components.

The common biodiesel characterization techniques are gas chromatography-mass spectroscopy (GC-MS) and Fourier transform infrared (FTIR). The evaluation process for biodiesel yield involves the utilization of a gas chromatography-flame ionization detector (GC-FID) technique to analyze and measure the amount of various methyl esters present. Biodiesel is further analyzed to ensure its physicochemical properties are within the range of stipulated standards according to American Society for Testing and Materials (ASTM) and European standards (EN). The biodiesel standard fuel properties include kinematic viscosity, flash point, pour point, cold flow, density, acid value, and so on.

4.7.1 Gas Chromatography

Gas chromatography (GC) analysis generally is used for the quantification and identification of certain components contained in various compounds. In biodiesel production, GC is used to identify and quantify the fatty acid composition present in a biodiesel feedstock. The distribution and composition of different fatty acids in a feedstock provide information about the type of methyl ester that will be produced when such feedstock is used. Therefore, when different types of feedstock are used for biodiesel production, the produced fatty acid methyl ester would contain various fatty acid compositions corresponding to the composition of the feedstock as presented in Table 4.7.

Islam et al. (2015) presented the procedure for analysis of the fatty acid methyl ester profile through GC. The biodiesel sample was analyzed using a Shimadzu GC-14B GC attached with a flame ionization detector (FID) and a capillary column Rtx-65 with a dimension of (30 m × 0.5 mm × 0.25 μm). Before the GC analysis, about 0.1 g of methyl heptadecanoate was dissolved in 100 milliliters (mL) of hexane for use as the internal standard for methyl ester analysis. About 0.2 g of biodiesel sample was dissolved in 10 mL of standard solution and 0.5 microliters (μL) of the sample was injected into the GC under the flow of helium as a carrier gas. The oven injector and detector temperature were set at 240°C and 280°C, respectively (Islam et al. 2013). The content of biodiesel was then calculated according to an EN accepted procedure (EN 14103) as shown in Eqs. 4.1 through 4.3 (Islam et al. 2013).

To estimate the biodiesel yield, the response factors (R_f) for each compound is calculated using the corresponding standard compound based on Eq. 4.1 (Islam et al. 2015).

$$R_f = \left(\frac{A_{is}}{A_{rs}}\right) \times \left(\frac{C_{rs}}{C_{is}}\right) \tag{4.1}$$

Where R_f is the response factor, A_{is} is the area of internal standard, C_{is} is the concentration of the internal standard, A_{rs} is the area of standard references, and C_{rs} is the concentration of internal standard references. The methyl ester is calculated using E. 4.2 (Islam et al. 2015).

TABLE 4.7
Fatty Acid Compositions of Biodiesel Produced from Selected Feedstocks

Feedstock	Lauric Acid (wt %)	Mystric Acid (wt %)	Palmitic Acid (wt %)	Palmitoleic Acid (wt %)	Stearic Acid (wt %)	Oleic Acid (wt %)	Linoleic Acid (wt %)	References
Crude palm oil	0.16	0.99	43.03	0.19	4.31	39.48	10.82	Melero et al. (2010)
Palm fatty acid distillate	–	1.93	45.68	–	4.25	40.19	7.90	Lokman et al. (2014)
Jatropha oil	–	–	20.16	1.32	7.22	39.77	31.53	Taufiq-yap et al. (2014)
Waste cooking oil	–	0.71	60.10	–	10.80	27.20	1.14	Mansir et al. (2017a)

$$\text{Methyl ester} = \frac{C_{iss} \times A_{if} \times R_f}{A_{iss}} \quad (4.2)$$

Where C_{iss} is the concentration of internal standard in the sample, A_{iss} is the area of internal standard in the sample, and A_{if} is the area of individual FAME compound in the sample. Finally, the biodiesel yield is calculated using Eq. 4.3 (Islam et al. 2015).

$$\text{Biodiesel yield}(\%) = \frac{\text{Total amount of methyl ester}(\text{mol})}{3 \times \text{Charged amount of triglycerols}(\text{mol})} \times 100 \quad (4.3)$$

4.7.2 Fourier Transform Infrared Spectroscopy

FTIR analysis is performed to find out the chemical functional groups present either in a solid, liquid, or gaseous sample. During the sample analysis, the sample usually is filled in the sample cup holder and inserted into the instrument for scanning. The sample usually is scanned at a wave number in the range of 400–4,000 cm^{-1}. The absorption frequency spectra of sample is recorded and plotted as transmittance (%) against wave number (cm^{-1}). The identification of functional groups from the FTIR spectra of biodiesel is based on the positions of adsorbed peaks. The peaks between 500 and 600 cm^{-1} represent the various inorganic compounds found in the oil sample. The peak at wave number of 700–750 cm^{-1} usually is portraying the presence of aromatic compounds in the biodiesel sample (Dutta et al. 2014). The stretching vibration of C—O ester groups appeared between 850 and 1,200 cm^{-1}. Moreover, other peaks related to the C—O stretching vibration of alcohol groups also are assigned at 1,200 cm^{-1}. The peaks at 1,200–1,400 cm^{-1} are mostly assigned to vending and vibrations of CH_2 and CH_3 aliphatic groups. The peaks at the wave number of 1,700–1,800 cm^{-1} are assigned to C=O stretching and vibrations of carboxylic acids and ester. The peaks at 2,850–2,950 cm^{-1} are assigned to stretching and vibrations of aliphatic C—H in CH_2 and terminal CH_3, respectively. The O–H group bending and vibration, which indicates the presence of water molecules, occur at either 1,650 or 3,700 cm^{-1} (Farooq et al. 2013).

4.8 CONCLUSIONS

Biodiesel production through the catalytic routes could be achieved through either a homogeneous catalyst system or a heterogeneous catalyst system. Nevertheless, a homogeneous catalyst system is suitable for high-grade oils, usually the edible oils. Using such catalysts on low-quality feedstocks such as PFAD, waste cooking oil, grease, and so one would result in the formation of soap emulsions, thus drastically reducing the overall biodiesel yield at the end of the reaction. A heterogeneous catalyst system can work on low-grade feedstocks for biodiesel production and is, therefore, more sustainable compared to a conventional system. The biodiesel is subjected to various characterizations such as GC-FID and FTIR to ensure its purity and composition after production.

ACKNOWLEDGMENTS

The authors heartily acknowledge the financial support from the Universiti Putra Malaysia through the research group project GP-IPB/2016/9490400.

REFERENCES

Amani, H., Z. Ahmad, and B. H. Hameed. 2014. Highly active alumina-supported Cs – Zr mixed oxide catalysts for low-temperature transesterification of waste cooking oil. *Applied Catalysis A: General* 487:16–25.

Avhad, M. R., and J. M. Marchetti. 2016. Innovation in solid heterogeneous catalysis for the generation of economically viable and ecofriendly biodiesel: A review. *Catalysis Reviews* 58:157–208.

Bhuiya, M. M. K., M. G. Rasul, M. M. K. Khan, N. Ashwath, A. K. Azad, and M. A. Hazrat. 2014. Second generation biodiesel: Potential alternative to-edible oil-derived biodiesel. *Energy Procedia* 61:1969–1972.

Cabanelas, I. T. D., Z. Arbib, F. A. Chinalia, C. O. Souza, J. A. Perales, P. F. Almeida, J. I. Druzian, and I. A. Nascimento. 2013. From waste to energy: Microalgae production in waste water and glycerol. *Applied Energy* 109:283–290.

Canakci, M., and J. V. Gerpen. 2004. A pilot plant to produce biodiesel from high free fatty acid feedstocks. *Transactions of the American Society of Agricultural Engineers* 46:945–955.

Chen, G., and B. Fang. 2011. Preparation of solid acid catalyst from glucose-starch mixture for biodiesel production. *Bioresource Technology* 102:2635–2640.

Cheng, J. 2009. *Biomass to Renewable Energy Processes*. Boca Raton, FL: CRC Press.

Correia, L. M., R. M. A. Saboya, J. A. de Sousa Campelo, N. Cecilia, C. L. Rodríguez- Castellón, E. Cavalcante Jr., and R. S. Vieira. 2014. Characterization of calcium oxide catalysts from natural sources and their application in the transesterification of sun-flower oil. *Bioresource Technology* 151:207–213.

Dehkordi, A. M., and M. Ghasemi. 2012. Transesterification of waste cooking oil to biodiesel using Ca and Zr mixed oxides as heterogeneous base catalysts. *Fuel Processing Technology* 97:45–51.

Demirbas, A. 2003. Biodiesel fuels from vegetable oils via catalytic and non-catalytic supercritical alcohol transesterification and other methods: A survey. *Energy Conversion and Management* 44:2093–2109.

Demirbas, A., and H. Kara. 2006. New options for conversion of vegetable oils to alternative fuels. *Energy Source Part A. Recovery Utilization and Environment Effects* 28:619–626.

Dutta, R., U. Sarkar, and A. Mukherjee. 2014. Extraction of oil from *Crotalaria juncea* seeds in a modified soxhlet apparatus: Physical and chemical characterization of a prospective bio-fuel. *Fuel* 116:794–802.

Escobar, J. C., E. S. Lora, O. J. Venturini, E. E. Yáñez, E. F. Castillo, and O. Almazan. 2009. Biofuels: Environment, technology and food security. *Renewable and Sustainable Energy Reviews* 13:1275–1287.

FAO. 2014. Food and Agriculture Organization. Oil Seeds Market Summary Trade and Markets Division, Food Outlook. Food and Agriculture Organization. http://www.fao.org/fileadmin/templates/est/COMM_M (Retrieved April 20, 2014).

Farooq, M., A. Ramli, and D. Subbarao. 2013. Biodiesel production from waste cooking oil using bifunctional heterogeneous solid catalysts. *Journal of Cleaner Production* 59:131–140.

Felizardo, P., M. J. N. Correia, I. Raposo, J. F. Mendes, R. Berkemeier, and J. M. Bordado. 2006. Production of biodiesel from waste frying oils. *Waste Management* 26:487–494.

Galadima, A. and O. Muraza. 2014. Biodiesel production from algae by using heterogeneous catalysts: A critical review. *Energy* 78:72–83.

Islam, A., Y. H. Taufiq-Yap, C. M. C. Chan, P. Ravindra, and E. S. Chan. 2013. Transesterification of palm oil using KF and $NaNO_3$ catalysts supported on spherical millimetric γ-Al_2O_3. *Renewable Energy* 59:23–29.

Islam, A., Y. H. Taufiq-Yap, P. Ravindra, and H. T. Siow. 2015. Biodiesel synthesis over millimetric γ-Al_2O_3/KI catalyst. *Energy* 89:965–973.

Issariyakul, T., and A. K. Dalai. 2014. Biodiesel from vegetable oils. *Renewable and Sustainable Energy Reviews* 31:446–471.

Lee, A. F., J. A. Bennett, J. C. Manayil, and K. Wilson. 2014. Heterogeneous catalysis for sustainable biodiesel production via esterification and transesterification. *Chemical Society Reviews* 22:7887–7916.

Lee, H. V, J. C. Juan, and Y. H. Tau. 2015. Preparation and application of binary acid-base Cao-La_2O_3 catalyst for biodiesel production. *Renewable Energy* 74:124–132.

Lokman, I. M., U. Rashid, R. Yunus, and Y. H. Taufiq-Yap. 2014. Carbohydrate-derived solid acid catalysts for biodiesel production from low-cost feedstocks: A review. *Catalysis Reviews: Science and Engineering* 56:187–219.

Lotero, E., Y. Liu, D. E. Lopez, K. Suwannakarn, D. A. Bruce, and J. G. Goodwin. 2005. Synthesis of biodiesel via acid catalysis. *Industrial and Engineering Chemistry Research* 44:5353–5363.

Mahesh, S. E., A. Ramanathan, K. M. Meera, S. Begum, and A. Narayanan. 2015. Biodiesel production from waste cooking oil using KBr impregnated CaO as catalyst. *Energy Conversion and Management* 91:442–450.

Mansir, N., S. H. Teo, M. L. Ibrahim, and Y. H. Taufiq-Yap. 2017a. Synthesis and application of waste egg shell derived CaO supported W-Mo mixed oxide catalysts for FAME production from waste cooking oil: Effect of stoichiometry. *Energy Conversion and Management* 151:216–26.

Mansir, N., S. H. Teo, U. Rashid, and Y. H. Taufiq-Yap. 2018. Efficient waste *Gallus domesticus* shell derived calcium-based catalyst for biodiesel production. *Fuel* 211:67–75.

Mansir, N., S. H. Teo, U. Rashid, M. I. Saimana, Y. P. Tan, G. A. Alsultan, and Y. H. Taufiq-Yap. 2017b. Modified waste egg shell derived bifunctional catalyst for biodiesel production from high FFA waste cooking oil. A review. *Renewable and Sustainable Energy Reviews* 82:3645–3655.

Mansir, N., Y. H. Taufiq-Yap, U. Rashid, and M. L. Ibrahim. 2017c. Investigation of heterogeneous solid acid catalyst performance on low grade feedstocks for biodiesel production: A review. *Energy Conversion and Management* 141:171–182.

Mata, T. M., A. A. Martins, and N. S. Caetano. 2010. Microalgae for biodiesel production and other applications: A review. *Renewable and Sustainable Energy Reviews* 14:217–232.

Melero, J.A., L. F. Bautista, G. Morales, J. Iglesias, and R. Sánchez-Vázquez. 2010. Biodiesel production from crude palm oil using sulfonic acid-modified mesostructured catalysts. *Chemical Engineering Journal* 161:323–331.

Muhammad, Y., W. M. A. W. Daud, and A. R. A. Aziz. 2014. Activity of solid acid catalysts for biodiesel production: A critical review. *Applied Catalysis A: General* 470:140–161.

Muppaneni, T., H. K. Reddy, S. Ponnusamy, P. D. Patil, Y. Sun, P. Dailey, and S. Deng. 2013. Optimization of biodiesel production from palm oil under supercritical ethanol conditions using hexane as co-solvent: A response surface methodology approach. *Fuel* 107:633–640.

Nakajima, K., and M. Hara. 2012. Amorphous carbon with SO_3H groups as a solid Brønsted acid catalyst. *American Chemical Society Catalysis* 2:1296–1304.

Navajas, A., T. Issariyakul, G. Arzamendi, L. M. Gandía, and A. K. Dalai. 2013. Development of eggshell derived catalyst for transesterification of used cooking oil for biodiesel production. *Asia-Pacific Journal of Chemical Engineering* 8:742–748.

Nigam P. S., and A. Singh. 2011. Production of liquid biofuels from renewable sources. *Progress in Energy and Combustion Science* 37:52–68.

Nizah, M. F. R., Y. H. Taufiq-Yap, U. Rashid, S. H. Teo, Z. A. S. Nur, and A. Islam. 2014. Production of biodiesel from non-edible *Jatropha curcas* oil via transesterification using Bi_2O_3-La_2O_3 catalyst. *Energy Conversion and Management* 88:1257–1262.

Ramachandran, K., T. Suganya, N. N. Gandhi, and S. Renganathan. 2013. Recent developments for biodiesel production by ultrasonic assist transesterification using different heterogeneous catalyst: A review. *Renewable and Sustainable Energy Reviews* 22:410–418.

Rathore, V., and G. Madras. 2007. Synthesis of biodiesel from edible and non-edible oils in supercritical alcohols and enzymatic synthesis in supercritical carbon dioxide. *Fuel* 86:2650–2659.

Sarin, R., M. Sharma, S. Sinharay, and R. K. Malhotra. 2007. Jatropha–palm biodiesel blends: An optimum mix for Asia. *Fuel* 86:1365–1371.

Shin, H. Y., S. H. An, R. Sheikh, Y. H. Park, and S. Y. Bae. 2012. Transesterification of used vegetable oils with a Cs-doped heteropolyacid catalyst in supercritical methanol. *Fuel* 96:572–578.

Shu, Q., J. Gao, Z. Nawaz, Y. Liao, D. Wang, and J. Wang. 2010. Synthesis of biodiesel from waste cooking oil with large amounts of free fatty acids using a carbon based solid acid catalyst. *Applied Energy* 87:2589–2596.

Singh, A., and S. D. Fernando. 2008. Transesterification of soybean oil using heterogeneous catalysts. *Energy and Fuels* 22:2067–2069.

Tan, Y. H., M. O. Abdullah, C. Nolasco-Hipolito, and Y. H. Taufiq-Yap. 2015. Waste ostrich- and chicken-eggshells as heterogeneous base catalyst for biodiesel production from used cooking oil: Catalyst characterization and biodiesel yield performance. *Applied Energy* 160:58–70.

Taufiq-Yap, Y. H., S. Hwa, U. Rashid, A. Islam, and M. Zobir. 2014. Transesterification of *Jatropha curcas* crude oil to biodiesel on calcium lanthanum mixed oxide catalyst: Effect of stoichiometric composition. *Energy Conversion and Management* 88:1290–1296.

Uzun, B. B., M. Kılıç, N. Özbay, A. E. Pütün, and E. Pütün. 2012. Biodiesel production from waste frying oils: Optimization of reaction parameters and determination of fuel properties. *Energy* 44:347–351.

Weckhuysen, B. M. G. Mestl, M. P. Rosynek, T. R. Krawietz, J. F. Haw, and J. H. Lunsford. 1998. Destructive adsorption of carbon tetrachloride on alkaline earth metal oxides. *Journal of Physical Chemistry B* 102:3773–3778.

Wilson, K., and J. H. Clark. 2000. Solid acids and their use as environmentally friendly catalysts in organic synthesis. *Pure and Applied Chemistry* 72:1313–1319.

Wong, Y. C., Y. P. Tan, I. Ramli, and H. S. Tee. 2015. Biodiesel production via transesterification of palm oil by using CaO–CeO_2 mixed oxide catalysts. *Fuel* 162:288–293.

5 Recent Advances in Hydrogen Production through Bi-Reforming of Biogas

Tan Ji Siang, Doan Pham Minh, Sharanjit Singh, Herma Dina Setiabudi, and Dai-Viet N. Vo

CONTENTS

5.1 Introduction 71
5.2 Thermodynamic Aspects for Bi-Reforming of Methane 74
5.3 Catalysts Used in Bi-Reforming of Methane 78
5.4 Effects of Catalyst Supports 81
5.5 Effects of Promoters 82
5.6 Effects of Process Variables on Methane Bi-Reforming Reaction 83
5.6.1 Influence of Gas Hourly Space Velocity 83
5.6.2 Influence of Reaction Temperature 84
5.6.3 Influence of Feedstock Composition 84
5.7 Mechanisms and Kinetics of Methane Bi-Reforming 86
5.8 Conclusions and Outlook 88
Acknowledgments 89
References 89

5.1 INTRODUCTION

The global energy requirement has been significantly increased for several decades because of rising global population and economic development. Currently, about 12 million ton of oil (85 million barrels), 22 million ton of coal, and 10 million cubic meters of natural gas are being consumed per day for satisfying about 82% of the world's energy consumption (Goeppert et al. 2014). In addition, based on the report of World Energy Outlook 2017 proposed by the International Energy Agency (2017), the global population was projected to increase from 7.4 billion in 2016 to 9.1 billion in 2040. Thus, it would further result in a substantial growth of energy demand in near future. In the report, the International Energy Agency (2017) also revealed that the rising global energy demand and depletion of non-renewable fossil fuels would lead to an increase in oil prices from \$83 per barrel in 2016 to an estimated \$111 per barrel by 2040. In addition, the increasing global energy consumption resulting from an upgraded living standard and a growing population triggered significant emissions of greenhouse gases (GHG) into the atmosphere, thus posing a threat to the environment and leading to the undesirable effect of climate change. Hence, from long-term economic and environmental perspectives, exploring an alternative renewable energy source with low carbon content is a necessary step to gradually reduce the dependency on petroleum resources.

Hydrogen (H_2) is regarded as an alternative and promising option to replace petroleum products since 1 kg of H_2 gas possesses a high energy density of 142 megajoules (MJ), which is equal to 3.1 kg of gasoline (Liu et al. 2009). In addition, it offers a versatile building block to various H-containing chemical products such as ammonia, urea, and fertilizers, as well as feedstock for fuel cells (Wood et al. 2012). Besides, its combustion only yields water so H_2 usage is free from environmental pollution (Wood et al. 2012). In fact, it is speculated that the consumption of coal and crude oil will reduce to 36.7% and 40.5%, respectively, by 2030 if hydrogen production technologies receive sufficient financial support for generation by industry (Hay et al. 2013). Thus, numerous research programs such as the Italy-Markal model of Italy (Contaldi et al. 2008), Swiss MARKAL model of Switzerland (Schulz et al. 2008), scenario-based model of Germany (Fischedick et al. 2005), HySociety Project of European Union (Wietschel et al. 2006), and Danish energy system of Denmark (Sørensen et al. 2004)) have been conducted to study the economics of hydrogen as an energy carrier in combined energy systems in various countries.

Hydrogen production can be easily achieved by separating H_2 from syngas (a mixture of H_2 and CO), which also can be produced from various synthesis routes. In fact, any hydrocarbon-containing compounds (e.g., methane (or biogas), propane, ethanol, glycerol, and naphtha) can be utilized as feedstocks to produce syngas (Abdullah et al. 2017; Bahari et al. 2017; Sudhakaran et al. 2017; Tavanarad et al. 2018). However, in large-scale syngas production, methane obtained from natural gas has been considered the most feasible and economical feedstock owing to its abundant availability, low cost, and comparatively established technology for the natural gas production system, as well as its wide supply network (Abdullah et al. 2017; Siang et al. 2018). Among syngas synthesis routes, reforming processes are the most conventional and cost-effective technologies for generating syngas from methane in industrial applications (Abdullah et al. 2017). The operating conditions, strengths, and weaknesses of the recent advanced reforming approaches (e.g., steam reforming of methane (SRM), partial oxidation of methane (POM), dry reforming of methane (DRM), and bi-reforming of methane (BRM) (also known as combined steam and dry reforming of methane, CSDRM)) are compared in Table 5.1.

Currently, SRM (cf. Eq. 5.1) is the commercially dominant and economically viable reforming process for syngas production from hydrocarbons. In the SRM process, steam reacts with CH_4 endothermically to yield H_2 and CO using heterogeneous catalysts (i.e., Ni- or Co-based catalysts (Abdullah et al. 2017). According to Fan et al. (2016), the industrial SRM process has approximately contributed to about 75% of global hydrogen production. However, this method releases an excessive amount of undesirable carbon dioxide (CO_2) greenhouse gas during processing and induces an unfavorable H_2/CO ratio (of above 3) for downstream processes (Abdullah et al. 2017). In the case of DRM (see Eq. 5.2), both unfavorable greenhouse gases (i.e., methane (CH_4) and CO_2) are consumed to yield the valuable syngas product but the resulting H_2/CO ratio of below unity owing to presence of the parallel reverse water-gas shift (RWGS) reaction is inappropriate for Fischer-Tropsch synthesis (FTS) requiring the practical and stoichiometric H_2/CO ratio of 2 (Olah et al. 2015). Since FTS is an outstanding and important industrial process for the generation of synthetic fuels to substitute for fossil fuels, the implementation of stand-alone SRM or DRM approaches could impose the installation of ancillary purification and separation units in the downstream process for adjusting H_2/CO ratio to 2 (Jabbour et al. 2017). Even though the POM (cf. Eq. 5.3) possesses numerous advantages including considerably short residence time, generation of desirable H_2/CO ratio of 2, and great CH_4 conversion, local hot spots on catalyst bed due to its exothermicity is the major setback leading to difficulty in the industrial operation and possible explosion risks (Asencios and Assaf 2013).

$$CH_4 + H_2O \rightarrow CO + 3H_2 \left(\Delta_r H^o_{298K} = +206 \text{ kJ/mol}\right) \quad (5.1)$$

$$CH_4 + CO_2 \rightarrow 2CO + 2H_2 \left(\Delta_r H^o_{298K} = +247 \text{ kJ/mol}\right) \quad (5.2)$$

$$CH_4 + 0.5O_2 \rightarrow CO + 2H_2 \left(\Delta_r H^o_{298K} = -22.6 \text{ kJ/mol}\right) \quad (5.3)$$

TABLE 5.1
Comparison of Methane Reforming Technologies for Syngas Production

			Conventional Operating Conditions			
Technology	**Benefits**	**Drawbacks**	**Temperature (°C)**	**Feedstock Ratio**	**Pressure (bar)**	**H_2/CO Ratio**
Steam reforming of methane	• Industrially established technology with large-scale production • High efficiency and no oxygen required • Highest H_2 selectivity	• High greenhouse gas (CO_2) emissions • Unfavorable H_2/CO ratio (>3) for downstream processes • Endothermic nature and severe heat duty	700–1000	CH_4/H_2O = 1/1	3–25	>3
Dry reforming of methane	• Efficient greenhouse gases utilization • Favorable H_2/CO ratio for long-chain hydrocarbon production via FTS • Green process	• Quick catalyst deterioration because of carbon deposit and metal sintering • Moderate reaction temperature requirement	650–850	CH_4/CO_2 = 1/1	1	<1
Partial oxidation of methane	• No requirement of feedstock desulfurization • High reactant conversions • Short residence time • A fast and exothermic reaction	• Hot spots or hot zones formation on catalyst bed owing to exothermic nature • Requirement of cryogenic unit for oxygen separation • High explosion risk	950–1100	CH_4/O_2 = 2/1	100	2
Bi-reforming of methane	• Negligible carbon deposition and prolonged catalyst lifespan • Flexible H_2/CO ratios via easy adjustment of feedstock composition • Environmentally friendly	• High reaction temperature requirement • Unavailable large-scale production	500–1000	$CH_4/H_2O/CO_2$ = 3/2/1	1–20	~2

Source: Li, D. et al., *Chem. Rev.*, 116, 11529–11653,2016; Jang, W.J. et al., *Appl. Energ.*, 173, 80–91, 2016; Abdullah, B. et al., *J. Clean. Prod.*, 162, 170–185, 2017.

For these reasons, the BRM (see Eq. 5.4) has recently emerged as a promising technique and feasible substitution for other conventional reforming processes in syngas and/or hydrogen generation because it can provide flexibly adjusted H_2/CO ratios including the practical and ideal value of 2 by simply manipulating reactant feed composition without the need for an auxiliary separation processes (Olah et al. 2012; Jabbour et al. 2017). Therefore, the resulting syngas (with $H_2/CO = 2$) from the BRM process could be directly fed to FTS to yield synthetic fuels with no required feed pretreatment. In addition, BRM exhibits high stability owing to coexistent H_2O and

CO_2 oxidizing agents in the feedstock. In fact, these oxidants could substantially gasify carbonaceous deposits (Olah et al. 2012; Siang et al. 2018) and hence prolong the catalyst lifespan.

$$3CH_4 + 2H_2O + CO_2 \rightarrow 4CO + 8H_2 \left(\Delta_r H^o_{298K} = +659 \text{ kJ/mol}\right) \quad (5.4)$$

BRM also is regarded as the future of reforming technology since it has the potential to replace the industrially recognized SRM process because it can effectively utilize biogas generated from anaerobic degradation of biomass including food waste, municipal sludge, agricultural waste, and landfill gases (Olah et al. 2015; Chen et al. 2017). Generally, the biogas derived from landfill and agricultural gases consists of CH_4 (55%–70%), CO_2 (30%–45%), N_2 (0%–15%), H_2O (1%–5%), O_2 (0%–3%), and small amounts of contaminants (i.e., NH_3 and H_2S) (Awe et al. 2017). Therefore, numerous biogas upgrading technologies including cryogenic separation, water scrubbing, pressure swing absorption, physical and chemical absorptions, membrane technology, hydrate formation, and biological approaches have been developed recently for biogas purification (Weiland 2010; Starr et al. 2012; Sun et al. 2015). In addition, continuing research attempts have been devoted to enhance the efficiency and lessen the investment and maintenance costs of biogas upgrading processes (Sun et al. 2015). Hence, considering the evolution in biogas upgrading market and associated technologies, hydrogen or syngas production through bi-reforming of biogas could be the most promising route for producing an alternative energy source in the near future. In fact, during the past two decades, the BRM reaction has been explored extensively for understanding the role of heterogeneous catalysts and operating parameters on catalytic performance. Thus, this chapter comprehensively reviews the current progress and advancements of the BRM process regarding thermodynamic aspects, process variables, catalyst use, and mechanisms.

5.2 THERMODYNAMIC ASPECTS FOR BI-REFORMING OF METHANE

As previously presented, CH_4 and CO_2 are the main components of biogas. Thus, bi-reforming of biogas or methane refers to the reaction between CH_4, CO_2, and H_2O. The ideal chemical reaction of this process is expressed in Eq. 5.5. However, other chemical reactions can occur from a given mixture of CH_4, CO_2, and H_2O as shown in the following equations:

$$3CH_4 + 2H_2O + CO_2 \rightarrow 4CO + 8H_2 \left(\Delta_r H^o_{298K} = +659 \text{ kJ/mol}\right) \quad (5.5)$$

$$CH_4 + CO_2 \rightarrow 2CO + 2H_2 \left(\Delta_r H^o_{298K} = +247 \text{ kJ/mol}\right) \quad (5.6)$$

$$2CO \rightarrow C + CO_2 \left(\Delta_r H^o_{298K} = -172 \text{ kJ/mol}\right) \quad (5.7)$$

$$CO + H_2O \rightarrow CO_2 + H_2 \left(\Delta_r H^o_{298K} = -41 \text{ kJ/mol}\right) \quad (5.8)$$

$$CH_4 \rightarrow C + 2H_2 \left(\Delta_r H^o_{298K} = +75 \text{ kJ/mol}\right) \quad (5.9)$$

$$CH_4 + H_2O \rightarrow CO + 3H_2 \left(\Delta_r H^o_{298K} = +206 \text{ kJ/mol}\right) \quad (5.10)$$

$$C_s + H_2O \rightarrow CO + H_2 \left(\Delta_r H^o_{298K} = +131 \text{ kJ/mol}\right) \quad (5.11)$$

$$CH_4 + 2H_2O \rightarrow CO_2 + 4H_2 \left(\Delta_r H^o_{298K} = +165 \text{ kJ/mol}\right) \quad (5.12)$$

For determining the thermodynamic equilibrium of CH_4, CO_2, and H_2O mixtures, FactSage software was used for the calculation. The principle of this calculation is based on the minimization of Gibbs free energy (Bale et al. 2010). Standard enthalpies of reaction ($\Delta_r H°_{298K}$) are calculated from standard enthalpies of formation (Lide 2003). The thermodynamic calculation gives the theoretical thermodynamic equilibriums under well-defined conditions of temperature and pressure of a system. It allows a predetermination about the behavior of the system before other experimental approaches.

As shown in Eq. 5.5, the ideal reaction of bi-reforming implies the molar ratio of $CH_4/CO_2/H_2O = 3/1/2$. However, a large excess of oxidants (CO_2 and H_2O) is generally needed to reach high methane conversion and to limit coke formation. Thus, in this section, the following initial mixture of CH_4, CO_2, and H_2O is studied at the atmospheric pressure (1 bar):

1. An equimolar initial mixture of CH_4, CO_2, and H_2O with 3, 1, and 2 moles of CH_4, CO_2, and H_2O, respectively.
2. Three other initial mixtures containing 3 moles of CH_4, 1 to 2.5 moles of CO_2, and 2 to 3 moles of H_2O. Hence, the molar ratio of oxidants (CO_2 and H_2O) to methane of these mixtures is higher than the stoichiometry of Eq. 5.5.

The conversion of CH_4 (X_{CH_4}), CO_2 (X_{CO_2}) and coke selectivity (Sel_{coke}) are computed according to the following equations where n_i is the initial amount, n_t is the amount at a given amount of time t.

$$X_{CH_4} = \frac{n_i^{CH_4} - n_t^{CH_4}}{n_i^{CH_4}} \times 100\% \tag{5.13}$$

$$X_{CO_2} = \frac{n_i^{CO_2} - n_t^{CO_2}}{n_i^{CO_2}} \times 100\% \tag{5.14}$$

$$Sel_{Coke} = \frac{n_C^t}{(n_i^{CH_4} - n_t^{CH_4}) + (n_i^{CO_2} - n_t^{CO_2})} \times 100\% \tag{5.15}$$

Figure 5.1 shows the methane conversion profiles as a function of temperature under atmospheric pressure. For the equimolar mixture according to Eq. 5.5, methane conversion is lower than 83% below 700°C (Curve 1 in Figure 5.1a). Then, increasing the temperature from 700°C to 1000°C allows reaching higher methane conversion, but this last conversation cannot be complete even at 1000°C, as highlighted by Curve 1 in Figure 5.1b. To increase methane conversion, an increasing amount of oxidants (i.e., CO_2 and H_2O in this case) can be used. As observed in Figure 5.1, increasing either H_2O (Curve 2) or CO_2 (Curve 3) amount, or both CO_2 and H_2O (Curve 4) allows generally increasing methane conversion, except for Curve 2 below around 720°C. In fact, in the case of Curve 2, the water-gas shift (WGS) reaction is favored for CO and H_2 amounts, which limits the consumption of CH_4 in other equilibriums. Regarding Figure 5.1b, at high temperature of 700°C–1000°C, increasing oxidant amounts strongly favors methane conversion, which reaches up to 99% at 850°C for Curves 2, 3, and 4. However, for the energy balance of the global process, the increasing oxidant amounts must be carefully adjusted to limit heat loss due to unreacted CO_2 and H_2O streams. A heat recovery unit can be envisaged for energy recovery downstream of the reforming reactor.

Figure 5.2 illustrates the CO_2 conversion as a function of reaction temperature for different initial mixtures. In all cases, CO_2 conversion decreases up to around 560°C because of the WGS reaction (Eq. 5.8), which produces CO_2. Particularly for the initial mixture corresponding to Curve 2, CO_2 conversion as shown in Eq. 5.14 is negative around 500°C–640°C. This result is because WGS is strongly favored by large amounts of water in the initial mixture. Thus, the equilibrium CO_2 amount

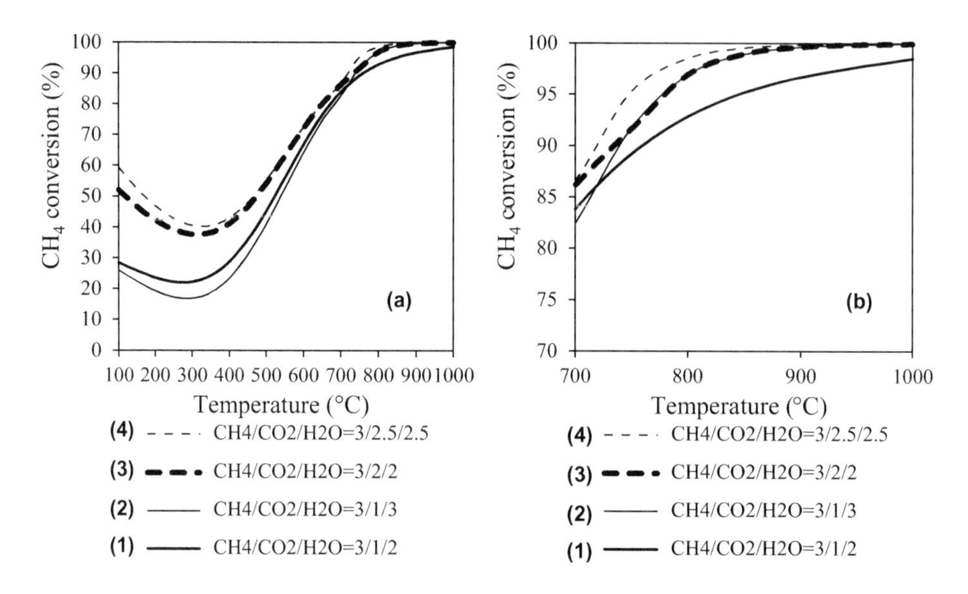

FIGURE 5.1 CH_4 conversion from different initial mixture of CH_4, CO_2, and H_2O. (a) Effect of initial $CH_4/CO_2/H_2O$ ratios on CH_4 conversion. (b) Magnified figure for CH_4 conversion at 700°C–1000°C and various initial $CH_4/CO_2/H_2O$ ratios.

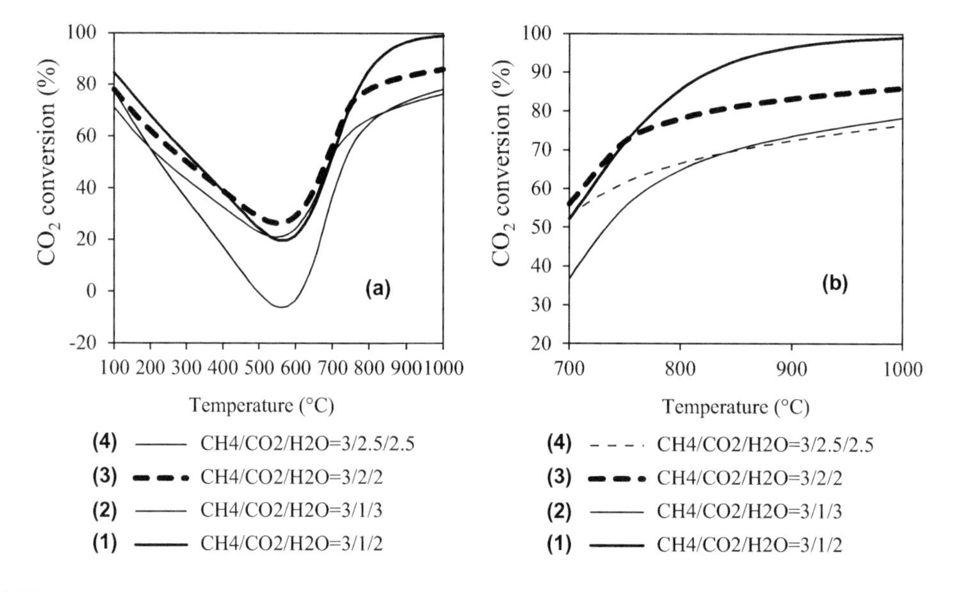

FIGURE 5.2 CO_2 conversion from different initial mixture of CH_4, CO_2, and H_2O. (a) Effect of initial $CH_4/CO_2/H_2O$ ratios on CO_2 conversion. (b) Magnified figure for CO_2 conversion at 700°C–1000°C and various initial $CH_4/CO_2/H_2O$ ratios.

is superior to the initial CO_2 amount within this temperature range, leading to the negative CO_2 conversion. Above 560°C, CO_2 conversion increases because of its participation at other equilibriums which are favored by high temperatures.

Regarding Figure 5.2b for high temperatures (700°C–1000°C), Curve 1 shows the highest conversion, which can be explained by the lowest initial molar ratio of oxidants (CO_2 and H_2O) to CH_4 of this mixture. Curve 2 and 4 show the lowest CO_2 conversion due to the large initial molar ratio of oxidants (CO_2 and H_2O) to CH_4, as well as large amounts of water in these mixtures, which favor

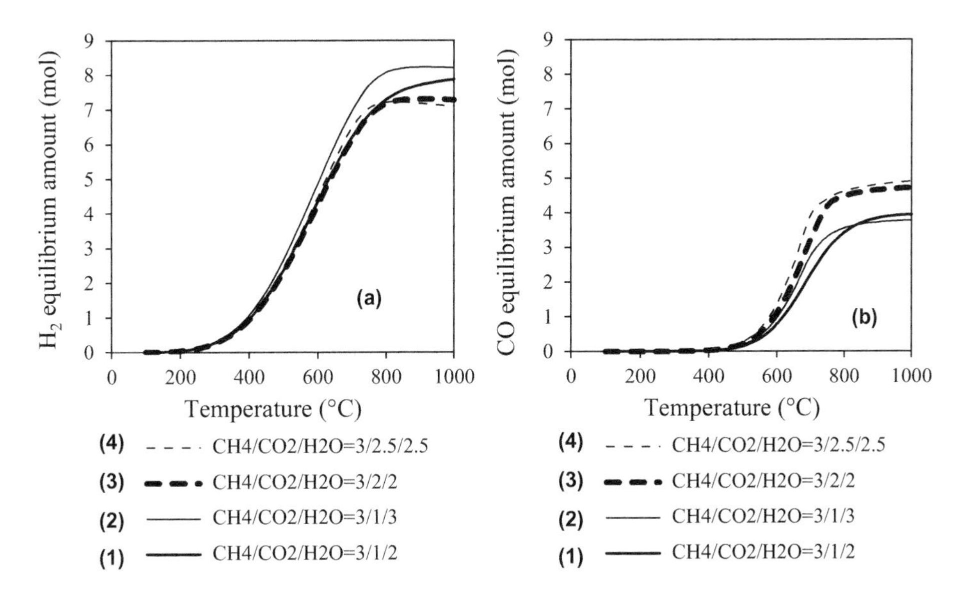

FIGURE 5.3 (a) H_2 and (b) CO amounts from different initial mixture of CH_4, CO_2, and H_2O.

WGS reaction. For Curves 2, 3, and 4, high residual CO_2 amounts are present in the product streams and can be separated and recycled.

Figure 5.3 shows the H_2 and CO amounts and Figure 5.4 shows H_2 to CO molar ratio and coke selectivity. The remainder of the initial mixtures contain 3 moles of CH_4, 1 to 2.5 moles of CO_2, and 2 to 3 moles of H_2O. In Figure 5.3a, H_2 equilibrium amount is similar below 700°C. Above 700°C, H_2 equilibrium amount is higher in the case of Curve 2, which corresponds to the highest amount of water in the initial mixture of CH_4, CO_2, and H_2O. A high water amount in the initial reaction mixture favors WGS reaction leading to a high H_2 equilibrium amount. For Curves 3 and 4 above 750°C, H_2 equilibrium amount is lower compared to Curves 1 and 2 because of the reverse WGS reaction which consumes H_2.

Figure 5.3b presents CO equilibrium amount. There is no CO formation until around 500°C because of the WGS reaction (Eq. 5.8), wherein CO reacts with steam, as well as the Boudouard reaction (Eq. 5.7) wherein CO decomposes into solid carbon (coke) and CO_2. The production of CO is preferred at high temperature (>700°C) and with high oxidant (CO_2 and H_2O) amounts in the initial reaction mixture.

Figure 5.4a shows H_2 to CO molar ratio as a function of the reaction temperature. The very high values of this ratio below 700°C is not significant because of the very low amount of CO within this temperature range as shown in Figure 5.3b. Above 700°C (in the insert of Figure 5.4a), the H_2/CO molar ratio varies between around 1.5 to around 2.8, depending on the composition of the initial mixtures. For Curves 1 and 2, H_2/CO ratio is relatively stable around 2 to 2.2. For Curves 3 and 4, this ratio is stable around 1.5. This ratio is directly due to the impacts of CO_2 and H_2O amounts in the initial reaction mixtures on the equilibriums of the system, as analyzed in Figure 5.3.

Figure 5.4b shows the selectivity into solid carbon (coke), according to Eq. 5.15. Coke is strongly favored at low temperature by the Boudouard reaction which is strongly endothermic. Coke selectivity starts to decrease around 400°C. At high temperature (above 700°C), coke selectivity is highly limited and depends strongly on the composition of the initial reaction mixtures. In general, increasing the oxidants-to-methane ratio (i.e., (CO_2 + H_2O)/CH_4 ratio) and increasing the water content allow decreasing coke selectivity at high temperature. For example, at 700°C, coke selectivity is zero with Curve 4 corresponding to the highest ratio of oxidants to CH_4, 3.7% with Curve 2 containing the highest water amount, 13.7% with Curve 3 containing the same oxidants to CH_4 ratio but

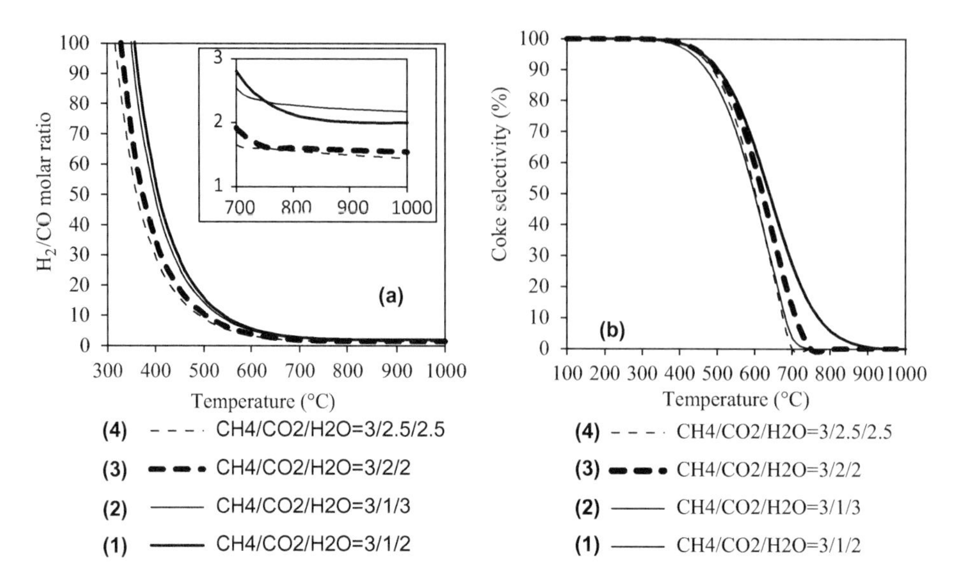

FIGURE 5.4 (a) H_2/CO molar ratio and (b) coke selectivity from different initial mixture of CH_4, CO_2, and H_2O.

more CO_2 than H_2O compared to Curve 2, and 27.3% with Curve 1, which has the lowest ratio of oxidants to CH_4. Thus, the composition of the initial mixture can be efficiently used to control solid carbon formation. As previously stated, a compromise between coke selectivity and energy balance must be established. Large amounts of unreacted CO_2 and H_2O lead to heat lost and so impact the global energy efficiency of the reforming process.

As a partial conclusion, the thermodynamic equilibrium of the syngas production from the bi-reforming of biogas can be predicted by FactSage software. This process is preferred at high reaction temperature (>700°C), high oxidants to methane ratio, and high steam content in the initial reaction mixture. These parameters are of crucial importance for controlling the methane conversion, H_2/CO molar ratio, and coke selectivity. Attention must be paid on the global energy efficiency of the reforming process. Heat loss by the unreacted CO_2 and H_2O must be minimized.

5.3 CATALYSTS USED IN BI-REFORMING OF METHANE

Many studies have reported the evolution of metal-based catalysts for the BRM reaction owing to its crucial role in stability and activity. The performance of a catalyst for BRM depends on the features of the support, type, and size of the particles used, and the degree of metal-support interaction. The common catalysts widely used in BRM are noble metal-based catalysts, usually Ru-based catalyst (Soria et al. 2011), and transition metal-based catalysts including Co-based catalyst (Olah et al. 2015) and Ni-based catalysts (Roh et al. 2009; Olah et al. 2012; Li et al. 2015 Li et al., 2016a). Noble metals are evidently very active and more coke resistance to reforming reaction than the transition catalysts. However, noble metals possess limited availability and are costly. Thus, transition metals have attracted considerable attention in recent years.

Ni-based catalysts have been broadly tested for BRM owing to their availability, low price, and great activity. However, the high coke formation leads to the fast deactivation and constitutes a major drawback for these catalysts to be applied in the industrial sector. Therefore, researchers have conducted several studies of the type of supports owing to the crucial role of support materials in properties and catalytic activity of the catalyst. In addition, attempts to improve catalytic activity have been carried out by the addition of secondary or tertiary elements as a promoter for catalyst. The performance of promoted and supported catalysts recently reported in literature is listed in Table 5.2.

TABLE 5.2
Performance of Selected Notable Catalysts Used in the BRM Reaction

Catalysts	$CH_4/H_2O/CO_2$ Ratio	T (°C)	GHSV (L g_{cat}^{-1} h^{-1})	Initial Performance			TOS (h)	Final Performance			D^a (%)	References
				CH_4 Conversion (%)	CO_2 Conversion (%)	H_2/CO Ratio		CH_4 Conversion (%)	CO_2 Conversion (%)	H_2/CO Ratio		
Types of supports												
Ni/MgO	1/0.8/0.4	830	60	71.3	73.8	2.0	320	70.8	73.4	2.0	1.0	Olah et al. (2012)
Ni/CeO_2	1/0.8/0.4	800	26.5	55.7	–	–	20	47.4	–	–	14.9	Roh et al. (2009)
Ni/ZrO_2	1/0.8/0.4	850	60	92.3	71.4	1.98	20	82.4	54.3	2.0	11.0	Li et al. (2015)
Ni/Mo_2C	1/0.25/0.05	950	–	70.5	99.8	0.96	300	5.8	25.8	0.46	91.8	Brush et al. (2016)
Ni/Mo_2C	1/0.375/0.375	950	–	57.6	99.8	1.26	250	7.9	13.0	1.40	86.3	Brush et al. (2016)
Ni/MgO-Al_2O_3 (Mg/Al=0.5)	1/0.8/0.4	700	1060	85.0	–	–	15	78.0	–	–	8.2	Koo et al. (2009)
Ni/MgO-Al_2O_3 (Mg/Al=3.5)	1/0.8/0.4	700	1060	65.0	–	–	15	60.0	–	–	7.7	Koo et al. (2009)
Ni/$Ce_{0.8}Zr_{0.2}O_2$	1/0.8/0.4	800	26.5	90.8	–	–	20	81.7	–	–	10.0	Roh et al. (2009)
Ni/$MgAl_2O_4$	1/1/0.4	850	5	77.4	35.9	–	24	75.6	32.0	–	2.3	Bae et al. (2011)
Ni/Al_2O_3 (mesoporous)	1/0.8/0.4	800	69	80.9	81.9	2.03	40	82.5	77.0	2.09	–	Jabbour et al. (2016)
Ni/SBA-15	1/0.5/0.5	800	36	67.3	–	0.93	10	64.5	–	1.17	4.2	Siang et al. (2017)
Types of promoters												
Ni/MgO/SBA-15	1/0.75/0.5	850	27	98.6	92.3	1.74	600	95.6	76.0	1.82	3.0	Huang et al. (2008)
Ni/MgO-Al_2O_3	1/1.2/0.4	900	4680	92.3	62.0	–	5	89.6	64.9	–	2.9	Koo et al. (2015)
Ni-Ca/Al_2O_3	1/0.8/0.4	800	138	64.0	58.0	2.09	40	65.0	58.0	2.05	–	Jabbour et al. (2017)

(*Continued*)

TABLE 5.2 (*Continued*)
Performance of Selected Notable Catalysts Used in the BRM Reaction

				Initial Performance				Final Performance				
Catalysts	$CH_4/H_2O/CO_2$ Ratio	T (°C)	GHSV (L g_{cat}^{-1} h^{-1})	CH_4 Conversion (%)	CO_2 Conversion (%)	H_2/CO Ratio	TOS (h)	CH_4 Conversion (%)	CO_2 Conversion (%)	H_2/CO Ratio	D[a] (%)	References
Ni-Mg/Al_2O_3	1/0.8/0.4	800	138	79.0	73.0	2.10	40	80.0	74.0	2.10	–	Jabbour et al. (2017)
Ru/$ZnLaAlO_4$	1/0.75/0.5	800	12	94.5	62.0	2.10	–	–	–	–	–	Khani et al. (2016)
Pt//$ZnLaAlO_4$	1/0.75/0.5	800	12	87.5	78.2	1.90	–	–	–	–	–	Khani et al. (2016)
Ni-La/$MgAl_2O_4$	1/1.2/0.4	900	4680	77.7	54.8	3.10	5	68.1	44.7	3.10	12.4	Park et al. (2015a)
Mo_2C-Ni/ZrO_2	1/0.8/0.4	850	60	97.7	79.2	1.95	20	96.5	73.2	1.89	1.3	Li et al. (2015)
Ni/Ce-SBA-15	1/0.5/0.5	800	36	77.7	–	1.60	10	71.3	–	1.65	8.2	Siang et al. (2017)
Co-Pt/Al_2O_3	1/0.2/1	800	–	99.5	73.7	–	–	–	–	–	–	Itkulova et al. (2014)
Ni-Ce/$MgAl_2O_4$ (Ce/Ni=0.25)	1/0.8/0.4	700	530	81.3	65.9	2.10	20	–	–	–	–	Koo et al. (2014)
Ni-Ce/$MgAl_2O_4$ (Ce/Ni=1.0)	1/0.8/0.4	700	530	74.9	52.3	2.20	20	–	–	–	–	Koo et al. (2014)
Ni/Ce-SBA-15	1/0.5/0.5	800	36	77.7	–	1.60	10	71.3	–	1.65	8.2	Siang et al. (2017)
Ni-Ce-Fe/Al_2O_3	1/1.3/0.8	900	24.6	90.3	65.6	–	50	87.8	64.0	–	2.8	Li et al. (2016b)

[a] Degree of catalyst deactivation, D (%) = [1 – (final CH_4 conversion/initial CH_4 conversion)] × 100%. TOS: time-on-stream.

5.4 EFFECTS OF CATALYST SUPPORTS

Support materials play a crucial role in developing heterogeneous catalysts with homogeneous metal dispersion and resistance for coke formation. Thus, a well-suited support is required by considering its physical and chemical natures including the surface area, pore characteristic, surface basicity, redox properties, thermal stability, and oxygen storage to promote the formation of metal-support interaction, well incorporation of active metal particles, sintering, and coke resistances.

Numerous studies have reported the performance of BRM over nickel (Ni) supported on metal oxides, metal carbides, hydrotalcite-like materials, and mesoporous materials, as summarized in Table 5.2. Metal oxides have attracted considerable attention in BRM reaction owing to their interesting textural properties (outstanding chemical and thermal stability, high oxygen mobilization, basicity, and redox attributes). As reported in the literature, NiO/MgO (Olah et al. 2012), Ni/CeO_2 (Roh et al. 2009), and Ni/ZrO_2 (Li et al. 2015) have good activity for bi-reforming reaction with CO_2 conversion of 71.3%, 55.7% and 92.3%, under 830°C, 800°C and 850°C, respectively. The degree of catalyst deactivation is 1% (time-over-stream (TOS) = 320 h), 14.9% (TOS = 20 h), and 11.0% (TOS = 20 h) for NiO/MgO Ni/CeO_2 and Ni/ZrO_2, respectively. Apart from single oxide supports, mixed oxides support such as MgO-Al_2O_3 was also reported in other studies (Koo et al. 2009) for BRM reaction with the CH_4 conversion of 85% and 65% for Mg/Al ratios of 0.5 and 3.5, respectively, under the reaction temperature of 850°C.

The potential of a metal carbide-based catalyst also has been explored in literature for BRM. This type of support has been widely explored since 1973 when Levy and Boudart (1973) declared that tungsten carbide (WC) demonstrated comparable activity to platinum (Pt). Along with WC, molybdenum carbide (Mo_2C) catalyst has been one of the primary carbides reported on the reforming reactions owing to its comparable activities to iridium (Ir) and ruthenium (Ru) catalysts (Claridge et al. 1998). The interesting features of the Mo_2C as the catalyst in reforming reaction is related to its capability to inhibit coke deposition without requiring an excess of oxidant because of its unusual mechanism that consists of competing carburization and oxidation reactions (LaMont and Thomson 2005). Brush et al. (2016) explored the performance of Ni/Mo_2C for bi-reforming reaction. They found that the Ni/Mo_2C catalyst can catalyze a BRM reaction (CH_4 conversion = 70.5%, CO_2 conversion = 99.8%) with no signs of coking or whiskering despite conducting the reaction with excessive methane at high temperature.

Apart from metal oxides and metal carbide, vast attention has been devoted to $Ce_{1-x}Zr_xO_2$ materials, attributed to their interesting properties, such as thermal stability, abundant oxygen storage capacity, and redox virtue because of the partial substitution of Ce^{4+} by Zr^{4+} in the CeO_2 lattice (Kašpar et al. 1999). Roh et al. (2009) reported enhanced performance of Ni/$Ce_{0.8}Zr_{0.2}O_2$ as compared to Ni/Al_2O_3 and Ni/CeO_2. Ni-$Ce_{0.8}Zr_{0.2}O_2$ showed better catalytic performance and stability with the percentage CH_4 conversion of 90.8% and 10% degree of catalyst deactivation at the same reaction situation. The key success for a favorable catalytic performance of Ni-$Ce_{0.8}Zr_{0.2}O_2$ was related to the homogeneously dispersed nano-sized NiO crystallites and nano-crystalline property of cubic $Ce_{0.8}Zr_{0.2}$ support leading to close Ni and support contact and better Ni distribution.

Moreover, recent study on the BRM reaction shows that catalysts dispersed on mesoporous materials such as Ni/SBA-15 (Siang et al. 2017) and Ni/Al_2O_3 (mesoporous) (Jabbour et al. 2016) possess outstanding results with respect to reactant conversion and stability because of the exceptional structural characteristics of these supports. Jabbour et al. (2016) explored the performance of mesoporous Ni/Al_2O_3 for the BRM reaction. In their study, mesoporous Ni/Al_2O_3 was compared with silica-based catalysts (Ni/CeliteS and Ni/SBA-15) to inspect the influence of the support for properties and catalytic activities. The result showed that the re-oxidation of the Ni^0 active phase led to deactivation of the silica-based catalysts while the metallic Ni^0 form was conserved in Al_2O_3-supported catalysts. For the impact of a mesoporous structure, Ni/Al_2O_3 (mesoporous) was compared with Ni/Al_2O_3(non-porous). The result revealed that Ni/Al_2O_3 (mesoporous) exhibited great activity and stability owing to the reinforced Ni-support interaction and well-scattered Ni^0

nanoparticles in the ordered Al_2O_3 framework, indicating the importance of ordered-structured material in the dispersion of Ni particles. According to the available literature summarized in Table 5.2, the types of support materials have a profound effect on metal particles dispersion, which directly affects the catalytic activity of a BRM reaction. Therefore, an appropriate selection of the support is required for efficient BRM.

5.5 EFFECTS OF PROMOTERS

As widely reported in literature, incorporating other appropriate metals as promoters can lessen carbon deposition on active sites and hence improve the total performance of a catalyst for BRM. In brief, the promoter can boost the catalytic activity of Ni-based catalysts by easing NiO reduction, rising amounts of basic sites, and facilitating the formation of low temperature active sites. The current promoters used for BRM are summarized in Table 5.2.

Several studies have reported the modification of Ni-based catalysts using basic modifiers, such as alkaline earth or alkali metal oxides. This type of modifier promotes the adsorption or activation of CO_2 and its following reaction with nearby carbonaceous deposits, and thus resulting in CO formation ($CO_2 + C \rightarrow 2CO$) (Huang et al. 2008). As described by Huang et al. (2008), incorporating MgO on Ni/SBA-15 ameliorated the activity and stability of catalyst whereby only a 3% degree of deactivation was detected after 600 hours. The exceptional catalytic performance of Ni/MgO/SBA-15 was related to homogeneous Ni^0 species dispersion through a nickel-magnesia solid solutions formation, and increasing CO_2 adsorption affinity, which could suppress deposited carbon formation and hinder deactivation. The result observed in Huang et al. (2008) concur with findings reported by Koo et al. (2015) for Ni/MgO-Al_2O_3 catalysts. They found that the introduction of MgO on Ni-Al_2O_3 successfully formed a nano-sized catalyst with high coke resistance as a result of the strong metal-support interaction with the introduction of MgO. Moreover, Jabbour et al. (2017) examined Ni-Ca/Al_2O_3 and Ni-Mg/Al_2O_3 performance for the BRM reaction. In their study, Mg^{2+} and Ca^{2+} containing salts (which are inexpensive and highly available were elected as additives) added to their capability to generate basic natures which are important in the BRM reaction. They found that the magnesium (Mg) or calcium (Ca) did not change the structural features of the Ni/Al_2O_3 but effectively played a positive role in the escaping side reactions (formation of carbon nanotubes) and resulted in high selectivity and a stable H_2/CO ratio.

Apart from alkali or alkaline earth metal oxides modification, the modification of Ni-based catalyst using noble metals could be considered a worthy attempt to enhance the catalytic performance, particularly in the stability and coke resistance of the catalyst. Several studies (Itkulova et al. 2014; Khani et al. 2016) found that the stability and activity of the Ni-based catalysts could be significantly enhanced by introducing a small quantity of noble metal as promoter. Khani et al. (2016) examined the introduction of Ru on $LaZnAlO_4$ for the BRM reaction. They found that the Ru addition facilitated the dispersion and reducibility of metal, enhanced sintering-resistibility, and inhibited coke formation. The introduction of Ru increased the CH_4 conversion from 92% to 94.5%, thus indicating the positive role of Ru in the BRM reaction over Ru/$ZnLaAlO_4$.

Itkulova et al. (2014) explored the addition of Pt on Co/Al_2O_3 for the BRM reaction. The results showed that Pt-Co/Al_2O_3 has great stability and carbon resistance because of Pt's positive role in the formation and stabilization of finely dispersed and reduced bimetallic nano-particles. In addition, this catalyst is able to produce syngas at a relatively low temperature (700°C–750°C). However, owing to the limitation of noble metals related to the high cost and poor availability, several types of secondary elements have been explored in recent years for BRM.

Rare earth metals such as La_2O_3 and/or CeO_2 have attracted considerable attention owing to their beneficial characteristics. Park et al. (2015a) explored the influence of lanthanum (La) as a promoter for the $MgAl_2O_4$ catalyst for the BRM reaction. The results indicated that Ni-La/$MgAl_2O_4$ has better stability and activity as compared to $MgAl_2O_4$ due to its relatively large surface area and Ni dispersion. This observation mostly was because of the positive role of La in inhibiting Ni particles from agglomerating because of enhanced metal-support interaction. In addition, Koo et al. (2014) found that the combination of Ni-based catalysts with active oxygen carriers such as CeO_2 produced an efficient catalyst for BRM because of the strong interaction between support and metal along with predominant active oxygen transfer through close contact with Ni-Ce. A similar result was observed by Siang et al. (2017) for Ni/Ce-SBA-15. They found that the introduction of cesium (Ce) on Ni supported on a mesoporous material (Ni/SBA-15) demonstrated better activity and stability than Ni/SBA-15 as a CeO_2 phase, which possessed great redox property and oxygen storage capacity, and was integrated into mesoporous SBA-15 framework.

In addition to rare earth metals, non-precious transition-metal carbides have gained significant attention lately owing to their outstanding bulk and surface properties. Li et al. (2015) reported that addition of low-content Mo_2C in Ni/ZrO_2 exhibits superior activity and stability during the BRM reaction owing to an improvement in Ni dispersion and the coke morphologies modification resulting from the transformation in the Ni-ZrO_2 interactions. Other than La and Mo_2C, the role of boron (B) as a promoter was investigated by Siang et al. (2018). In their study, B-promoted Ni/SBA-15 was compared with the Ni/SBA-15 for the BRM reaction. They found that the addition of B suppressed graphitic carbon formation and caused a four times reduction of carbonaceous deposition.

In certain cases, the addition of tertiary metals is needed to increase the performance of a catalyst. Li et al. (2016b) explored the influence of Ce and iron (Fe) as promoters for Ni/Al_2O_3. In their study, the properties and catalytic performance of Ni–Ce–Fe/Al_2O_3 were compared to Ni/Al_2O_3. They found that Ce and Fe additions significantly improved Ni particles dispersion with strong metal-support interaction and enhanced resistance for coke formation. CH_4 and CO_2 conversions grew nearly 20% and 16%, respectively, compared to Ni/Al_2O_3. Interestingly, the weight loss of spent catalyst (50 h TOS) was almost negligible indicating the high resistance of catalyst for coke formation.

5.6 EFFECTS OF PROCESS VARIABLES ON METHANE BI-REFORMING REACTION

In recent years, numerous studies about the role of process variables on the catalytic BRM performance have been extensively investigated to ascertain the dependency degree of catalytic activity and selectivity from process variables (involving reactant feed composition, temperature, and gas hourly space velocity (GHSV)). The exploration of operating variables is crucial for catalyst design and development (Al-Nakoua and El-Naas 2012; Ryi et al. 2014), optimization of BRM operating conditions (Olah et al. 2015; Karemore et al. 2016), and the derivation of intrinsic BRM kinetics (Park et al. 2015b; Jang et al. 2016).

5.6.1 Influence of Gas Hourly Space Velocity

As an important operating parameter in heterogeneous catalytic reaction, gas hourly space velocity (GHSV) has been investigated extensively for BRM to avoid present mass and heat transfer limitations and hence obtaining inherent kinetics of a catalyst as well as optimizing catalytic performance. Huang et al. (2008) examined the influence of GHSV on BRM at 850°C and CH_4:CO_2:H_2O of 2:1:1.5 over 3%MgO-10%Ni/SBA-15. They noticed that as GHSV was smaller than 27 L g_{cat}^{-1} h^{-1}, the dependency of CH_4 and CO_2 conversions on GHSV was not obvious. However, as GHSV increased from 27 to 54 L g_{cat}^{-1} h^{-1}, conversion of CO_2 was dramatically dropped from 93.4% to 64.3% while a slight decline in CH_4 conversion was evident, implying that GHSV has more significant influence on DRM than SRM.

Li and Veen (2018) reported that an appreciable enhancement in reactant conversions (of about 81% and 87% for CO_2 and CH_4, respectively) was achieved with reducing GHSV over Mg-Al oxide-supported Ni catalyst from 13.6 × 10^4 to 2.8 × 10^4 h^{-1} during a BRM reaction. They also found that both CH_4 and CO_2 conversions reached near to equilibrium conversions for CH_4 of 94% and CO_2 of 77% at a GHSV value of 2.8 × 10^4 h^{-1} while the ratio of H_2 to CO was unchanged and ranged from 1.99 to 2.01. Thus, they deduced that varying GHSV did not alter or significantly affect the equilibrium between SRM and DRM side reactions. However, decreasing GHSV could enhance these reactions.

Kim et al. (2017) also observed similar findings in their study of Ni/M-Al_2O_3 catalysts (with M being samarium (Sm), Mg, or Ce promoters) for BRM (see Table 5.3) at different GHSV values (ranging within 36 to 72 L g_{cat}^{-1} h^{-1}). Regardless of catalyst types, they found that increasing GHSV could drop CH_4 and CO_2 conversions approximately 13.9% and 31.9%, respectively. Nevertheless, varying GHSV did not affect the sequence of CH_4 conversion; Ni/Sm-Al_2O_3 > Ni/Al_2O_3 > Ni/Ce-Al_2O_3 > Ni/Mg-Al_2O_3. In addition, all catalysts exhibited the highest CH_4 conversion close to equilibrium conversion values (97.3%) at low GHSV = 36 L g_{cat}^{-1} h^{-1}.

5.6.2 Influence of Reaction Temperature

As BRM is an extremely endothermic reaction, a BRM catalytic performance could be favored with a raise in reaction temperature. However, active metal sintering and carbon deposition resulting from endothermic methane decomposition may be induced by harsh reaction temperature. Carbonaceous deposition and metal sintering have been widely recognized as main factors triggering catalyst deactivation in a BRM reaction (Al-Nakoua and El-Naas 2012; Ryi et al. 2014; Jabbour et al. 2017). Hence, reaction temperature is a vital process variable requiring meticulous investigation for manipulating the magnitude of undesirable parallel side reactions and simultaneously enhancing the performance of primary reforming processes.

In the studies of reaction temperature effect on BRM reactions over various catalysts (Soria et al. 2011; Ryi et al. 2014; Karemore et al. 2016), a rise in temperature from 400°C to 800°C reportedly enhanced CH_4 and CO_2 conversions by up to 81% but reduced the H_2/CO ratio by about 67.3% (cf. Table 5.3) because of the BRM endothermic nature. However, at a low temperature of 500°C, Karemore et al. (2016) found that the predominant WGS side reaction generated an excessive amount of intermediate CO_2 during BRM and thus strongly suppressed CO_2 reactant consumption. Therefore, it could yield negative CO_2 conversion (–19.4% as seen in Table 5.3). Karemore et al. (2016) also reported that irrespective of reaction temperature within 500°C–700°C, CH_4 conversion was greater than CO_2 conversion because of the dominant occurrence of WGS and SRM reactions.

Ryi et al. (2014) examined the BRM process with the use of a catalytic Ni membrane at a varying reaction temperature from 650°C to 750°C. They found that CH_4 preferentially reacted with H_2O instead of CO_2 oxidants in BRM at a low reaction temperature because of lower H_2O stability. The preferred reaction between CH_4 and H_2O could lead to the dominance of SRM. However, when reaction temperature increased from 650°C to 750°C, CO_2 conversion reportedly improved from 80.6% to 94.5% due to the predominance of DRM, RWGS, and CO_2 gasification reactions. Therefore, a substantial decrease in ratio of H_2/CO from 7.5 to 5.3 (cf. Table 5.3) was experienced for the BRM reaction (Ryi et al. 2014).

5.6.3 Influence of Feedstock Composition

The BRM reaction is an integrated process of DRM and SRM reactions to yield syngas (H_2/CO = 2) at a temperature range of 700°C–850°C. One of the main benefits from implementing a BRM reaction is the capability to provide flexible adjustment for H_2/CO ratios used in downstream processes by the regulation of feedstock composition. For this reason, examining the influence of feedstock composition on BRM performance for tuning H_2/CO ratios is significantly essential for various industrial applications. Primarily, the feed ratio of oxidants (i.e., H_2O/CO_2 ratio) played a pivotal role in controlling the H_2/CO ratio and carbon hindrance (Soria et al. 2011).

TABLE 5.3
Effects of Process Variables on Performance of Selected Catalysts Used in the BRM Reaction

Catalyst	Operating Conditions			CH_4 Conversion (%)	CO_2 Conversion (%)	H_2/CO Ratio	References
	$CH_4/H_2O/CO_2$ Ratio	Temperature (°C)	GHSV ($L\ g_{cat}^{-1}\ h^{-1}$)				
Ni/Al_2O_3	3/1–2/2–1	800	2.5	84.1–83.0	85.7–78.2	1.48–2.20	Karemore et al.
	3/2/1	500–800	2.5	16.0–84.0	(−19.4)–79.0	6.70–2.19	(2016)
Ni/Mo_2C	1/0.625–0.125/0.125–0.625	700	–	77.6–82.8	97.5–99.3	3.73–1.08	Brush et al. (2016)
$Ni/Sm\text{-}Al_2O_3$	1/1.2/0.38	800	36–72	97.1–90.0	42.8–34.2	–	Kim et al. (2017)
$Ni/Ce\text{-}Al_2O_3$				96.4–83.4	40.5–27.6	–	
$Ni/Mg\text{-}Al_2O_3$				94.2–81.1	37.1–26.9	–	
$Ru/ZrO_2\text{-}La_2O_3$	1/0.1–0.5/1	500	400	18.2–24.4	17.0–9.8	0.93–0.90	Soria et al. (2011)
	1/0.5/1	400–500	400	7.7–24.4	0–9.8	3.69–1.90	
Mg-promoted Ni/	2/2–0.5/0.5–2	850	27	97.7–98.9	94.1–84.0	1.61–1.85	Huang et al. (2008)
SBA-15	2/1.5/1	700–850	27	68.1–98.4	65.0–91.9	–	
	2/1.5/1	850	10–60	99.2–93.7	93.4–64.3	1.70–1.91	
$Ni/MgAl_2O_4$	3/0–3/1.2	775	8.6×10^{4} [a]	35.5–90.5	93.2–50.0	2.25–1.00	Li and Veen (2018)
	3/2.2/1.2	700–800	8.6×10^{4} [a]	52.0–78.0	39.0–68.0	2.22–1.95	
	3/2.2/1.2	775	$(2.8–13.6) \times 10^{4}$ [a]	86.2–69.5	80.9–60.3	2.01–1.99	
Ni membrane	1/2.7–1.5/0.3–1.5	800	–	96.0–92.9	–	7.50–1.80	Ryi et al. (2014)
	1/2.7/0.3	650–750	–	80.6–94.5	–	7.50–5.30	

[a] The unit of GHSV is h^{-1}.

Brush et al. (2016) synthesized the molybdenum carbide supported nickel (Ni/Mo_2C) catalyst by carburization of $NiMoO_4/MoO_3$ material. The resulting catalyst was subjected for the BRM reaction at a constant temperature of 950°C and varying H_2O/CO_2 ratios (0%–30 %), whereas the composition of CH_4 was kept consistent at 40%. They found that the H_2/CO ratio can be altered within 0.91–3.0 by adjusting H_2O/CO_2 oxidant ratios. Irrespective of oxidant compositions, the catalyst was deactivated after certain time intervals. However, both whisker or graphitic carbons were not found on spent catalyst surface, as normally anticipated from nickel-based catalysts during a BRM reaction. Therefore, they claimed that the catalytic deactivation occurred with time-on-stream due to the re-oxidation of molybdenum carbide to its respective MoO_2 oxide.

In another report, a series of MgO-promoted Ni/SBA-15 catalysts were tested for the BRM reaction (Huang et al. 2008). The initial experiments found that a 3% MgO- promoted catalyst had the greatest catalytic performance compared to its counterparts. Then, this catalyst was further tested with different CO_2/H_2O ratios. The H_2/CO ratio was reportedly increased from 1.61 to 2.78 by increasing H_2O composition in the feedstock. The stability tests were also carried out over 3%MgO promoted and unpromoted 10%Ni/SBA-15 catalysts for 620 hours. The activity of the 10%Ni/SBA-15 catalyst was dropped after 120 hours, whereas the catalyst promoted with MgO showed comparatively better stability due to the synergistic effect of promoter and oxidant composition. However, its catalytic activity was also decreased after 400 hours. Most importantly, they also reported a drop in the performance of 3%MgO-Ni/SBA-15 because of metallic Ni^0 oxidation to NiO species. Similarly, Ryi et al. (2014) prepared a catalytic nickel membrane and tested at various oxidant mixtures for three different temperatures (650°C–750°C) and a short interval of 4 hours. They reported that H_2/CO ratio and CH_4 conversion was increased to 11% and 84.2%, respectively, by decreasing the CO_2/H_2O ratio at temperature of 923 K. Moreover, negligible carbon formation was observed in post-reaction characterization. Therefore, it can be concluded from these findings that the major reason for catalytic deactivation is the re-oxidation of active metal sites to the corresponding inactive metal oxides.

Karemore et al. (2016) examined the impact of oxidants (H_2O and CO_2) and reaction temperature on H_2/CO ratio for a BRM reaction on the Ni/Al_2O_3 catalyst. The results for the BRM reaction revealed that the H_2/CO ratio was decreased from 6.73 to 2.2 by increasing the temperature from 500°C to 800°C. In contrast, H_2/CO ratio was reduced from 2.20 to 1.48 by increasing the CO_2/H_2O ratio from 0.5 to 2. Recently, the effect of varying CH_4/H_2O ratios was investigated for the BRM reaction on $Ni/MgAl_2O_4$ catalyst by Li and Veen (2018). The results revealed that the CH_4 conversion was improved by almost three-fold (from 35.5% to 90.5%), whereas CO_2 conversion was dropped to 50% by increasing CH_4/H_2O ratios from 3/0 to 3/3. In contrast, H_2/CO ratio was relatively enhanced from 1.5 to 2.4 by increasing H_2O composition. Hence, it can be deduced from these observations that the ratio of H_2/CO can be easily tuned by varying the amounts of oxidants in reactant mixtures. However, the impact of CH_4 partial pressure on the catalytic performance and carbon formation is nascent on the ground of the BRM reaction studies.

5.7 MECHANISMS AND KINETICS OF METHANE BI-REFORMING

The kinetic studies are the main concerns in both academic and industrial realms to establish an adequate reaction rate model primarily derived from the intrinsically mechanistic reaction steps. The kinetic model also must assure the best fit to experimental data and potentially capture reaction rates of reactants and formation rate of products. Studying the mechanisms and kinetics of a BRM reaction is crucial since the mechanistic-derived kinetic model could benefit from the optimization of a BRM process, industrial BRM reactor design, and catalyst synthesis. However, in comparison with other common reforming reactions, such as SRM, DRM, and POM, kinetic investigation for BRM is comparatively fewer. Thus, the kinetic research of the BRM process could be a noteworthy emphasis in the near future.

To understand the inherently mechanistic pathways of BRM, Qin et al. (1996) conducted the in-situ isotope experiment using the $^{13}CO_2$ reactant for BRM on an MgO supported rhodium (Rh) catalyst.

They found that the CH_x species ($0 \leq x \leq 3$) arising from CH_4 dissociation were more active than surface carbon produced from CO or CO_2 decomposition and suggested that SRM and DRM could consume the same type of intermediate (i.e., adsorbed oxygen on catalyst surface, O_{ad} originating from CO_2 or H_2O decomposition) to yield a CO product since these reactions evidently occurred at the same time. In addition, they found that SRM and DRM reactions could occur concurrently and possess relatively similar reaction pathways. Hence, Qin et al. (1996) proposed a Langmuir-Hinshelwood (LH) BRM mechanism (as summarized in Table 5.4), wherein surface reaction between adsorbed species is the rate determining step. Jabbour et al. (2017) also suggested an analogous mechanistic pathway for BRM over a Mg-doped Ni/Al_2O_3 catalyst. In addition, they found that the basic property of catalyst remarkably facilitated the dissociation rates of H_2O and CO_2 to yield surface activated O_{ad} species which could react with nearby adsorbed carbonaceous species. Thus, it could hinder the polymerization of carbon to coke formation and improve catalytic activity.

To the best of our knowledge, there are no reported kinetic BRM models derived from inherently mechanistic steps. Most attempts for expressing reactant consumption and product formation rates were based on the established kinetic models of SRM and DRM reactions. Challiwala et al. (2017) conducted a kinetic assessment for BRM using the kinetic rate expressions of stand-alone SRM and DRM Langmuir-Hinshelwood-Hougen-Watson (LHHW) mechanisms proposed by Xu and Froment (1989) and Verykios (2003), respectively. Abashar (2004) and Shahkarami and Fatemi (2015) also used similar SRM and DRM kinetic rate expressions to perform mathematical modelling for BRM process optimization and reactor design without proposing BRM kinetic models. Instead of deriving kinetic model from mechanistic pathways, Park et al. (2015b) used a power-law model (see Eq. 5.16) to capture CH_4 consumption rate and estimate the corresponding reactant reaction order for BRM over unpromoted and Co- or La-promoted Ni/Al_2O_3 catalysts. BRM runs with various feedstock composition ($CH_4/CO_2/H_2O/Ar$) under temperature ranging from 600°C to 850°C were conducted for deriving the power-law model where r_{CH4} and k_{CH4} are CH_4 reaction rate and CH_4 rate constant in this order.

$$r_{CH_4} = k_{CH_4}(P_{CH_4})^{\alpha}(P_{CO_2})^{\beta}(P_{H_2O})^{\gamma} \quad (5.16)$$

The partial pressure of CH_4, CO_2, and H_2O is denoted as P_{CH_4}, P_{CO_2}, and P_{H_2O}, respectively, while α, β, and γ are the corresponding reaction orders for CH_4, CO_2, and H_2O reactants. As summarized in Table 5.5, depending on catalyst types, the reaction orders of CH_4 were computed within 1.35 to 1.39, whereas reaction orders for other reactants (i.e., H_2O and CO_2) were close to zero. The apparent

TABLE 5.4
BRM Mechanistic Steps over Rh/MgO Catalyst Proposed

Mechanistic steps	
CH_4 activation	$CH_4 + 2X \rightarrow CH_3 - X + H - X$ [a]
	$CH_3 - X + 2X \rightarrow CH - X + 2H - X$
	$CH - X + X \rightarrow C - X + H - X$
H_2O and CO_2 dissociation	$H_2O + 3X \rightarrow O - X + 2H - X$
	$CO_2 + 2X \rightarrow O - X + CO - X$
Surface reaction of adsorbed species	$CH_x - X + O - X + (x-1)X \rightarrow CO - X + xH - X$
H_2 and CO formation steps	$CO - X \rightarrow CO + X$
	$2H - X \rightarrow H_2 + 2H$

Source: Qin, D. et al., *J. Catal.*, 159, 140–149, 1996.

[a] *X* denotes the available Rh active sites.

TABLE 5.5
Estimated Associated Parameters from the Power-Law Model for BRM over Selected Ni-based Catalysts

Catalysts	Reaction Temperature (°C)	Apparent Activation Energy, E_a(kJ mol^{-1})	Pre-exponential Factor, A × 10^{10}	Reaction Order of CH_4	CH_4 rate Constant, k_{CH_4}
15%Ni/Al_2O_3	600–850	82.7	2.96	1.36	20.48
5%La-15%Ni/Al_2O_3	600–850	85.2	7.21	1.39	20.98
1%Co-5%La-15%Ni/Al_2O_3	600–850	93.8	13.50	1.35	23.15
3%Co-5%La-15%Ni/Al_2O_3	600–850	99.4	29.80	1.39	26.61

activation energy, E_a, for CH_4 consumption was also estimated with the values of 82.7–99.4 kJ mol^{-1} for different types of catalysts. In addition, Park et al. (2015b) noticed that the apparent activation energies increased with cobalt (Co) promotion in the order: 15%Ni/Al_2O_3 < 5%La-15%Ni/Al_2O_3 < 1%Co-5%La-15%Ni/Al_2O_3 < 3%Co-5%La-15%Ni/Al_2O_3 catalysts. The increase in apparent activation energy with Co addition was ascribed to the dynamic redox cycle in which metallic Co was initially oxidized to Co-O and these cobalt oxide species were subsequently reduced back to metallic Co form by carbonaceous species on catalyst surface (Ruckenstein and Wang 2002; Park et al. 2015b). Hence, they deduced that the latter step (the reduction of Co-O to metallic Co species) was the rate-determining step for the bi-reforming of methane over Co-doped catalysts leading to a higher apparent activation energy.

5.8 CONCLUSIONS AND OUTLOOK

Noble catalysts have been widely used in the BRM. Despite their excellent catalytic performance, precious metal-based catalysts are not the best-suited candidates for industrial purpose due to price and availability concerns. Therefore, the current research is focused on the Ni-based catalysts because of their activity, availability, and stability, which is also as an extension based on the commercial catalysts for SRM and DRM reactions. However, the Ni-based catalyst has high carbonaceous formation and thus leading to the rapid deactivation. Owing to a major drawback of Ni-based catalyst for application in the industrial sector, the modification of the catalyst by varying the support materials and promoters could enhance the performance of catalyst and hence result in higher conversion and selectivity.

Based on this review, appropriate support selection must be done considering its textural and chemical attributes to boost the active metal particles dispersion, facilitate the metal-support interaction, minimize metal sintering, stimulate the reduction of the catalyst, and most importantly to reduce or resist the carbonaceous species formation. Meanwhile, the addition of suitable promoters with the appropriate crystallites size deposited onto the support can enhance metal-support interaction. Therefore, it could avoid metal particles agglomeration during a BRM reaction. It is proposed that developing the synergistic effects of suitable metals composition with proper selection of the support can successfully synthesize a favorable catalyst for the BRM reaction. Moreover, the research on a catalyst that is stable at bi-reforming operating conditions and particularly more practical in industrial scale is highly needed for commercialization.

The operating conditions including GHSV, reaction temperature, and feedstock composition have demonstrated an essential role in BRM catalytic performance. Thus, it is crucial to conduct the optimization of process parameters for BRM or establish an accurate correlation between process variables and catalytic performance in future studies. In addition, considering the limited BRM kinetic models and inadequate knowledge about BRM mechanism, advanced characterization techniques including in situ adsorption spectroscopy (of CO_2 and CH_4 reactants on catalyst surface), temperature-programmed desorption, and chemisorption must be performed for a thorough understanding of BRM surface reaction steps for deriving mechanistic-based kinetic models.

ACKNOWLEDGMENTS

The financial sponsorship offered by Universiti Malaysia Pahang (UMP Research Grant Scheme: RDU170326) for carrying out this work is appreciatively acknowledged.

REFERENCES

Abashar, M. E. E. 2004. Coupling of steam and dry reforming of methane in catalytic fluidized bed membrane reactors. *International Journal of Hydrogen Energy* 29:799–808.

Abdullah, B., N. A. A. Ghani, and D.-V. N. Vo. 2017. Recent advances in dry reforming of methane over Ni-based catalysts. *Journal of Cleaner Production* 162:170–185.

Al-Nakoua, M. A. and M. H. El-Naas. 2012. Combined steam and dry reforming of methane in narrow channel reactors. *International Journal of Hydrogen Energy* 37:7538–7544.

Asencios, Y. J. and E. M. Assaf. 2013. Combination of dry reforming and partial oxidation of methane on NiO–MgO–ZrO_2 catalyst: Effect of nickel content. *Fuel Processing Technology* 106:247–252.

Awe, O. W., Y. Zhao, A. Nzihou, D. P. Minh, and N. Lyczko. 2017. A review of biogas utilisation, purification and upgrading technologies. *Waste and Biomass Valorization* 8:267–283.

Bae, J. W., A. R. Kim, S. C. Baek, and K. W. Jun. 2011. The role of CeO_2–ZrO_2 distribution on the Ni/$MgAl_2O_4$ catalyst during the combined steam and CO_2 reforming of methane. *Reaction Kinetics, Mechanisms and Catalysis* 104:377–388.

Bahari, M. B., N. H. H. Phuc, F. Alenazey, K. B. Vu, N. Ainirazali, and D.-V. N. Vo. 2017. Catalytic performance of La-Ni/Al_2O_3 catalyst for CO_2 reforming of ethanol. *Catalysis Today* 291:67–75.

Bale, C. W., E. Bélisle, P. Chartrand, S. A. Decterov, G. Eriksson, A. E. Gheribi, K. Hack, I. H. Jung, Y. B. Kang, J. Melançon, A. D. Pelton, S. Petersen, C. Robelin, J. Sangster, and M. A. Van Ende. 2010. FactSage thermochemical software and databases. *Calphad* 54:35–53.

Brush, A., E. J. Evans, G. M. Mullen, K. Jarvis, and C. B. Mullins. 2016. Tunable Syngas ratio via bi-reforming over coke-resistant Ni/Mo_2C catalyst. *Fuel Processing Technology* 153:111–120.

Challiwala, M. S., M. M. Ghouri, P. Linke, M. M. El-Halwagi, and N. O. Elbashir. 2017. A combined thermo-kinetic analysis of various methane reforming technologies: Comparison with dry reforming. *Journal of CO_2 Utilization* 17:99–111.

Chen, X., J. Jiang, K. Li, S. Tian, and F. Yan. 2017. Energy-efficient biogas reforming process to produce syngas: The enhanced methane conversion by O_2. *Applied Energy* 185:687–697.

Claridge, J. B., A. P. York, A. J. Brungs, C. Marquez-Alvarez, J. Sloan, S. C. Tsang, and M. L. Green. 1998. New catalysts for the conversion of methane to synthesis gas: Molybdenum and tungsten carbide. *Journal of Catalysis* 180:85–100.

Contaldi, M., F. Gracceva, and A. Mattucci. 2008. Hydrogen perspectives in Italy: Analysis of possible deployment scenarios. *International Journal of Hydrogen Energy* 33:1630–1642.

Fan, J., L. Zhu, P. Jiang, L. Li, and H. Liu. 2016. Comparative exergy analysis of chemical looping combustion thermally coupled and conventional steam methane reforming for hydrogen production. *Journal of Cleaner Production* 131:247–258.

Fischedick, M., J. Nitsch, and S. Ramesohl. 2005. The role of hydrogen for the long-term development of sustainable energy systems: A case study for Germany. *Solar Energy* 78:678–686.

Goeppert, A., M. Czaun, J. P. Jones, G. S. Prakash, and G. A. Olah. 2014. Recycling of carbon dioxide to methanol and derived products-closing the loop. *Chemical Society Reviews* 43:7995–8048.

Hay, J. X. W., T. Y. Wu, and J. C. Juan. 2013. Biohydrogen production through photo fermentation or dark fermentation using waste as a substrate: Overview, economics, and future prospects of hydrogen usage. *Biofuels, Bioproducts and Biorefining* 7:334–352.

Huang, B., X. Li, S. Ji, B. Lang, F. Habimana, and C. Li. 2008. Effect of MgO promoter on Ni-based SBA-15 catalysts for combined steam and carbon dioxide reforming of methane. *Journal of Natural Gas Chemistry* 17:225–231.

Internal Energy Agency. 2017. World energy outlook 2017. Chapter 1 (February): 33–61. http://www.iea.org/media/weowebsite/2017/Chap1_WEO2017.pdf.

Itkulova, S. S., G. D. Zakumbaeva, Y. Y. Nurmakanov, A. A. Mukazhanova, and A. K. Yermaganbetova. 2014. Syngas production by bi-reforming of methane over Co-based alumina-supported catalysts. *Catalysis Today* 228:194–198.

Jabbour, K., N. El Hassan, A. Davidson, S. Casale, and P. Massiani. 2016. Factors affecting the long-term stability of mesoporous nickel-based catalysts in combined steam and dry reforming of methane. *Catalysis Science & Technology* 6:4616–4631.

Jabbour, K., P. Massiani, A. Davidson, S. Casale, and N. El Hassan. 2017. Ordered mesoporous "one-pot" synthesized Ni-Mg (Ca)-Al_2O_3 as effective and remarkably stable catalysts for combined steam and dry reforming of methane (CSDRM). *Applied Catalysis B: Environmental* 201:527–542.

Jang, W. J., D. W. Jeong, J. O. Shim, H. M. Kim, H. S. Roh, I. H. Son, and S. J. Lee. 2016. Combined steam and carbon dioxide reforming of methane and side reactions: Thermodynamic equilibrium analysis and experimental application. *Applied Energy* 173:80–91.

Karemore, A. L., P. D. Vaidya, R. Sinha, and P. Chugh. 2016. On the dry and mixed reforming of methane over Ni/Al_2O_3–Influence of reaction variables on syngas production. *International Journal of Hydrogen Energy* 41:22963–22975.

Kašpar, J., P. Fornasiero, and M. Graziani, M. 1999. Use of CeO_2-based oxides in the three-way catalysis. *Catalysis Today* 50:285–298.

Khani, Y., Z. Shariatinia, and F. Bahadoran. 2016. High catalytic activity and stability of $ZnLaAlO_4$ supported Ni, Pt and Ru nanocatalysts applied in the dry, steam and combined dry-steam reforming of methane. *Chemical Engineering Journal* 299:353–366.

Kim, A. R., H. Y. Lee, J. M. Cho, J. H. Choi, and J. W. Bae. 2017. Ni/M-Al_2O_3 (M= Sm, Ce or Mg) for combined steam and CO_2 reforming of CH_4 from coke oven gas. *Journal of CO_2Utilization* 21:211–218.

Koo, K. Y., H. S. Roh, U. H. Jung, D. J. Seo, Y. S. Seo, and W. L. Yoon. 2009. Combined H_2O and CO_2 reforming of CH_4 over nano-sized Ni/MgO-Al_2O_3 catalysts for synthesis gas production for gas to liquid (GTL): Effect of Mg/Al mixed ratio on coke formation. *Catalysis Today* 146:166–171.

Koo, K. Y., J. H. Lee, U. H. Jung, S. H. Kim, and W. L. Yoon. 2015. Combined H_2O and CO_2 reforming of coke oven gas over Ca-promoted Ni/$MgAl_2O_4$ catalyst for direct reduced iron production. *Fuel* 153:303–309.

Koo, K. Y., S. H. Lee, U. H. Jung, S. H. Roh, and W. L. Yoon. 2014. Syngas production via combined steam and carbon dioxide reforming of methane over Ni–Ce/$MgAl_2O_4$ catalysts with enhanced coke resistance. *Fuel Processing Technology* 119:151–157.

LaMont, D. C. and W. J. Thomson. 2005. Dry reforming kinetics over a bulk molybdenum carbide catalyst. *Chemical Engineering Science* 60:3553–3559.

Levy, R. B. and M. Boudart. 1973. Platinum-like behavior of tungsten carbide in surface catalysis. *Science* 181:547–549.

Li, D., X. Li, and J. Gong. 2016a. Catalytic reforming of oxygenates: State of the art and future prospects. *Chemical Reviews* 116:11529–11653.

Li, M. and A. C. V. Veen. 2018. Coupled reforming of methane to syngas ($2H_2$-CO) over Mg-Al oxide supported Ni catalyst. *Applied Catalysis A: General* 550:176–183.

Li, P., Y. H. Park, D. J. Moon, N. C. Park, and Y. C. Kim. 2016b. Carbon deposition onto Ni-Based catalysts for combined steam/CO_2 reforming of methane. *Journal of Nanoscience and Nanotechnology* 16:1562–1566.

Li, W., Z. Zhao, P. Ren, and G. Wang. 2015. Effect of molybdenum carbide concentration on the Ni/ZrO_2 catalysts for steam-CO_2 bi-reforming of methane. *RSC Advances* 5:100865–100872.

Lide, D. R. 2003. *Handbook of Chemistry and Physics*, 84th ed. Boca Raton, FL: CRC Press/Taylor & Francis.

Liu, K., C. Song, and V. Subramani. 2009. *Hydrogen and Syngas Production and Purification Technologies.* New Jersey, NJ: John Wiley & Sons.

Olah, G. A., A. Goeppert, M. Czaun, and G. S. Prakash. 2012. Bi-reforming of methane from any source with steam and carbon dioxide exclusively to metgas (CO–$2H_2$) for methanol and hydrocarbon synthesis. *Journal of the American Chemical Society* 135:648–650.

Olah, G. A., A. Goeppert, M. Czaun, T. Mathew, R. B. May, and G. S. Prakash. 2015. Single step bi-reforming and oxidative bi-reforming of methane (natural gas) with steam and carbon dioxide to metgas (CO-$2H_2$) for methanol synthesis: Self-sufficient effective and exclusive oxygenation of methane to methanol with oxygen. *Journal of the American Chemical Society* 137:8720–8729.

Park, J. E., K. Y. Koo, U. H. Jung, J. H., Lee, H. S. Roh, and W. L. Yoon. 2015a. Syngas production by combined steam and CO_2 reforming of coke oven gas over highly sinter-stable La-promoted Ni/$MgAl_2O_4$ catalyst. *International Journal of Hydrogen Energy* 40:13909–13917.

Park, M. H., B. K. Choi, Y. H. Park, D. J. Moon, N. C. Park, and Y. C. Kim. 2015b. Kinetics for steam and CO_2 reforming of methane over Ni/La/Al_2O_3 catalyst. *Journal of Nanoscience and Nanotechnology* 15:5255–5258.

Qin, D., J. Lapszewicz, and X. Jiang. 1996. Comparison of partial oxidation and steam-CO_2 mixed reforming of CH_4 to Syngas on MgO-Supported metals. *Journal of Catalysis* 159:140–149.

Roh, H. S., K. Y. Koo, and W. L. Yoon. 2009. Combined reforming of methane over co-precipitated Ni–CeO_2, Ni–ZrO_2 and Ni–$Ce_{0.8}Zr_{0.2}O_2$ catalysts to produce synthesis gas for gas to liquid (GTL) process. *Catalysis Today* 146:71–75.

Ruckenstein, E., and H. Y. Wang. 2002. Carbon deposition and catalytic deactivation during CO_2 reforming of CH_4 over Co/γ-Al_2O_3 catalysts. *Journal of Catalysis* 205:289–293.

Ryi, S. K., S. W. Lee, J. W. Park, D. K. Oh, J. S. Park, and S. S. Kim. 2014. Combined steam and CO_2 reforming of methane using catalytic nickel membrane for gas to liquid (GTL) process. *Catalysis Today* 236:49–56.

Schulz, T. F., S. Kypreos, L. Barreto, and A. Wokaun. 2008. Intermediate steps towards the 2000 W society in Switzerland: An energy–economic scenario analysis. *Energy Policy* 36:1303–1317.

Shahkarami, P., and S. Fatemi. 2015. Mathematical modeling and optimization of combined steam and dry reforming of methane process in catalytic fluidized bed membrane reactor. *Chemical Engineering Communications* 202:774–786.

Siang, T. J., H. T. Danh, S. Singh, Q. D. Truong, H. D. Setiabudi, and D.-V. N. Vo. 2017 Syngas production from combined steam and carbon dioxide reforming of methane over Ce-modified silica-supported nickel catalysts. *Chemical Engineering Transactions* 56:1129–1134.

Siang, T. J., T. L. Pham, N. Van Cuong, P. T. Phuong, N. H. H. Phuc, Q. D. Truong, and D.-V. N. Vo. 2018. Combined steam and CO_2 reforming of methane for syngas production over carbon-resistant boron-promoted Ni/SBA-15 catalysts. *Microporous and Mesoporous Materials* 262:122–132.

Sørensen, B., A. H. Peterson, C. Juhl, H. Ravn, C. Søndergren, P. Simonsen, K. Jørgensen, L. H. Nielsen, H. V. Larsen, P. E. Morthorst, L. Schleisner, F. Sørensen, and T. E. Pedersen. 2004. Hydrogen as an energy carrier: scenarios for future use of hydrogen in the Danish energy system. *International Journal of Hydrogen Energy* 29:23–32.

Soria, M. A., C. Mateos-Pedrero, A. Guerrero-Ruiz, and I. Rodríguez-Ramos. 2011. Thermodynamic and experimental study of combined dry and steam reforming of methane on Ru/ZrO2-La_2O_3 catalyst at low temperature. *International Journal of Hydrogen Energy* 36:15212–15220.

Starr, K., X. Gabarrell, G. Villalba, L. Talens, and L. Lombardi. 2012. Life cycle assessment of biogas upgrading technologies. *Waste Management* 32:991–999.

Sudhakaran, M. S. P., L. Sultana, M. M. Hossain, J. Pawlat, J. Diatczyk, V. Brüser, S. Reuter, and Y. S. Mok. 2017. Iron–ceria spinel ($FeCe_2O_4$) catalyst for dry reforming of propane to inhibit carbon formation. *Journal of Industrial and Engineering Chemistry* 61:142–151.

Sun, Q., H. Li, J. Yan, L. Liu, Z. Yu, and X. Yu. 2015. Selection of appropriate biogas upgrading technology-A review of biogas cleaning, upgrading and utilisation. *Renewable and Sustainable Energy Reviews* 51:521–532.

Tavanarad, M., F. Meshkani, and M. Rezaei. 2018. Production of syngas via glycerol dry reforming on Ni catalysts supported on mesoporous nanocrystalline Al_2O_3. *Journal of CO_2 Utilization* 24:298–305.

Verykios, X. E. 2003. Catalytic dry reforming of natural gas for the production of chemicals and hydrogen. *International Journal of Hydrogen Energy* 28:1045–1063.

Weiland, P. 2010. Biogas production: Current state and perspectives. *Applied microbiology and biotechnology* 85:849–860.

Wietschel, M., U. Hasenauer, and A. de Groot. 2006. Development of European hydrogen infrastructure scenarios – CO_2 reduction potential and infrastructure investment. *Energy Policy* 34:1284–1298.

Wood, D. A., C. Nwaoha, and B. F. Towler. 2012. Gas-to-liquids (GTL): A review of an industry offering several routes for monetizing natural gas. *Journal of Natural Gas Science and Engineering* 9:196–208.

Xu, J., and G. F. Froment. 1989. Methane steam reforming, methanation and water-gas shift: I. Intrinsic kinetics. *AIChE Journal* 35:88–96.

6 Effects of Mesoporous Supports and Metals on Steam Reforming of Alcohols

Richard Y. Abrokwah, William Dade, Sri Lanka Owen, Vishwanath Deshmane, Mahbubur Rahman, and Debasish Kuila

CONTENTS

6.1 Introduction 93
6.2 Experimental 95
6.2.1 Materials 95
6.2.2 Experimental Procedure 95
6.2.3 Characterization of Catalysts 95
6.2.4 Testing of Catalysts 96
6.3 Results and Discussion 98
6.3.1 X-ray Diffraction Studies of Powder Catalysts 98
6.3.2 Textural Property Analysis of Catalysts 99
6.3.3 H_2-TPR Studies: Reduction Behavior of M-TiO_2 and M-MCM-41 Catalysts 100
6.4 Catalysts Performance Test for Steam Reforming Activity 101
6.4.1 SRM Activity Tests 101
6.4.2 Effect of Different Metals Supported on Mesoporous Silica, MCM-41 for SRM Studies 103
6.4.3 Glycerol Stream Reforming Activity Test 104
6.4.4 GSR Catalyst Stability Studies 105
6.5 Conclusions 106
Acknowledgments 106
References 106

6.1 INTRODUCTION

The world's continuously growing energy needs necessitates the search for more efficient alternative energy sources to supplement traditional fossil fuels. The salient concerns of the Intergovernmental Panel on Climate Change (IPCC) are to reduce particulate emissions, hydrofluorocarbons (HFCs), perfluorocarbons (PFCs), sulfur, and nitrogen oxides among other pollutants. The IPCC envisages that the ecosystems, water resources, coastlines, and human health could be detrimentally compromised if fossil fuel explorations are not abated. One major source of energy that has not been utilized much is hydrogen. In industry, hydrogen is a raw material for the manufacture of fertilizer, drugs, and plastics as well as hydrogenation of fats and oils. Hydrogen also is in high demand in the proton exchange membrane fuel cells (PEMFC) to power modern fuel cell electric vehicles (Hoffmann 2001; D'Souza et al. 2006; Deshmane and Adewuyi 2012). Steam reforming of alcohols (methanol, ethanol, and glycerol) remains a promising route to hydrothermally extract hydrogen from these alcohols. Methanol reforming is more popular because it can be performed at low temperature. However, glycerol and ethanol require high temperatures for reforming due to the presence of C–C bonds that require much higher energy to break. Nonetheless, the heightened interest in biodiesel production, which yields a significant 10% of glycerol as by-product, has renewed the use of glycerol as a feedstock for hydrogen production. Besides its abundant availability, it has a high hydrogen throughput to glycerol ratio (1 mole glycerol:7 moles H_2).

Steam reforming of alcohols is an endothermic process that requires a catalyst to produce hydrogen. We and others (Park and Lee 2010; Deshmane et al. 2015a, 2015b) have shown that in addition to the optimum thermodynamic conditions, the choice of metal-support system plays an indispensable role in conversion of the feedstock, selectivity towards hydrogen, and the overall performance of the catalysts. Deshmane et al. (2015a) found that Cu-MCM-41, with 15% Cu loading, showed excellent catalytic performance compared to other metals, and yielded ~90% methanol conversion, 100% H_2 selectivity, and a carbon monoxide (CO) selectivity of 0.8% for steam reforming of methanol (SRM) at 300°C and 2838 h^{-1} gas hourly space velocity (GHSV). Furthermore, the Cu-MCM-41 catalyst exhibited strong resistance to deactivation and maintained consistent performance over a time-on-stream (TOS) for 48 hours.

Deshmane et al. (2015a) observed that although Cu-MCM-41 is an excellent catalyst for H_2 production with minimal CO formation, Cu supported on TiO_2 is not. The SRM activity and performance of a Cu catalyst encapsulated in different supports (TiO_2, MCM-41, SBA-15, CeO_2, and ZrO_2) and that of different metals in M-MCM-41 (M = Cu, nickel (Ni), cobalt (Co), zinc (Zn), palladium (Pd), and tin (Sn)) are discussed in this paper. We also have used the knowledge of steam reforming activity of methanol and that reported for glycerol steam reforming (GSR) in literature in the design of catalysts for H_2 production from glycerol. Some of the results from our GSR work using MCM-41 support containing CeO_2 and TiO_2 are included in this chapter.

Among the metals tested as catalysts for GSR are Co (Zhang et al. 2007) Ni, (Adhikari et al. 2007; Douette et al. 2007; Zhang et al. 2007; Iriondo et al. 2009), iridium (Ir) (Adhikari et al. 2007; Zhang et al. 2007), Pd (Swami and Abraham 2006; Adhikari et al. 2007; Douette et al. 2007), ruthenium (Ru) (Hirai et al. 2005; Byrd et al. 2008), rhodium (Rh), and platinum (Pt) (Adhikari et al. 2007). Urasaki et al. (2005) suggested that perovskite-type oxides as catalytic support reduced the formation of carbon due to their high content of lattice oxygen. It has been reported that CeO_2 as a support exhibits good redox characteristics (high oxygen storage capacity) that is able to significantly diminish the residual CO formed in steam reforming reactions (SSRs) (Llorca et al. 2003; Ramírez-Cabrera et al. 2003). In an experiment conducted by Gonzalez and co-workers (González Vargas et al. 2013), structural analysis of Rh/CeO_2 catalyst revealed that CeO_2 prevented the highly dispersed Rh particles from sintering and thus maintained sufficient Rh/CeO_2 interfacial areas, which facilitated coke gasification through the high oxygen storage-release capacity (Cai et al. 2008). In addition, CeO_2 is known to efficiently catalyze the water-gas shift reaction (Zhang et al. 2007; Montini et al. 2010), enhancing CO elimination, thus preventing catalyst deactivation.

As in the case of CeO_2, TiO_2 has been found to be useful in catalytic applications. A photo catalysis study conducted by Wu et al. (2008) on TiO_2 revealed that the catalytic effect of TiO_2 is enhanced by the ease of reducibility of the O_2 moiety, which could improve hydrogen selectivity in SRRs. This effect was proven indeed in a previous work on thermal SRM carried out by our group, wherein a one-pot hydrothermal method was used to synthesize mesoporous titania having a higher surface area (Owen 2014). Impregnation of acidic supports like Al_2O_3 with basic oxides like La_2O_3 or CeO_2 also have been shown to minimize the acidic site density on the effective catalyst surface (Wen et al. 2008). Wu and Kawi (2009) showed that Ce^{4+} and Ti^{4+} could be introduced into the structure of MCM-41 and SBA-15 by a hydrothermal route, which strongly improved the reducibility of the Rh/Ce-MCM-41 and Rh/Ce-SBA-15 catalysts used for ethanol steam reforming studies. We recently elucidated the high thermal stability of amorphous MCM-41 and the strong electronic properties of the crystalline TiO_2 supports, respectively (Abrokwah et al. 2016). In this chapter, different metals immobilized on MCM-41 support modified with TiO_2 and CeO_2 (CeO_2-MCM-41 and TiO_2-MCM-41) are investigated for GSR activity. This hybrid support system can provide many benefits such as enhancement of the metal reducibility, reduction in sintering and coke formation as well as increased stability and better performance of the catalyst.

6.2 EXPERIMENTAL

6.2.1 Materials

The precursors, i.e., 98% titanium isopropoxide (TIPR), tetramethylorthosilicate, 99% (TMOS), tetraethylorthosilicate, 99% (TEOS), pluronic acid-P123, as well as ammonium hydroxide were obtained from Acros Organics, New Jersey, USA. Sigma-Aldrich, Missouri, provided hexadecyltrimethylammonium bromide, 99% (CTAB) and metal precursors such as $Cu(NO_3)_2.H_2O$ 98% $CoCl_2.6H_2O$, $ZnNO_3.6H_2O$ 98%, and $SnCl_2.2H_2O$ 98%. The other analytical grade chemicals such as $PdNO_3.H_2O$, ZrO_2, $Ni(NO_3)_2$, $Ce(NO_3)_4$, hydrogen fluoride (HF), anhydrous ethanol, and acetone were acquired from Fischer Scientific, New Jersey, USA. Deionized water was utilized for all experiments and we used the reagents without any additional purification.

6.2.2 Experimental Procedure

The MCM-41, CeO_2, ZrO_2, and TiO_2-supported catalysts were prepared based on the optimized (hydrothermal one-pot) procedure from our previous work (Deshmane et al. 2015a, 2015b). On the other hand, SBA-15 preparation followed a one-pot direct synthesis approach wherein neither a basic precipitation of the metal hydroxides nor catalyst washing was performed. In SBA-15 synthesis molar ratios of [1 TEOS: 0.081 CTAB: 41 H_2O: 7.5 ethanol: 0.01679 P123: 5.981 HCl] was used. The quantity of metal precursors used, as specified in the previous section, depended on the desired metal loading in the final support. The SBA-15 samples were specifically prepared by first stirring P123 in 2M HCl at 35°C until complete dissolution (designated *solution one*). A second solution (*solution two*) was prepared by dissolving CTAB in deionized water until a homogeneous mixture was obtained. *Solution two* was then gently poured into *solution one* and stirred continuously for 30 minutes to obtain a new *mixture A*. The metal precursors were dissolved in ethanol and added dropwise to *mixture A*. The limiting reagent-TEOS was then added dropwise to the *final mixture* and stirred for 20 hours at 35°C. The *final mixture* was aged at 80°C for 48 hours in an oven followed by air-drying for at least 24 hours. Afterwards the powder sample was oven-dried again for 24 hours at 98°C. Finally, the dried materials were calcined. The amorphous MCM-41 and SBA-15 samples were calcined in a stepwise fashion – 1°C/minute at 350°C for 8 hours, 1°C/minute at 450°C for 8 hours, and 1°C/minute at 550°C for 8 hours. The crystalline TiO_2, CeO_2, and ZrO_2 samples were calcined at a maximum temperature of 450°C for 5 hours.

6.2.3 Characterization of Catalysts

Small and wide-angle powder X-ray diffractions (XRD) were recorded using a D8 discover X-ray diffractometer from Bruker (Bruker Optics, Inc., Billerica, Massachusetts, USA) with a Position sensitive device (PSD) detector, using CuKα radiation generated at 40 mA and 40 kV at the scanning rate of 0.014°/s. The crystal sizes of the metal oxides were determined using Scherrer equation (Ingham and Toney 2014).

$$\tau = \frac{0.9\lambda}{\beta \cdot \cos\theta} \tag{6.1}$$

where, τ is the crystal size, λ (0.1541 nm) is the wavelength of the Cu Kα radiation, β is the full width half maximum, and θ is the Braggs diffraction angle. Surface area, pore-size, and pore volume of the catalysts were determined using a Quantachrome NOVA 2200e instrument (Quantachrome Instruments, Boynton Beach, Florida, USA). The surface area was calculated using the Brunauer-Emmett-Teller (BET) equation from the adsorption branch of the isotherm in a relative pressure (P/P_0) range of 0.07–0.3. The total pore volume was evaluated based on the amount of N_2 adsorbed at a $P/P_0 = 1$.

The non-local density functional theory (NLDFT) (based on the classical Kelvin equation) and Barrett-Joyner-Halenda (BJH) (based on modified Kevin equation) models are both commercially available and widely used for pore size and distribution computation. The elaborate review by Lowell et al. (2004) indicated that the BJH model is more accurate for heterogeneous surfaces (like those of crystalline materials) in the meso-macropore regime, whereas the NLDFT model is more precise for mixed micro-mesopore range materials having well-defined pore geometry and long-range ordered periodicity like MCM-41, MCM-48, and SBA-15. They affirmed that, unlike the NLDFT, the BJH model was based on several macroscopic thermodynamic assumptions; hence, the pore mode diameter and/or pore sizes determined by the BJH approach (for pores less than 10 nm) are usually 25% short of the actual values. Therefore, the pore size distribution and average pore sizes of the amorphous MCM-41 and SBA-15 samples were determined by the NLDFT method (Jaroniec and Solovyov 2006) and those of crystalline samples (TiO_2, CeO_2 and ZrO_2) by the BJH method (Barrett et al. 1951).

The H_2 TPR analysis was carried out with the AutoChem II 2920 Chemical Analyzer from Micromeritics Instrument Corp. (Norcross, Georgia, USA) equipped with a thermal conductivity detector (TCD). In a typical TPR method, the catalyst was purged with a 10% H_2/90% argon (Ar) gas mixture at flow rate of 50 milliters per minute (mL/min) while the temperature was ramped at 10°C/minute from room temperature to 1000°C and held for 25 minutes. Thermogravimetric analysis (TGA) and differential scanning calorimetry (DSC) were performed using the SDT Q600 V20.4 Build 14 system (TA Instruments, New Castle, Delaware, USA) at a heating rate of 10°C/minute with airflow rate of 50 mL/min. The elemental compositions were estimated using an inductively coupled plasma optical emission spectroscopy (ICP-OES) Agilent 710-ES spectrometer and an energy dispersive X-ray spectroscopy (EDX)-Zeiss EVO LS10 scanning electron microscopy (SEM) equipped with Oxford INCA X-act detector. Fourier transform infrared (FTIR) spectra were recorded using Shimadzu IR Prestige-21 FTIR 8300 spectrometer equipped with liquid-N_2–cooled mercury-cadmium-telluride (MCT) detector. The KBr pelletization method was employed for the sample preparation and the spectrum was recorded at room temperature in the range of 400–4,000 cm^{-1} at 4 cm^{-1} resolution. The catalyst morphology was studied with the Zeiss Libra 120 (Carl Zeiss NTS GmbH, Oberkochen, Germany) transmission electron microscope (TEM) operated at an accelerating voltage of 120 kilovolts (kV).

6.2.4 Testing of Catalysts

SRM and GSR experiments were carried out in a tubular stainless steel-packed bed reactor. The operating and reaction parameters are summarized in Table 6.1.

The catalyst/sand mixture was packed into the reactor and sealed at both ends with quartz wool. The middle of the reactor was fitted with a thermocouple to monitor temperature. Hitherto reforming reactions, the catalyst was reduced in-situ for 1 hour using 10% H_2/Ar at 350°C and 550°C for the crystalline and amorphous supports, respectively. The gaseous reformates and collected

TABLE 6.1
Operating Parameters and Reaction Conditions for Steam Reforming Reactions

	Reaction Conditions	
Operation Parameters	**GSR**	**SRM**
Temperature (°C)	450–700	200–350
Gas hour space velocity (hr^{-1} at standard temperature and pressure, STP)	2,200	2,838
Water-to-feed ratio	12:1	3:1
Carrier gas	N_2 used as internal standard at 50 mL/min	No internal standard
Amount of catalyst (g)	3	2
Amount of white (50–70 mesh) sand (g)	3	1
Tubular reactor dimensions	Internal diameter: 10 mm, length 70 mm	Internal diameter: 6.2 mm, length 70 mm

condensates were analyzed with Agilent 7890B gas chromatography system. The gaseous products were separated using Restek Shin Carbon (2 m × 2 mm × 1/8") packed column and analyzed on a TCD with Ar as a carrier gas. The condensate was separated on Agilent DB-1 (60 m × 250 × 1 μm) capillary column and analyzed on flame ionization detector (FID) using helium (He) as the carrier gas. Although the selectivity and conversion of methanol were obtained using methanol mass balance based on feed and condensate (Equations 6.2 through 6.4), the glycerol counterpart was analyzed based on Eqs. 6.5 through 6.7.

$$X_{Methanol} = \frac{CH_3OH \text{ moles converted}}{CH_3OH \text{ moles fed}} \times 100\% \tag{6.2}$$

$$S_{H_2} = \frac{H_2 \text{ moles in product}}{H_2 \text{ moles in product} + 2 \times CH_4 \text{ moles in product}} \times 100\% \tag{6.3}$$

$$S_{CO/CO_2/CH_4} = \frac{CO/CO_2/CH_4 \text{ moles in product}}{(CO_2 + CO + CH_4) \text{ moles in product}} \times 100\% \tag{6.4}$$

The glycerol conversion and selectivity of gaseous products were calculated based on the following formulas:

$$H_2 \text{ Selectivity} = \frac{H_2 \text{ moles in product}}{\text{Moles C atoms produced in gas phase}} \times \left(\frac{3}{7}\right) \times 100\% \tag{6.5}$$

$$\text{Selectivity of x} = \frac{\text{Moles of C atoms in x}}{\text{Moles of C atoms produced in gas phase}} \times 100\% \tag{6.6}$$

$$\text{Glycerol conversion} = \frac{\text{Glycerol in} - \text{Glycerol out}}{\text{Glycerol in}} \times 100\% \tag{6.7}$$

Gas chromatography-mass spectrometry (GC-MS) analysis of the glycerol condensate recorded negligible amounts of formic acid, acetone, and acetic acid, which were not considered in our calculations.

6.3 RESULTS AND DISCUSSION

6.3.1 X-ray Diffraction Studies of Powder Catalysts

Figure 6.1 represents the small-angle XRD and wide-angle XRD patterns of different Ni/Co-MCM-41 samples that were modified with ceria and titania. The small-angle XRD of both Co and Ni catalysts showed a peak at a 2θ angle of ~2° corresponding to the (100) reflective plane, characteristic of ordered mesoporous silica (Zhou et al. 2011; Goscianska et al. 2014). The fact that the (200) reflective plane peak expected around 4.5° is absent which suggests that the catalysts possessed predominantly short-moderate range ordered mesoporosity with limited or no long range ordered mesoporous (Loebick et al. 2010). The wide-angle XRD of Co/MCM-41 and Ni/MCM-41 was X-ray amorphous. However, appearance of distinct diffraction peaks after modification confirmed that addition of CeO_2 and TiO_2 significantly improved the crystallinity of the host support. The sharp diffraction peaks observed at ~37° for both Ni and Co were probably due to the anatase phase of TiO_2 and mixed cobalt oxide crystals (Pouretedal and Ahmadi 2012). The Ni catalysts displayed peaks at 2θ of 44°, and 63.5° attributed to the corresponding (200) and (220) planes of cubic NiO crystallites (JCPDS# 78-0643) (Rossetti et al. 2013; Liu et al. 2014).

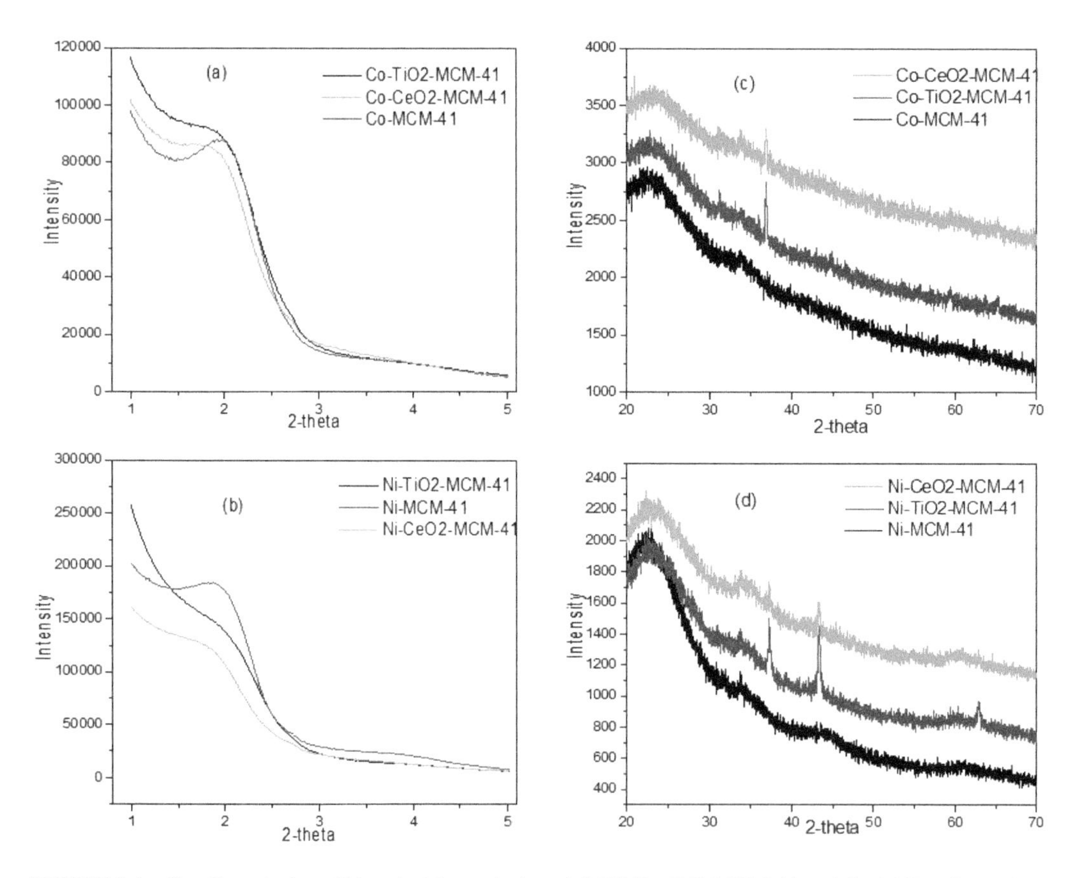

FIGURE 6.1 Small-angle (a and b) and wide-angle (c and d) XRD of Ni-MCM-41 and Co-MCM-41 catalysts modified with CeO_2 and TiO_2.

6.3.2 Textural Property Analysis of Catalysts

From Table 6.2, the actual metal loadings of different catalysts varied in the range of 8.35–11.31 weight percent (wt%), which is close to the intended/theoretical loading of 10 wt%. The actual ceria and titania loadings determined by EDX and ICP-OES were about 2.6–2.9 wt% and 4.5–4.9 wt%, respectively, compared to their theoretical loadings of 5 wt%. The lower weight percent of TiO_2 could be due to rapid condensation of the very reactive TIPR precursor resulting in some loss of the intended loading/material during the washing step of synthesis (Deshmane et al. 2015b).

The surface area of the MCM-41 support declined (by ~25%) as the metal loading was increased. This decline is expected as addition of metallic oxides to a mesoporous matrix tend to block some of the pores, thus decreasing the average porosity and consequently decreasing the surface area. Overall, the pore volume and pore sizes held steady upon metal incorporation and did not change much. All the catalysts exhibited type IV isotherms (figure not shown) signifying their mesoporosity.

It is worth mentioning from Table 6.3 that the surface area of 10 wt% Cu (catalysts used for SRM studies) supported on amorphous materials (SBA-15 and MCM-41) was about 3–7 times greater than 10 wt% Cu supported on crystalline materials-TiO_2, CeO_2, and ZrO_2. The pore diameters of all the catalysts were in the mesoporous regime with 10Cu/CeO_2 showing the largest ~10 nanometers (nm) and 10Cu/TiO_2 the least of 2.65 nm.

TABLE 6.2
Surface Areas, Pore Sizes, Pore Volumes, and Actual Metal Loadings of Different Catalysts before Steam Reforming Reactions

Catalyst	Surface Area (m^2/g)	Pore Size (nm)	Pore Volume (cm^3/g)	Actual Metal Loading (wt %) (EDX)	Actual CeO_2/TiO_2 Loading (wt %) (ICP-OES)
MCM-41	1039.2	3.31	0.75	0	0
10Ni-MCM-41	824.6	3.78	0.84	10.50	0
10Co-MCM-41	815.8	3.66	0.76	11.10	0
10Ni5TiO_2-MCM-41	787.3	3.78	0.82	9.45	2.64
10Co5TiO_2-MCM-41	754.9	3.66	0.69	11.31	2.94
10Ni5CeO_2-MCM-41	716.3	3.65	0.75	8.35	4.97
10Co5CeO_2-MCM-41	725.1	3.66	0.76	9.14	4.54
10-Sn-MCM-41	1009.3	3.33	0.77	8.45	0
10-Zn-MCM-41	800.3	3.50	0.78	8.99	0
10-Pd-MCM-41	797.6	2.90	0.81	8.95	0

TABLE 6.3
Textural Properties of Supported Copper Catalysts

Catalyst	Surface Area (m^2/g)	Pore Size (nm)	Pore Volume (cm^3/g)
10Cu/MCM-41	795.7	3.71	0.62
10Cu/SBA-15	875.1	4.82	0.85
10Cu/TiO_2	285.6	2.65	0.19
10Cu/CeO_2	121.7	10.21	0.53
10Cu/ZrO_2	124.2	3.64	0.16

6.3.3 H_2-TPR Studies: Reduction Behavior of M-TiO_2 and M-MCM-41 Catalysts

The TPR was performed to investigate the relative degree of reducibility of the metal oxides. The TPR profiles in Figure 6.2 show the contrast in H_2 consumption (activation) of the metal oxides when the support was changed. The reduction of bulk NiO crystallites that Richardson et al. (1992) reported that usually occurs in the range of 280°C–300°C was not observed in our study for Ni-TiO_2 and Ni-MCM41. We noticed reduction at a higher temperature (i.e., >600°C) that can be attributed to Ni-spinel polymorphs (Park and Lee 2010). Park and Lee (2010) drew the same conclusion for the reduction of $NiTiO_3$. Therefore, we suspect that the peak of Ni-MCM-41 (Figure 6.2a) between 600°C and 900°C is due to the formation of Ni-silicates/spinel- $NiSiO_3$/$NiSiO_4$ species. This observation suggested that NiO particles interacted more strongly with MCM-41 than TiO_2 support because oxides that interact strongly with the support tend to be reduced at higher temperatures. For Sn/TiO_2, the two peaks centered at 500°C and 650°C are attributed to a two-step reduction of bulk SnO_2 to metallic Sn ($Sn^{4+} \rightarrow Sn^{2+} \rightarrow Sn^0$) (Wang et al. 2012). Clearly, these two steps seem to be overlapped when the support was switched in the case of Sn/MCM41. Since the reduction temperature for bulk SnO_2 is normally above 750°C (Burch et al. 2000), reduction of SnO_2 species on both supports around 200°C–400°C represents the particles/oxides with expedient/optimum support interactions.

In Figure 6.2c, the H_2 consumption of 10Cu-TiO_2 suggested three types of reduction phases of CuO crystallites ($Cu^{2+} \rightarrow Cu^+ \rightarrow Cu^0$) (Tu et al. 2006). The first two peaks almost overlapping

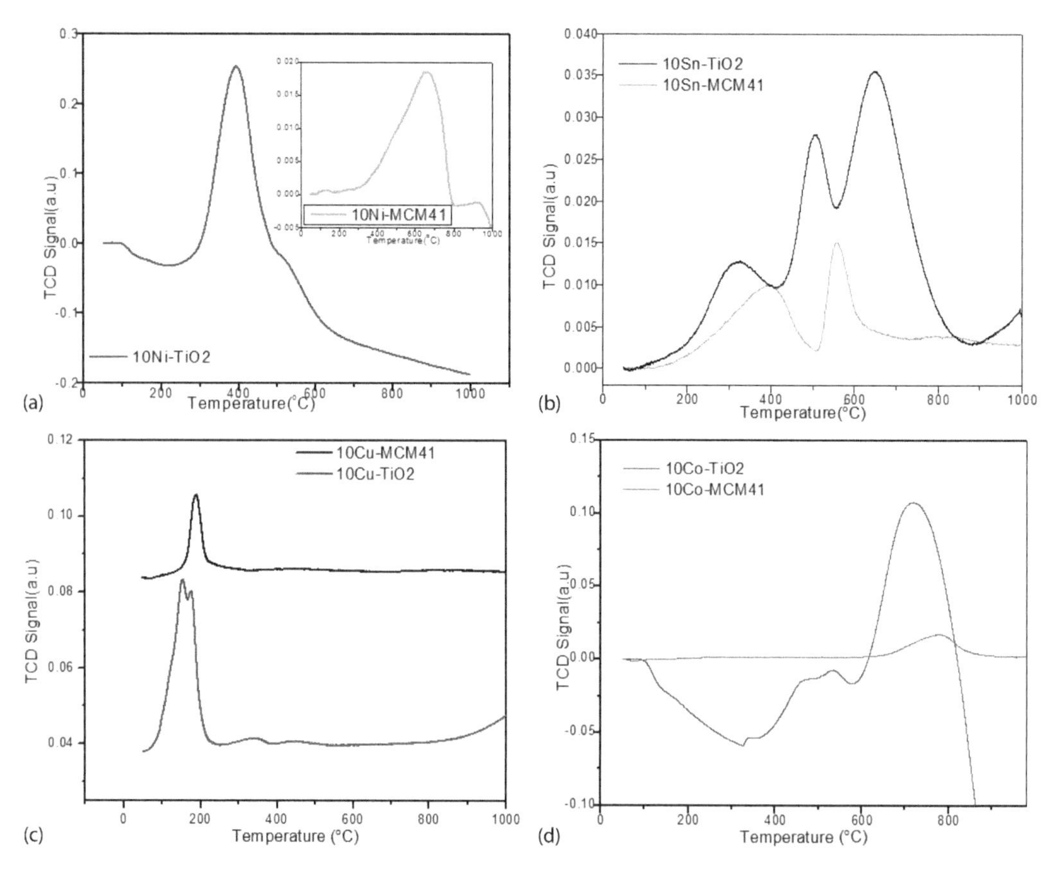

FIGURE 6.2 H_2-TPR profiles of (a) Ni-MCM-41 and Ni-TiO_2, (b) Sn-TiO_2 and Sn-MCM-41, (c) Cu-TiO_2 and Cu-MCM-41, and (d) Co-TiO_2 and Co-MCM-41.

occurred in the range of 70°C–250°C, and the third from 265°C to 380°C is assigned to the activation/reduction of CuO in the bulk phase (Tu et al. 2006). In contrast, the reduction of CuO-MCM41 was a one-step process from $Cu^{2+} \rightarrow Cu^0$ in the temperature range of 100°C–200°C. The reduction of Co-MCM-41 occurred in primarily one-step (600°C–800°C), whereas that of Co-TiO_2 consisted of several overlapping peaks. In the case of Co-MCM-41, Co_3O_4 oxides, which usually reduce between 400°C and 500°C, were not observed suggesting that CoO was most likely the dominant phase in the MCM-41 framework. The three-step reduction pattern of Co-TiO_2, with first H_2 consumption centered at ~450°C, the second at ~540°C, and the third at ~700°C correspond to the three-stage reduction of Co_3O_4 to Co^0 ($Co^{3+} \rightarrow Co^{2+} \rightarrow Co^0$) (Suriye et al. 2005; Yung et al. 2007).

The TPR profiles of Ni/Co-MCM-41, Co/Ni-CeO_2-MCM-41, and Co/Ni-TiO_2-MCM-41 (figure not shown) showed a doublet peak observed for Co/CeO_2-MCM reduction which agreed with reported observations by Vita et al. (2014). It is believed that due to the different binding strengths of Ce^{3+} and Ce^{4+}, the weaker interactions resulted in the reduction of cobalt oxides at 700°C and the stronger at 850°C. Other researchers have argued that the higher mobility of oxygen ions over ceria could be responsible for this two-stage reduction (Lucredio et al. 2011). The reduction of Ni^{2+} to Ni^0 reduction occurred in a wide temperature range of 600°C–680°C. The peak observed at 650°C for Ni/MCM-41 shifted to 600°C in Ni/TiO_2-MCM-41 and 680°C in Ni-CeO_2-MCM-41, thus showing the effect of ceria and titania as dopants in mesoporous silica (Cui et al. 2009).

6.4 CATALYSTS PERFORMANCE TEST FOR STEAM REFORMING ACTIVITY

6.4.1 SRM Activity Tests

The SRM activity of the copper catalysts with different supports was tested in the temperature range of 200°C–350°C at a GHSV of 2838 h^{-1} while the GSR was performed at 450°C–700°C at 2200 h^{-1} GHSV. Figure 6.3 illustrates the catalytic behavior of each catalyst for SRM showing the effect of reaction temperature on methanol conversion. Figure 6.4 shows the selectivity towards H_2, CO, CO_2, and CH_4. Conspicuously, the performance of the catalysts varied drastically when the support was changed from amorphous (MCM-41 and SBA-15) to crystalline (TiO_2, CeO_2, and ZrO_2).

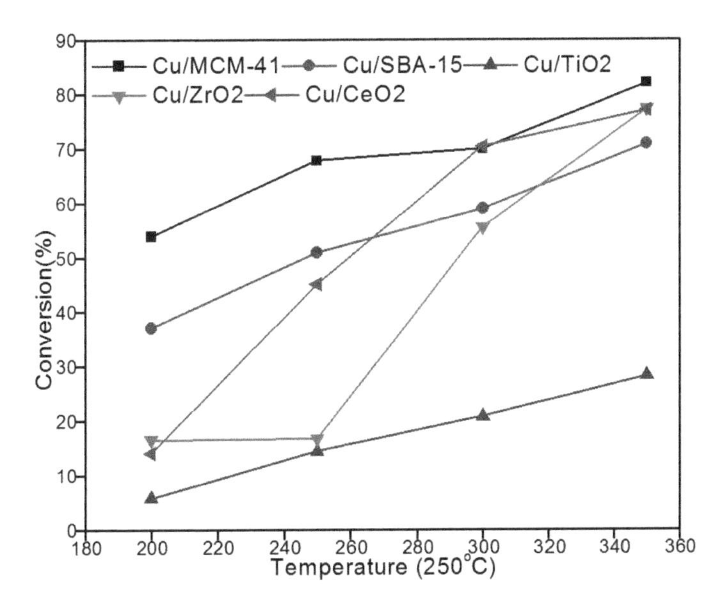

FIGURE 6.3 SRM activity of copper catalysts with different supports; reaction conditions: 200°C–350°C, 1 atm, 1:3 CH_3OH:H_2O molar ratio, and GHSV of 2,838 h^{-1} at STP.

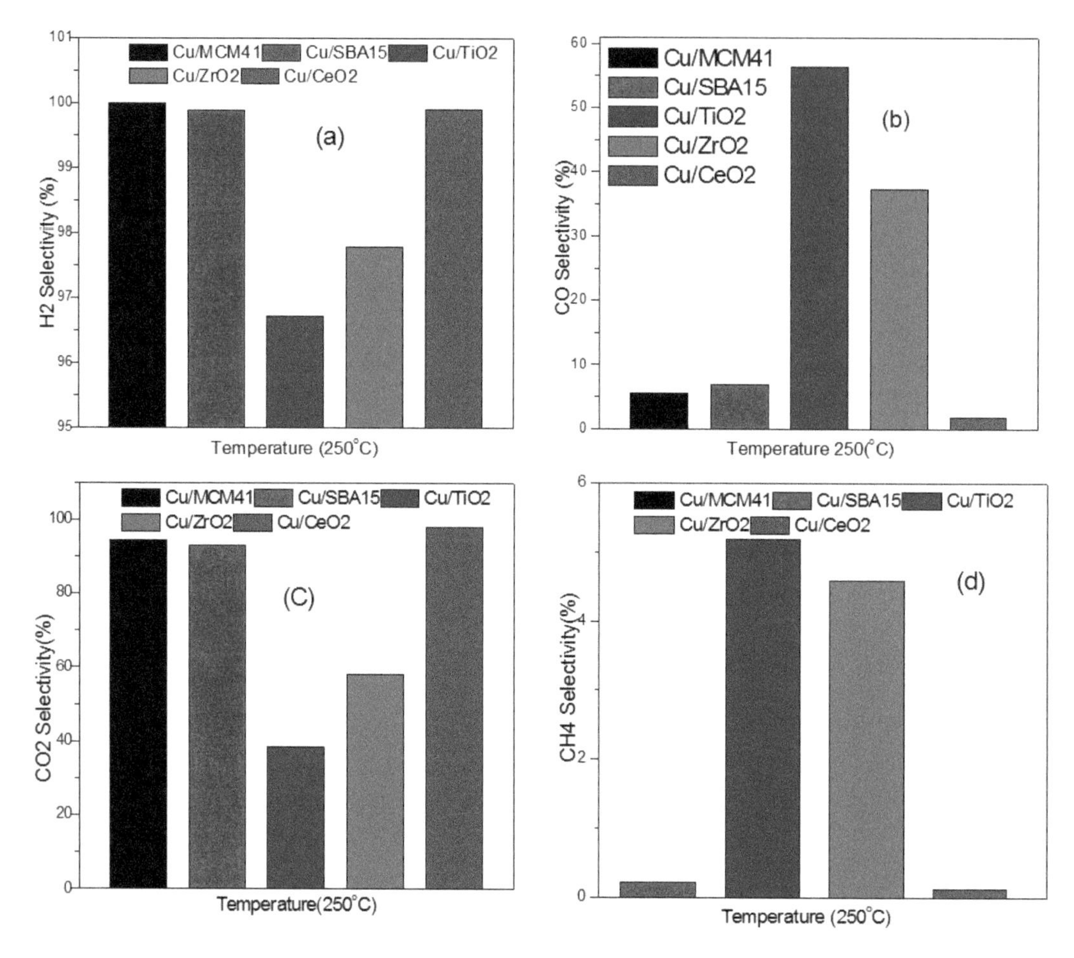

FIGURE 6.4 Effect of copper-support interactions on (a) H_2, (b) CO, (c) CO_2, and (d) CH_4 selectivity at 250°C, 1 atm, 1:3 CH_3OH:H_2O molar ratio and GHSV of 2,838 h^{-1} at STP.

For instance, at lower temperature (<250°C), the SRM activity of Cu was remarkably higher when immobilized on MCM-41 and SBA-15 compared to the TiO_2, CeO_2, and ZrO_2 supports.

This observation is evident around 200°C, wherein Cu/MCM-41 and Cu/SBA-15 yielded ~68% and 50% methanol conversion, while Cu/TiO_2, Cu/ZrO_2, Cu/CeO_2 produced 14%, 16%, and 45% conversions, respectively. However, after 250°C, the exponential change in methanol conversion by the crystalline materials, especially CeO_2 and ZrO_2, was drastic relative to MCM-41 and SBA-15 that increased steadily. Furthermore, at 350°C, CeO_2 and ZrO_2 showed better activity than SBA-15 support. This result could be because the crystallinity of the support increased with temperature, thus promoting many strong and favorable interactions between Cu atoms and the support. Overall, Cu/MCM-41 showed the best activity and Cu/TiO_2 the least. As shown in the TPR profiles, Cu/MCM-41 reduced easily to active Cu^0 in one-step compared to that observed for Cu/TiO_2, which occurred in more than two-steps. This difference in the ease of activation likely accounted for the slight activity dominance of Cu/MCM-41 over Cu/TiO_2.

Figure 6.4 compares the SRM selectivity of copper catalysts towards H_2, CO, CO_2, and CH_4 at the optimum performance temperature regarding the selectivity, conversion, and stability of 250°C, which was established in our previous work (Deshmane et al. 2015a, 2015b). All the catalyst showed excellent (>95%) H_2 selectivity. The CO selectivity showed significant disparities between the crystalline and amorphous catalysts. Cu/MCM-41 and Cu/SBA-15 exhibited less than 7% CO; compared to Cu/TiO_2 and Cu/ZrO_2 that showed quite high CO selectivity of ~56% and ~37%, respectively. Cu/CeO_2 yielded the lowest CO selectivity of 1.84% and the highest CO_2 selectivity of ~98%. Our results are consistent with that of CeO_2 reported in literature (Deeprasertkul et al. 2014). Deeprasertkul et al. (2014) inferred that CeO_2 has an inherent ability to activate water and retain oxygen in its crystal lattice. These properties enable CeO_2 to promote the water-gas shift reaction by further oxidizing any CO to CO_2. The availability of this excess oxygen also enabled the secondary reforming of any CH_4 formed during the reforming to CO_2. Overall, CH_4 selectivity by all the catalysts was less than 5%. Notably, Cu/MCM-41 produced no CH_4, while Cu/CeO_2 and Cu/SBA-15 produced CH_4 in negligible amounts (i.e., $<0.5\%$). We observed that as the CO_2 selectivity of Cu/MCM-41, Cu/SBA-15, and Cu/CeO_2 increased, the yield of CO decreased dramatically. This indicated that these catalysts strongly promoted the water-gas reaction as well as secondary reforming of methane. Conversely, in the case of Cu/TiO_2 and Cu/ZrO_2, as the CO_2 selectivity declined the CO yield augmented suggesting that Cu/TiO_2 and Cu/ZrO_2 enhanced the reverse water gas shift activity.

In our recent work (Deshmane et al. 2015b), the SRM activity for different metals (M-TiO_2) supported on polycrystalline titania was extensively studied. The performance of the catalysts followed the sequence of Pd-TiO_2 > Ni-TiO_2 > Zn-TiO_2 > Co-TiO_2 > Cu-TiO_2 > Sn-TiO_2; and the ability to lower CO selectivity was observed in the order of Zn-TiO_2 < Sn-TiO_2 < Co-TiO_2 < Cu-TiO_2 < Pd-TiO_2 < Ni-TiO_2. This diverse catalytic behavior was due to specific metal-support interactions controlling the reducibility and dispersion of metal particles in the TiO_2 matrix.

6.4.2 Effect of Different Metals Supported on Mesoporous Silica, MCM-41 for SRM Studies

Our quest to gain insights into how some selected metals (10 wt%) behaved on the same MCM-41 support yielded very intriguing results. Figure 6.5 shows a sharp contrast in the SRM activity and stability when the MCM-supported catalysts were run continuously for 40 hours (Abrokwah et al. 2016). Notably, although Pd, Sn, and Ni showed a methanol conversion increased monotonically for the first 10 hour, their activity declined steadily afterwards. Cu/MCM-41 exhibited the best activity

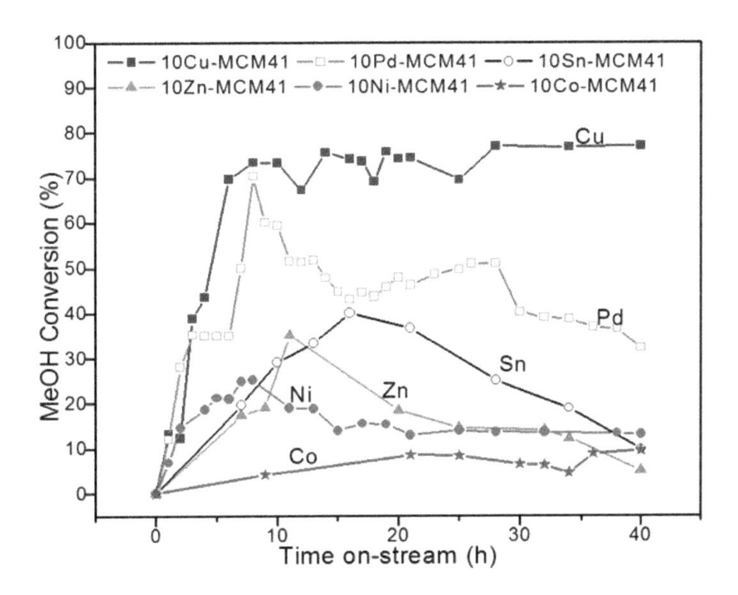

FIGURE 6.5 Catalysts activity versus time during SRM; reaction conditions: 300°C, 1 atm, methanol:water molar ratio of 1:3 and a GHSV of 2,838 h^{-1} at STP.

and stability by sustaining ~75% methanol steady state conversion throughout the 40 hours. Co/MCM-41 was stable but recorded very low conversion (i.e., <10%), making it an unsuitable SRM candidate under our experimental conditions.

6.4.3 Glycerol Stream Reforming Activity Test

Table 6.4 summarizes glycerol conversion and selectivity of different products obtained from GSR reactions. The glycerol conversion increased with increasing temperature consistent with the thermodynamically endothermic nature of GSR (Lucredio et al. 2011). The reaction was performed at 450°C, 550°C, 650, and 700°C. Based on the H_2 selectivity, the glycerol conversion and the ability of the catalysts to reduce CO under our reaction conditions, 650°C was selected as the optimum reaction temperature. H_2 selectivity increased with temperature and attained maximum at 650°C while the conversion decreased at temperatures less than 650°C. In general, the Ni catalysts exhibited slightly higher glycerol conversion (93.7%, 100%, and 100% for Ni/MCM-41, Ni/CeO_2-MCM-41, and Ni/TiO_2-MCM-41, respectively) compared to their Co counterparts (91.56%, 95.46%, and 94.12% for Co/MCM-41, Co/CeO_2-MCM, and Co/TiO_2-MCM-41, respectively, at 700°C) under the same reaction conditions.

TABLE 6.4
Summary of Glycerol Stream Reforming Catalyst Activity Tests Performed in the Range of 450°C–700°C

			Selectivity (%)			
Catalyst	**Temperature (°C)**	**Conversion (%)**	H_2	CO	CH_4	CO_2
Ni-MCM-41	450	53.0	42.0	67.9	8.3	23.8
	550	80.7	71.9	29.3	10.0	60.7
	650	91.9	81.0	21.5	7.1	71.4
	700	93.4	84.3	27.0	3.6	69.4
Co-MCM-41	450	4.7	24.0	74.8	10.2	15.1
	550	82.6	72.3	32.1	8.4	59.5
	650	84.2	70.0	42.4	7.1	50.4
	700	91.6	67.7	25.8	12.2	62.0
Ni-CeO_2-MCM-41	450	16.6	40.6	53.4	7.2	39.4
	550	85.2	77.7	11.5	9.1	79.4
	650	97.3	88.4	25.4	3.0	71.6
	700	100.0	85.4	30.3	1.3	68.3
Co-CeO_2-MCM-41	450	3.8	66.6	62.7	8.5	28.8
	550	77.8	80.7	20.8	4.2	74.9
	650	87.5	82.4	26.3	5.4	68.3
	700	95.5	68.0	24.6	11.2	64.2
Ni-TiO_2-MCM-41	450	19.8	52.8	72.5	7.2	20.3
	550	87.8	73.9	9.6	13.9	76.6
	650	98.9	86.0	18.0	3.5	78.5
	700	100.0	84.8	22.9	2.9	74.1
Co-TiO_2-MCM-41	450	5.2	22.9	67.9	3.3	28.8
	550	86.7	72.5	31.5	7.9	60.5
	650	90.5	80.7	26.6	3.9	69.5
	700	94.1	75.2	15.9	10.2	74.0

The H_2 selectivity was higher for Ni than Co catalysts. For example, Ni-MCM-41, Ni-CeO_2-MCM-41, and Ni-TiO_2-MCM-41 showed 81%, 88.4%, and 86% H_2 selectivity, respectively, compared to 70%, 82.4%, and 80.7% for the Co catalyst, respectively, at 650°C. The results also suggest that support modification with CeO_2 and TiO_2 improved glycerol conversions up to 6% and H_2 selectivity by 17% in Co and Ni catalysts. This improvement could be due to enhanced metal support interaction improving metal dispersion, thereby promoting secondary methane reforming and water-gas shift reaction. Unlike Co catalysts, Ni catalysts showed a general trend of decreasing methane selectivity with increasing temperature (Zhang et al. 2007; Adhikari et al. 2008). Overall, high CO selectivity, which occurred at 450°C, was accompanied by low H_2 selectivity and low glycerol conversions. This observation concurred with the report by Dou et al. (2009) that a large number of light hydrocarbons and oxygenates such as CO are produced during the GSR reaction.

6.4.4 GSR Catalyst Stability Studies

The GSR was conducted continuously for 40 hours at 650°C at GHSV of 2200 h^{-1} to ascertain the extent to which each catalytic system resisted deactivation (Figure 6.6). Ceria and titania containing catalysts significantly improved the GSR activity and stability of all the catalysts. Overall Ni catalysts performed better than Co catalysts. Particularly, CeO_2 and TiO_2 modified Ni catalysts did not show any loss of activity (i.e., ~100% glycerol conversion) for the entire 40 hours on stream. Co-TiO_2-MCM-41 was the least stable among the catalysts tested for GSR. Comparatively, addition of CeO_2 yielded better stability than TiO_2. This result stems from the ability of CeO_2 to diminish coking due to its oxygen holding capacity to either inhibit coke formation and/or facilitate the oxidative elimination of coke (Adhikari et al. 2007; Zhang et al. 2007; Montini et al. 2010). Although the uniquely poor performance of Co-TiO_2-MCM-41 is not fully understood, it may be attributed to the high acidity of TiO_2 with the ability to attract large coke deposition over the catalyst active sites (Claridge et al. 1993; Montini et al. 2010; Liu et al. 2014).

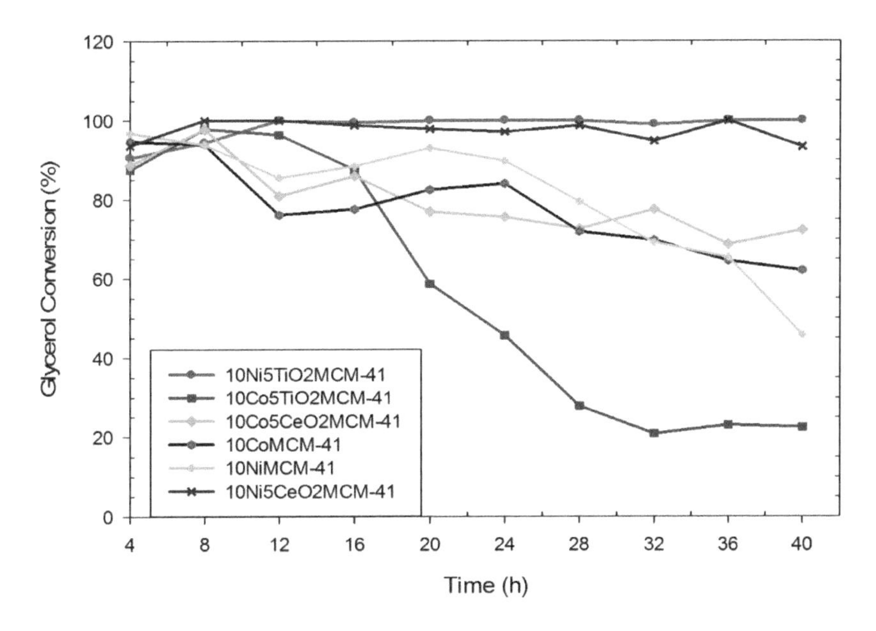

FIGURE 6.6 Glycerol conversion trend for catalysts conducted at 650°C with water:glycerol molar ratio of 12:1 and GHSV of 2,200 h^{-1}.

6.5 CONCLUSIONS

We have successfully synthesized various high surface area catalytic systems by a one-pot hydrothermal method and tested for SRM and GSR studies. Small angle XRD analyses showed that all catalysts possessed ordered mesoporous structure. The SRM reactions revealed that at lower temperature (<250°C), Cu interactions with the amorphous supports performed better than their crystalline analogs, such as TiO_2, ZrO_2, and CeO_2. Although 10% Cu-MCM41 showed the highest SRM activity with ~68% methanol conversion and ~6% CO selectivity, 10% $CuCeO_2$ yielded ~45% methanol conversion with the lowest ~1.84% CO selectivity. When the metals are varied in M-MCM-41 catalysts, we ascertained that although methanol conversion of Pd, Sn, and Ni increased monotonically for the first 10 hours, their activity declined steadily afterwards.

In the case of GSR studies, 650°C was identified to be the optimum performance temperature of all catalysts. In general, the Ni catalysts exhibited higher glycerol conversion and H_2 selectivity relative to the Co catalysts. In addition to its enhanced catalyst stability, the MCM-41 supports modified with CeO_2 and TiO_2 also improved glycerol conversions up to 6% and H_2 selectivity by 17% in Co and Ni catalysts. Except for Co/TiO_2-MCM-41, all the catalysts exhibited excellent stability and GSR activity after 40 hours on-stream. Our study demonstrated that Ni/TiO_2-MCM-41, Ni/CeO_2-MCM-41 and Co/CeO_2-MCM-41 could be potential candidates for hydrogen generation through hydrothermal reforming of alcohols. In addition, the type of support is also vital for the catalyst performance and product distribution.

ACKNOWLEDGMENTS

Our sincere gratitude goes to the National Science Foundation (NSF) for funding this research (Grant No. HRD-124215). We thank Dr. Shamsuddin Ilias and Dr. Lijun Wang for the use of their laboratories for surface analysis and TPR studies, respectively. The authors greatly appreciate the analytical support provided by Mr. James King (Chemistry Department) and Mr. Bryce Holmes (School of Agriculture and Environmental Sciences). Dr. Jag Sankar and Dr. Sergey Yarmolenko (Center for Advanced Materials and Smart Structures-CAMSS) at NCAT provided their XRD for our material characterization; and we applaud the assistance. The final acknowledgment goes to the Joint School of Nanoscience and Nanoengineering (JSNN) for the availability of their TEM and SEM-EDX elemental analyzers.

REFERENCES

Abrokwah, R. Y., V. G. Deshmane, and D. Kuila. 2016. Comparative performance of M-MCM-41 (M: Cu, Co, Ni, Pd, Zn and Sn) catalysts for steam reforming of methanol. *Journal of Molecular Catalysis A: Chemical* 425:10–20.

Adhikari, S., S. D. Fernando, and A. Haryanto. 2008. Hydrogen production from glycerin by steam reforming over nickel catalysts. *Renewable Energy* 33:1097–1100.

Adhikari, S., S. Fernando, S. R. Gwaltney, S. D. F. To, R. M. Bricka, P. H. Steele, and A. Haryanto. 2007. A thermodynamic analysis of hydrogen production by steam reforming of glycerol. *International Journal of Hydrogen Energy* 32:2875–2880.

Barrett, E. P., L. G. Joyner, and P. P. Halenda. 1951. The determination of pore volume and area distributions in porous substances. I. Computations from nitrogen isotherms. *Journal of the American Chemical Society* 73:373–380.

Burch, R., V. Caps, D. Gleeson, S. Nishiyama, and S. C. Tsang. 2000. Nanoscopic tin-oxygen linings on mesoporous silica as a novel catalyst for organic hydrogen transfer reaction. *Applied Catalysis A: General* 194:297–307.

Byrd, A. J., K. K. Pant, and R. B. Gupta. 2008. Hydrogen production from glycerol by reforming in supercritical water over Ru/Al_2O_3 catalyst. *Fuel* 87:2956–2960.

Cai, W., F. Wang, A. C. Van Veen, H. Provendier, C. Mirodatos, and W. Shen. 2008. Autothermal reforming of ethanol for hydrogen production over an Rh/CeO_2 catalyst. *Catalysis Today* 138:152–156.

Claridge, J. B., M. L. H. Green, S. C. Tsang, A. P. E. York, A. T. Ashcroft, and P. D. Battle. 1993. A study of carbon deposition on catalysts during the partial oxidation of methane to synthesis gas. *Catalysis Letters* 22:299–305.

Cui, Y., V. Galvita, L. Rihko-Struckmann, H. Lorenz, and K. Sundmacher. 2009. Steam reforming of glycerol: The experimental activity of La1–xCexNiO_3 catalyst in comparison to the thermodynamic reaction equilibrium. *Applied Catalysis B: Environmental* 90:29–37.

D'Souza, L., A. Suchopar, K. Zhu, D. Balyozova, M. Devadas, and R. M. Richards. 2006. Preparation of thermally stable high surface area mesoporous tetragonal ZrO_2 and Pt/ZrO_2: An active hydrogenation catalyst. *Microporous and Mesoporous Materials* 88:22–30.

Deeprasertkul, C., R. Longloilert, T. Chaisuwan, and S. Wongkasemjit. 2014. Impressive low reduction temperature of synthesized mesoporous ceria via nanocasting. *Materials Letters* 130:218–222.

Deshmane, V. G., and Y. G. Adewuyi. 2012. Synthesis of thermally stable, high surface area, nanocrystalline mesoporous tetragonal zirconium dioxide (ZrO_2): Effects of different process parameters. *Microporous and Mesoporous Materials* 148:88–100.

Deshmane, V. G., R. Y. Abrokwah, and D. Kuila. 2015a. Synthesis of stable Cu-MCM-41 nanocatalysts for H_2 production with high selectivity via steam reforming of methanol. *International Journal of Hydrogen Energy* 40:10439–10452.

Deshmane, V. G., S. L. Owen, R. Y. Abrokwah, and D. Kuila. 2015b. Mesoporous nanocrystalline TiO_2 supported metal (Cu, Co, Ni, Pd, Zn, and Sn) catalysts: Effect of metal-support interactions on steam reforming of methanol. *Journal of Molecular Catalysis A: Chemical* 408:202–213.

Dou, B., V. Dupont, G. Rickett, N. Blakeman, P. T. Williams, H. Chen, Y. Ding, and M. Ghadiri. 2009. Hydrogen production by sorption-enhanced steam reforming of glycerol. *Bioresource Technology* 100:3540–3547.

Douette, A. M. D., S. Q. Turn, W. Wang, and V. I. Keffer. 2007. Experimental investigation of hydrogen production from glycerin reforming. *Energy & Fuels* 21:3499–3504.

González Vargas, O. A., J. A. de los Reyes, Heredia, A. M. Castellanos, L. F. Chend, and J. A. Wang. 2013. Cerium incorporating into MCM-41 mesoporous materials for CO oxidation. *Materials Chemistry and Physics* 139:125–133.

Goscianska, J., A. Olejnik, and R. Pietrzak. 2014. Adsorption of L-phenylalanine on ordered mesoporous carbons prepared by hard template method. *Journal of the Taiwan Institute of Chemical Engineers* 45:347–353.

Hirai, T., N. O. Ikenaga, T. Miyake, and T. Suzuki. 2005. Production of hydrogen by steam reforming of glycerin on ruthenium catalyst. *Energy & Fuels* 19:1761–1762.

Hoffmann, P. 2001. *Tomorrow's Energy: Hydrogen, Fuel Cells, and the Prospects for a Cleaner Planet.* Cambridge, MA: MIT Press.

Ingham, B., and M. F. Toney. 2014. X-ray diffraction for characterizing metallic films. In *Metallic Films for Electronic, Optical and Magnetic Applications*, ed. K. Barmak, and K. Coffey, pp. 3–38. Philadelphia, PA: Woodhead Publishing Limited.

Iriondo, A., V. L. Barrio, J. F. Cambra, P. L. Arias, M. B. Güemez, R. M. Navarro, M. C. Sanchez-Sanchez, and J. L. G. Fierro. 2009. Influence of La_2O_3 modified support and Ni and Pt active phases on glycerol steam reforming to produce hydrogen. *Catalysis Communications* 10:1275–1278.

Jaroniec, M., and L. A. Solovyov. 2006. Improvement of the Kruk-Jaroniec-Sayari method for pore size analysis of ordered silicas with cylindrical mesopores. *Langmuir* 22:6757–6760.

Liu, L., X. Ma, and J. Li. 2014. Hydrogen production from ethanol steam reforming over Ni/SiO_2 catalysts: A comparative study of traditional preparation and microwave modification methods. *International Journal of Energy Research* 38:860–874.

Llorca, J., P. R. la Piscina, J.-A. Dalmon, J. Sales, and N. Homs. 2003. CO-free hydrogen from steam-reforming of bioethanol over ZnO-supported cobalt catalysts: Effect of the metallic precursor. *Applied Catalysis B: Environmental* 43:355–369.

Loebick, C. Z., S. Lee, S. Derrouiche, M. Schwab, Y. Chen, G. L. Haller, and L. Pfefferle. 2010. A novel synthesis route for bimetallic CoCr–MCM-41 catalysts with higher metal loadings. Their application in the high yield, selective synthesis of single-wall carbon nanotubes. *Journal of Catalysis* 271:358–369.

Lowell, S., J. E. Shields, M. A. Thomas, and M. Thommes. 2004. *Characterization of Porous Solids and Powders: Surface Area, Pore Size and Density.* Dordrecht, the Netherlands: Springer.

Lucredio, A. F., J. D. A. Bellido, A. Zawadzki, and E. M. Assaf. 2011. Co catalysts supported on SiO_2 and γ-Al_2O_3 applied to ethanol steam reforming: Effect of the solvent used in the catalyst preparation method. *Fuel* 90:1424–1430.

Montini, T., R. Singh, P. Das, B. Lorenzut, N. Bertero, P. Riello, A. Benedetti et al. 2010. Renewable H_2 from glycerol steam reforming: Effect of La_2O_3 and CeO_2 addition to Pt/Al_2O_3 catalysts. *ChemSusChem* 3:619–628.

Owen, S. L. S. 2014. Mesoporous TiO_2 encapsulated monometallic (Cu, Co, Ni, Pd, Sn, Zn) nanocatalysts for steam reforming of methanol. Master of Science, North Carolina A&T University.

Park, S.-J. and S.-Y. Lee. 2010. A study on hydrogen-storage behaviors of nickel-loaded mesoporous MCM-41. *Journal of Colloid and Interface Science* 346:194–198.

Pouretedal, H. R. and M. Ahmadi. 2012. Synthesis, characterization, and photocatalytic activity of MCM-41 and MCM-48 impregnated with CeO_2 nanoparticles. *International Nano Letters* 2:1–8.

Ramírez-Cabrera, E., N. Laosiripojana, A. Atkinson, and D. Chadwick. 2003. Methane conversion over Nb-doped ceria. *Catalysis Today* 78:433–438.

Richardson, J. T., B. Turk, M. Lei, K. Forster, and M. V. Twigg. 1992. Effects of promoter oxides on the reduction of nickel oxide. *Applied Catalysis A: General* 83:87–101.

Rossetti, I., A. Gallo, V. Dal Santo, C. L. Bianchi, V. Nichele, M. Signoretto, E. Finocchio, G. Ramis, and A. Di Michele. 2013. Nickel catalysts supported over TiO_2, SiO_2 and ZrO_2 for the steam reforming of glycerol. *ChemCatChem* 5:294–306.

Suriye, K., P. Praserthdam, and B. Jongsomjit. 2005. Impact of Ti^{3+} present in titania on characteristics and catalytic properties of the Co/TiO_2 catalyst. *Industrial & Engineering Chemistry Research* 44:6599–6604.

Swami, S. M. and M. A. Abraham. 2006. Integrated catalytic process for conversion of biomass to hydrogen. *Energy & Fuels* 20:2616–2622.

Tu, C. H., A. Q. Wang, M. Y. Zheng, X. D. Wang, and T. Zhang. 2006. Factors influencing the catalytic activity of SBA-15-supported copper nanoparticles in CO oxidation. *Applied Catalysis A: General* 297:40–47.

Urasaki, K., Y. Sekine, S. Kawabe, E. Kikuchi, and M. Matsukata. 2005. Catalytic activities and coking resistance of Ni/perovskites in steam reforming of methane. *Applied Catalysis A: General* 286:23–29.

Vita, A., L. Pino, F. Cipitì, M. Laganà, and V. Recupero. 2014. Biogas as renewable raw material for syngas production by tri-reforming process over $NiCeO_2$ catalysts: Optimal operative condition and effect of nickel content. *Fuel Processing Technology* 127:47–58.

Wang, Z., G. Zou, X. Luo, H. Liu, R. Gao, L. Chou, and X. Wang. 2012. Oxidative coupling of methane over $BaCl_2$-TiO_2-SnO_2 catalyst. *Journal of Natural Gas Chemistry* 21:49–55.

Wen, G., Y. Xu, H. Ma, Z. Xu, and Z. Tian. 2008. Production of hydrogen by aqueous-phase reforming of glycerol. *International Journal of Hydrogen Energy* 33:6657–6666.

Wu, W.-H., X. Sun, Y. P. Yu, J. Hu, L. Zhao, Q. Liu, Y. F. Zhao, and Y. M. Li. 2008. TiO_2 nanoparticles promote β-amyloid fibrillation in vitro. *Biochemical and Biophysical Research Communications* 373:315–318.

Wu, X., and S. Kawi. 2009. Rh/Ce-SBA-15: Active and stable catalyst for CO_2 reforming of ethanol to hydrogen. *Catalysis Today* 148:251–259.

Yung, M. M., E. M. Holmgreen, and U. S. Ozkan. 2007. Cobalt-based catalysts supported on titania and zirconia for the oxidation of nitric oxide to nitrogen dioxide. *Journal of Catalysis* 247:356–367.

Zhang, B., X. Tang, Y. Li, Y. Xu, and W. Shen. 2007. Hydrogen production from steam reforming of ethanol and glycerol over ceria-supported metal catalysts. *International Journal of Hydrogen Energy* 32:2367–2373.

Zhou, Y., F. N. Gu, L. Gao, J. Y. Yang, W. G. Lin, J. Yang, Y. Wang, and J. H. Zhu. 2011. 3D net-linked mesoporous silica monolith: New environmental adsorbent and catalyst. *Catalysis Today* 166:39–46.

7 Current Developments in the Production of Liquid Transportation Fuels through the Fischer-Tropsch Synthesis

Venu Babu Borugadda and Ajay K. Dalai

CONTENTS

7.1 Introduction 109
7.2 Integrated Route for Biomass-to-Gas and Gas-to-Liquid Conversion 110
  7.2.1 Collection and Pretreatment of Feedstock 111
  7.2.2 Gasification of Feedstock 112
  7.2.3 Syngas Cleaning and Conditioning 112
  7.2.4 Conversion of Syngas to Synthetic Fuels and Chemicals through the Fischer-Tropsch Process 113
7.3 Technological Developments in the Fischer-Tropsch Process 114
7.4 Chemistry of Fischer-Tropsch Process 114
7.5 Selection of Fischer-Tropsch Catalysts 117
7.6 Reactors Used in the Fischer-Tropsch Process 118
7.7 Conclusions and Future Perspectives 119
References 119

7.1 INTRODUCTION

Currently, there has been significant attention towards the production of liquid transportation fuels from renewable feedstocks as the price of fossil fuels continues to rise. Another reason for an interest shift towards green fuels is the massive amount of heat-trapping greenhouse gas (GHG) emissions into the atmosphere from the burning fossil fuels (Nanda et al. 2015a). Despite the ecological concerns, the present trend of energy consumption suggests that by 2025 the worldwide demand for petroleum would be 35% (NREL 2006). Globally, the overall usage of liquid transportation fuels could increase from nearly 54% in 2008 to 60% in 2035 (Figure 7.1). The fastest growing alternative energy source (i.e., from renewable wastes) is expected to be increasing at 8.2% per year, equivalent to 300 million barrels of oil equivalent per day (Mboe/d) between 2010 and 2030. Above all, the utilization of liquid transportation fuels continues to rise at an average rate of 1.6% annually (Veziroglu and Sahin 2008). Biofuels can potentially supplement this increasing demand for energy alongside the fossil fuels that could continue to dominate the transportation sector for the coming decades because of their well-established and robust refining and utilization infrastructures.

Biomass-derived liquid transportation fuels could play a significant role in cutting down the GHG emissions and the dependency on fossil resources by reducing their consumption. The first-generation biomass-derived fuels such as bioethanol and biodiesel can partially substitute the fossil-derived transportation fuels through blends in current vehicular engines (Nanda et al. 2014).

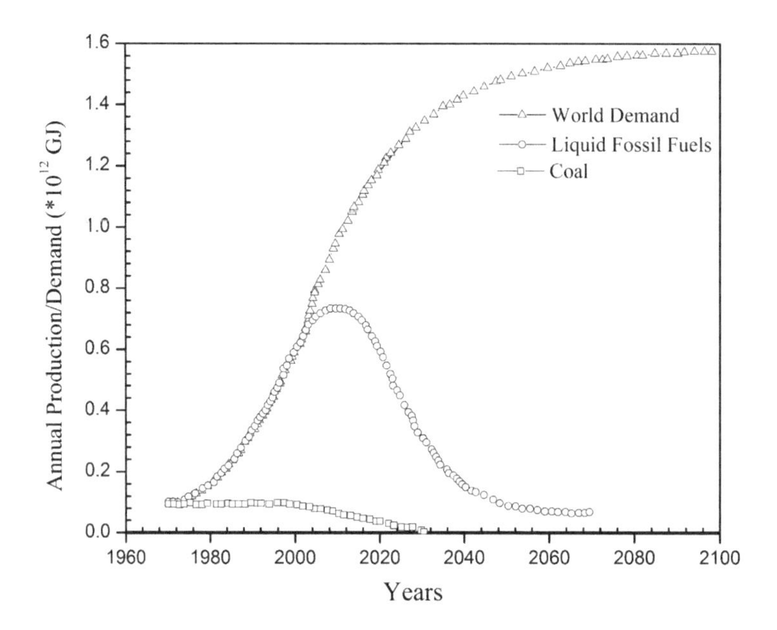

FIGURE 7.1 Estimated global energy demand and fossil-derived liquid transportation fuels production. (Data adapted from Conti, J. and Holtberg, P., *World Energy Outlook 2011*, Department of Energy, U.S. Energy Information Administration, Washington, DC, https://www.iea.org/publications/freepublications/publication/WEO2011_WEB.pdf, accessed on April 6, 2018; BP Energy Outlook, edition, https://www.bp.com/content/dam/bp/en/corporate/pdf/energy-economics/energy-outlook/bp-energy-outlook-2018.pdf, accessed on April 21, 2018.)

However, because of the food-versus-fuel issue and socio-environmental problems such as land availability, the high fertilizer requirement in feedstock cultivation and the low productivity per hectare limited the production of first-generation biofuels (Hamelinck and Faaij 2006). To overcome these limitations, the second-generation biofuels have emerged from non-edible feedstocks such as agricultural crop wastes, woody biomass, municipal solid waste (MSW), food waste, cattle manure, pulp and paper processing waste, and sewage sludge (Nanda et al. 2013, 2015b, 2016a, 2016b, 2017; Gong et al. 2017a, 2017b). The examples of second-generation biofuels include bioethanol, biobutanol, bio-oil, biodiesel, biogas (methane), and hydrogen-rich synthesis gas.

Several biochemical and thermochemical technologies categorized under biomass-to-liquid (BTL), biomass-to-gas (BTG), and gas-to-liquid (GTL) are available for the conversion of biomass to respective solid, liquid, and gaseous fuels. While BTL technologies include pyrolysis, liquefaction, and fermentation, BTG include gasification and anaerobic methanation (Nanda et al. 2014). The GTL technologies include Fischer-Tropsch (FT) synthesis and syngas fermentation that use syngas as a feedstock to produce liquid hydrocarbon fuels. This chapter is dedicated to reviewing the advanced research and development in the production of liquid transportation fuels through an integrated BTG and GTL technology, especially gasification of biomass followed by the FT technology. The process benefits of different gasification processes and the FT synthesis are discussed.

7.2 INTEGRATED ROUTE FOR BIOMASS-TO-GAS AND GAS-TO-LIQUID CONVERSION

The conversion of renewable biomass or petrochemical feedstocks into liquid fuels is a multi-step process through thermochemical synthesis methods such as gasification and the FT process. The detailed process flow consists of the following stages:

1. Collection and pretreatment of feedstock
2. Gasification of feedstock
3. Syngas cleaning and conditioning
4. Conversion of syngas to synthetic fuels and chemicals through the FT process

7.2.1 Collection and Pretreatment of Feedstock

In most of the BTG processes, agricultural crop residues (straw) and forestry biomass (wood) are considered as the feedstocks owing to their high cellulose, hemicellulose, and lignin contents. However, these plant residues contain significant amount of moisture that require drying and pretreatment, thereby increasing the overall process expenditures. Biomass can be dried using a rotary dryer, which is the most commonly used technique due to its low sensitivity to the biomass particle size and bigger processing capacity compared to the super-heated steam dryer. However, super-heated steam dryers are preferable to use because the generated steam and be used in other processes within the plant such as the FT process. In addition, super-heated steam dryers are safer than rotary driers with respect to fire hazard due to the absence of oxygen.

In terms of the commercial scale production, two kinds of gasifiers are commonly used for lignocellulosic biomass gasification: the circulating fluidized bed and the entrained flow gasifier (Figure 7.2). The entrained flow gasifier requires that the particle size of the wood chips be 1 millimeter (mm). Therefore, a biomass pulverizer (grinder or hammer mill) is placed next to the dryer to reduce the size of the biomass to 1 mm. However, a circulating fluidized bed gasifier can

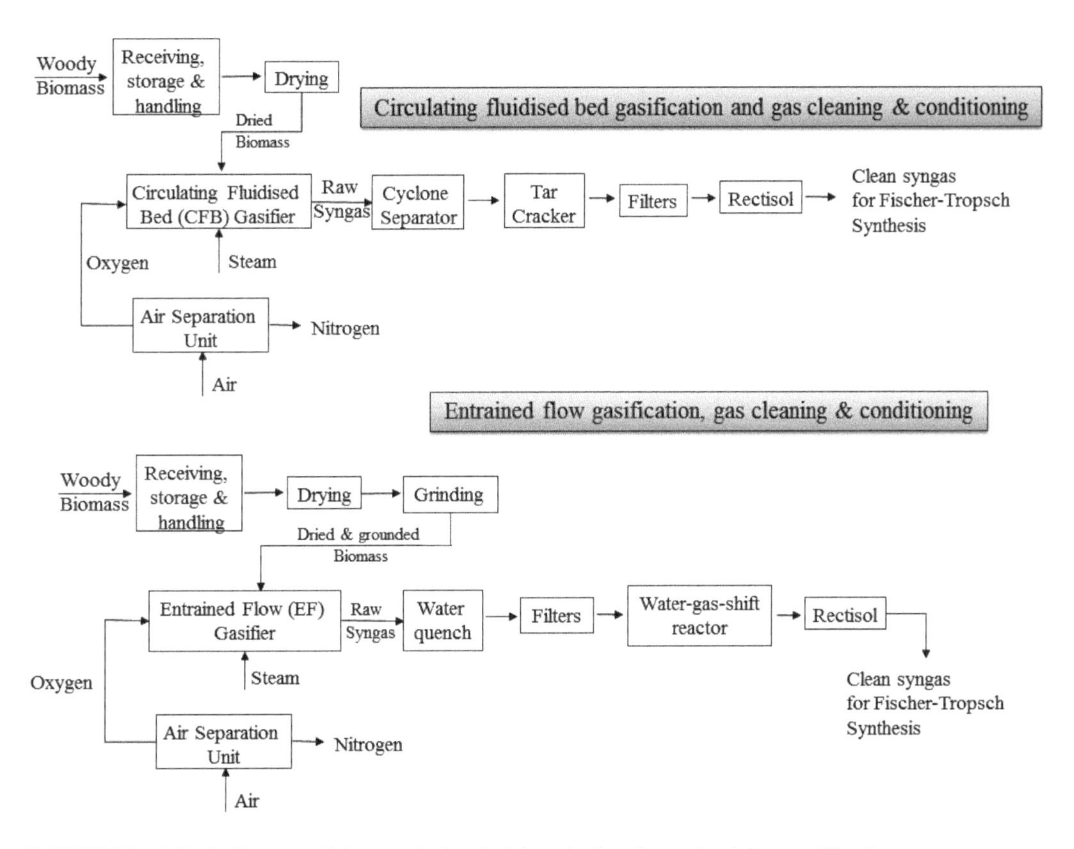

FIGURE 7.2 Block diagram of the circulating fluidized bed and entrained flow gasification processes.

handle varied biomass particle sizes. Hence, no intensive size reduction is needed in circulating fluidized bed gasification reactors (Drift et al. 2004; Swanson et al. 2010).

7.2.2 Gasification of Feedstock

As mentioned in the previous section, circulating fluidized bed and entrained flow gasifiers are most appropriate for commercial-scale BTG conversion due to the ease of scalability. Circulating fluidized bed gasifiers have their operating temperature between 700°C and 1100°C with a typical feeding rate of 15 tons per hour (t/h) of dry feed. On the other hand, entrained flow gasifiers have a typical operating temperature requirement of 1200°C–1400°C with a feeding rate of 20 dry t/h. The higher temperature in entrained flow reactors leads to very low methane and tar content, higher carbon conversion, and low gas cleaning requirements in comparison to other gasifiers (Drift et al. 2004; Cristales et al. 2015).

Currently, lock hoppers are the established pressurized technology for coal-fired gasifiers but one of the key drawbacks of this technology is a huge requirement of inert gas (either recovered CO_2 or usually N_2) and extra cost for the inert gas compression. Therefore, advanced developments are needed to overcome such a challenge in biomass feeding for pressurized gasifiers. In addition, many efforts have been made to develop alternative pressurized feed systems to address the issue of lock hopers, which includes screw, piston, and rotary feeders (Sara et al. 2016).

Another advancement in BTG technology is the use of water and a reaction medium for hydrothermal gasification. Hydrothermal gasification is advantageous over conventional thermochemical gasification because it can use high-moisture containing feedstocks such as algae, municipal solid wastes, sewage sludge, and industrial effluents. The use of high-moisture containing feedstocks in hydrothermal gasification has significant cost-saving benefits due to the elimination of the biomass drying or pretreatment step (Reddy et al. 2016). Hydrothermal gasification can be classified into subcritical and supercritical water gasification. Subcritical water deploys water near or below its critical temperature and pressure (i.e., $T_c \leq 374°C$ and $P_c \leq 22.1$ MPa) (Reddy et al. 2014). Supercritical water uses water above its critical temperature and pressure (i.e., $T_c \geq 374°C$ and $P_c \geq 22.1$ MPa). In hydrothermal gasification, water acts as a reactant, catalyst, and a green solvent to efficiently degrade highly recalcitrant biomasses to hydrogen-rich syngas (Rana et al. 2017).

7.2.3 Syngas Cleaning and Conditioning

In the process of developing successful BTG technology, syngas cleaning is considered as one of the major challenges. Based on the mode of operation, the nature of the catalyst, and type of the FT reactor, the acceptable levels of impurities may vary. Alkali compounds, particulates, tars, nitrogen, chloride, and sulfur-based compounds need to be removed from the syngas before it enters the FT reactor. Cyclone separators and filters are used to remove particulates and alkali compounds. As many authors suggested, a water quencher is used next to the entrained flow reactor for syngas cooling and removal of particulates (Figure 7.2) (Drift and Boerrigter 2006; Phillips et al. 2011). In the case of a circulating fluidized bed gasifier, the heat exchanger is used to cool the syngas and to recover the steam generated for tar cracking. Furthermore, a water quencher also eliminates ammonia from the syngas. However, for the gasifiers coupled with tar cracking, maximum nitrogenous compounds are cracked in the tar reformer.

Owing to the presence of sulfur in certain biomasses such as municipal solid wastes and sewage sludge, the formation of hydrogen sulfide (H_2S) and carbonyl sulfide (COS) can be occur. To remove the sulfur compounds, the Rectisol process has been proven as a promising technique for large-scale FT processes and coal gasification units. The Rectisol process can bring down the concentration of H_2S and CO_2 to as low as 0.1 and 2 ppm, respectively. Furthermore, the Rectisol process also can eliminate trace compounds such as HCl and HCN (Zhang et al. 2012).

The simulation studies with entrained flow gasifiers have shown that the molar ratio of H_2/CO is less than the required optimal ratio ($H_2/CO = 2$) for FT synthesis. Thus, the installation of the water-gas shift reactor after the Rectisol unit could make up the required H_2/CO ratio. Contrastingly, water-gas shift reactors are not needed for the circulating fluidized bed gasifiers since the H_2/CO ratio of the tar cracker is nearly 2. The H_2/CO ratio of 2 is attributed to the light hydrocarbons and tars that are almost fully removed in the tar cracker. Nonetheless, this ratio completely depends on the chemical composition of the raw syngas from the gasifier (Dimitriou et al. 2018).

7.2.4 Conversion of Syngas to Synthetic Fuels and Chemicals through the Fischer-Tropsch Process

To produce liquid transportation fuels and some base chemicals from syngas, different technologies have been developed such as Fischer-Tropsch synthesis, methanol-to-gasoline (MTG) synthesis, and Topsoe integrated gasoline synthesis (TIGAS) (Figure 7.3). However, the major focus of this

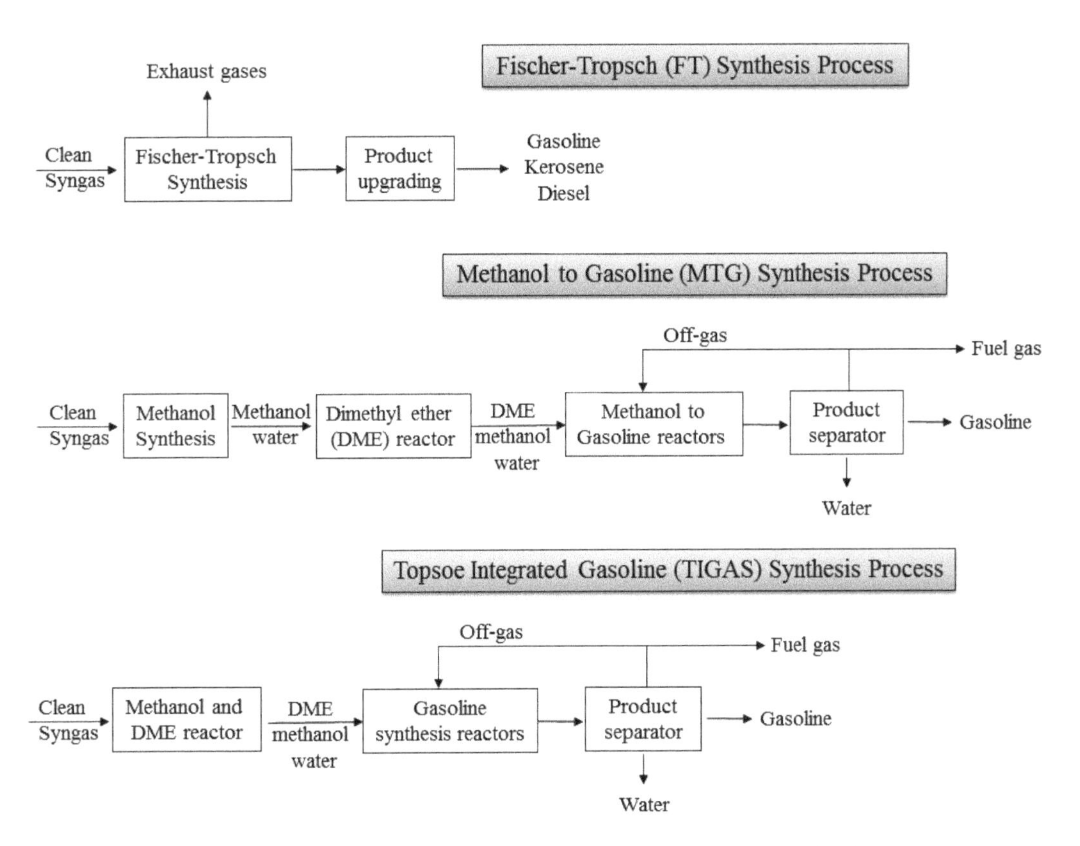

FIGURE 7.3 Block diagram for the Fischer-Tropsch (FT), methanol to gasoline (MTG), and Topsoe integrated gasoline (TIGAS) synthesis processes.

chapter is on the production of liquid hydrocarbon fuels from the FT process. The FT synthesis is a process for catalytic conversion of H_2 and CO into a wide variety of organic compounds by varying the carbon chain lengths typically from C_1 to C_{100}.

7.3 TECHNOLOGICAL DEVELOPMENTS IN THE FISCHER-TROPSCH PROCESS

The Fischer-Tropsch (FT) process is a fully established industrial practice for the conversion of synthesis gas to liquid hydrocarbon fuels by varying the carbon chain length. The graphical representation of the FT route as a GTL technology is shown in Figure 7.4.

Franz Fischer, Hans Tropsch, and Helmut Pichler developed the Fischer-Tropsch process in 1923 at the Kaiser Wilheim Institute during the reactions involving syngas with cobalt catalyst to produce various liquid fuels (e.g., gasoline, alcohols, diesel, middle, and heavy distillate oils) (Fischer and Tropsch 1923). The Ruhrchemie atmospheric fixed bed reactor was the first industrial FT reactor established in 1935. It operated under medium pressure (0.5–1.5 MPa) at 180°C–200°C reaction temperature with Co as a catalyst to produce diesel, gasoline lubrication oils, and other chemicals. Later, many commercial-scale FT plants were established, and these plants used synthesis gas generated from coal gasification or the methane reforming reaction (Ail and Dasappa 2016). Table 7.1 lists some of the BTG and GTL demonstration-scale plants worldwide.

7.4 CHEMISTRY OF FISCHER-TROPSCH PROCESS

The carbonaceous materials (coal or biomass) are gasified first to produce a syngas, followed by conversion of the produced syngas in a FT reactor to obtain the desired hydrocarbons of different chain lengths. The FT reaction chemistry is the critical step because it involves a catalytic chemical reaction for converting a mixture of CO and H_2 to hydrocarbon constituents of gasoline, diesel, wax, olefins, and alcohols based on the following equations. In these equations, n is the average length of the hydrocarbon chain (up to 50). All the reactions are exothermic, and the products are a mixture of different hydrocarbons, although paraffins and olefins are the main products (Stelmachowski and Nowicki 2003).

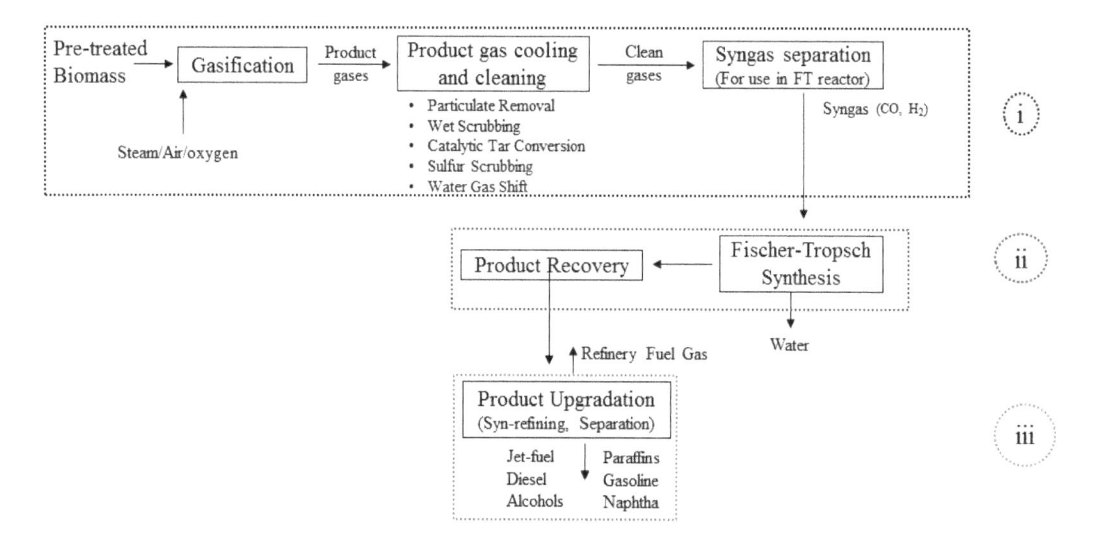

FIGURE 7.4 Flow diagram of the typical Fischer-Tropsch synthesis. (i) syngas production through gasification of biomass, (ii) syngas conversion to hydrocarbons through FT synthesis, and (iii) refining and upgradation of FT products.

TABLE 7.1
List of Selected BTG and GTL Demonstration-Scale Plants Worldwide

Company	Project	Gasification Technology	Capacity	Feedstocks	Products
British Airways, UK, 2015	GreenSky	Solena's Plasma gasification	500,000 t/y	Food waste, plant waste and municipal solid waste	Jet fuel, naphtha, and electricity
ThyssenKrupp Uhde and Five French partners	BioTfueL	Entrained flow	€112.7 M	Lignocellulosic biomass (straw, forest waste, energy crops, etc.)	Diesel and jetfuel
CEA and Air Liquide	SYNDIESE-BtS	Entrained flow	23000 t/y	Forestry and agricultural residues	Diesel, kerosene, and naphtha
Falcrum Bioenergy Inc.	Sierra BioFuels	TRI steam reformer	175,000 t/y	Municipal solid waste	Jet fuel, diesel and electricity
Cool Planet Energy Systems	Project Genesis	Thermo-mechanical fractionation	10 M gallons/year	Pine chips and other cellulosic material	Gasoline, jet fuel and diesel
	Red Rock Biofuels	TRI steam reformer	136,000 t/y	Forest and sawmill residues, agricultural residues	Jet fuel, naphtha, and diesel
	UPM Stracel facility	Gasification and Fischer-Tropsch	105000 tons	Woody biomass	Diesel, naphtha and power generation
	AJos BtL	Gasification and Fischer-Tropsch	1.8 M t/y	Woody biomass	Diesel and naphtha
Linde Engineering Dresden GmbH	Finnish Forest BtL Oy	Carbo-V biomass gasification	1.5 M t/y	Woody biomass	Diesel and naphtha
Neste Oil and Stora Enso	NSE Biofuels Oy	Gasification and Fischer-Tropsch	200,000 t/a	Woody biomass	Diesel

Source: ETIP Bioenergy, Discontinued BtL Projects, http://www.etipbioenergy.eu/value-chains/conversion-technologies/advanced-technologies/biomass-to-liquids/discontinued-btl-projects, accessed on April 15, 2018.

Paraffin synthesis (saturated hydrocarbons)

$$nCO + (2n+1)H_2 \rightarrow C_nH_{2n+2} + nH_2O \tag{7.1}$$

Olefin synthesis (unsaturated hydrocarbons)

$$nCO + 2nH_2 \rightarrow C_nH_{2n} + nH_2O \tag{7.2}$$

Alcohol synthesis

$$nCO + 2nH_2 \rightarrow C_nH_{2n+1}OH + (n-1)H_2O \tag{7.3}$$

Water-gas shift reaction

$$CO + H_2O \rightarrow CO_2 + H_2 \tag{7.4}$$

The water-gas shift reaction is also one of the significant reactions, which is dependent on the catalyst and process conditions. During the FT reaction, syngas is adsorbed on the catalyst surface to form a chain initiator through the surface polymerization reaction, which proceeds by chain propagation, termination, and desorption of the FT products. The reactants (H_2 and CO) and the FT products also are oxidized or reduced on the catalyst surface by carbide formation and coke deposition through the Boudouard reaction. The best H_2/CO ratio in the synthesis gas for the FT reaction is around 2:1. Therefore, water-gas shift reactors are integrated to regulate the H_2/CO proportion in synthesis gas prior to the FT reactor for high output (Liu 2009).

During the FT reaction, a wide range of hydrocarbons are generated, the composition and quality of which depend on the reaction conditions (i.e., temperature, pressure, H_2/CO ratio, type of catalyst, and catalyst loading). The formulation of several FT products is driven by the kinetic and mechanistic factors in addition to thermodynamic considerations (Dry 2004). The stability of the FT reaction is achieved over the rearrangement of the catalyst under the influence of reactants, products, and intermediates. FT synthesis involves certain polymerization reactions such as initiation, propagation, and termination (Flory 1936). The initiation step is the genesis of the hydrocarbon chain monomer building block from the adsorbed CO and H_2 on the surface of the catalyst. The propagation step involves the addition of these monomer building blocks to increase the hydrocarbon chains. Finally, termination is the interruption of growing hydrocarbon chains from the surface of the catalyst (Steen and Schulz 1999). The FT polymerization model was extended for chain branching by Friedel and Andreson (1950). In the FT process, product distribution follows the Anderson-Schultz-Flory (ASF) hydrocarbon chain length statistics as shown in the following equation.

$$\log\left(\frac{wi}{i}\right) = i\log\alpha + \log\left(\frac{(1-\alpha)^2}{\alpha}\right) \tag{7.5}$$

where wi is the weight fraction of FT product, i is the chain length of the produced hydrocarbons, and α is known as the chain growth probability. The ASF model accompanies the rate of consumption of the syngas (CO and H_2). Besides, ASF also enables FT kinetics to predict the formation rate of FT products according to the carbon number. Furthermore, ASF can convert the experimental FT product distributions into kinetic data, which determines the product selectivity such as:

1. Chain growth probability as a function of carbon number
2. Chain branching probability as a function of carbon number
3. Olefinicity per carbon number

Depending on the FT reaction conditions and type of the catalyst used, the FT product distribution varies according to the carbon number such as C_1–C_2 (synthetic natural gas), C_3–C_4 (liquefied petroleum gas), C_5–C_7 (light petroleum), C_8–C_{10} (heavy petroleum), C_{11}–C_{20} (middle distillates), C_{11}–C_{12} (kerosene), C_{13}–C_{20} (diesel), C_{21}–C_{30} (soft wax), and C_{31}–C_{60} (hard wax). Furthermore, the chain growth probability and chain branching probability as a function of carbon number can be explained with the help of individual mechanisms such as the alkyl mechanism, the alkenyl mechanism, the enol mechanism, and the CO-insertion mechanism. Among all these mechanisms, the most widely accepted mechanism is the alkyl mechanism for hydrocarbon chain growth in FT process. The detailed schematic representation of these mechanisms can be found elsewhere (Steynberg and Dry 2004; Fontenelle and Fernandes 2011) and the summary of the FT reaction mechanisms are shown in Table 7.2.

TABLE 7.2
Summary of the Fischer-Tropsch Reaction Mechanisms

Type of Mechanism	Class of Monomer	Chain Initiator	Products
Alkyl	$M{=}CH_2$	$M\text{-}CH_3$	α-Olefins, n-paraffins
Alkenyl	$M{=}CH_2$	$M\text{-}CH{=}CH_2$	α-Olefins
Enol	$M{=}C(H)(OH)$	$M{=}C(CH_3)(OH)$	Aldehydes, alcohols, and α-olefins
CO-insertion	M=CO	$M{=}CH_3$	Alcohols, aldehydes, α-olefins, n-paraffins

7.5 SELECTION OF FISCHER-TROPSCH CATALYSTS

The FT product distribution can be varied according to the reaction conditions and the type of reactor. Nevertheless, the key to attain the desired product distribution is based on the activity and selectivity of the catalyst. An ideal FT catalyst has a rich H_2 activity in order to catalyze the hydrogenation of CO into complex hydrocarbons. All metallic elements of group VIII (transition metals) can adsorb CO chemically and further dissociate it. There are four transition metals having adequate activity for commercial application (i.e., iron (Fe), cobalt (Co), nickel (Ni), and ruthenium (Ru)) (Yang et al. 2015).

Most of the catalysts used in FT processes can produce high-molecular weight linear alkanes and diesel fuels. Nickel inclines towards the production of methane, which is undesirable together with low olefin formation despite its higher activities for commercial production. In addition, under high-pressure conditions, nickel functions with less activity owing to the propensity of producing oxygenates (i.e. < volatile carbonyl compounds). Furthermore, the high cost of the ruthenium compared to cobalt limits its commercial use. Cobalt-based catalyst have relatively less water-gas shift activity and higher cost compared to iron-based catalyst. Iron has more water-gas shift activity to produce H_2, hence iron is more appropriate for the conversion of syngas with lower H_2/CO ratio.

Iron-based catalysts also can operate at high-temperature (300°C–350°C) and low-temperature (220°C–270°C) systems. Therefore, iron-based catalysts are more commonly used to produce liquid fuels from syngas through the FT process. However, cobalt-based catalysts are more suitable for low-temperature regimes (190°C–240°C). Moreover, at identical operating conditions, cobalt-based catalysts have a higher tendency to acquire hydrocarbons with higher molecular weight than iron-based catalyst. High temperatures promote product selectivity towards lower hydrocarbons and higher hydrogenated products with improved branching in the product molecules and increased by-product contents such as ketones and aromatics. Hence, the two major commercial catalysts are based on either iron or cobalt (Hao et al. 2008).

By considering the assortment of activity on unsupported metals, iron-based catalysts have abrupt catalyst lifetime. Nonetheless, high pressure (1.5 MPa) can improve the product distribution by two-fold and catalyst lifetime by many folds (Sandeep and Dasappa 2014). Iron-based catalysts are considered as the best option for gasoline production and α-olefins at high-temperature (Sandeep and Dasappa 2014). On the other hand, cobalt catalyst is more active compared to iron-based catalysts towards the formation of diesel fuels or waxes at low-temperature FT conditions. Furthermore, cobalt-based catalysts have higher stability and product conversion, higher productivity and good aging at low-temperatures. In the mechanistic point of view, as soon as CO is adsorbed on the catalyst surface, the surface carbon is formed, which reacts with both Co and Fe to form carbides. However, iron carbides are more stable than that of cobalt (Paterson et al. 2017). Further, the optimal H_2/CO ratio for iron- and cobalt-based catalysts are 0.5–2.5 and 2–2.3, respectively (Ail and Dasappa 2016). Table 7.3 compares the significant features of iron and cobalt-based catalysts used in FT process (Ail and Dasappa 2016).

In addition to the FT processing conditions, the mode of preparation of the catalyst also directly affects the structure, morphology, diffusion and aging effects. During the preparation of the catalysts, the following goals should be considered:

1. Proper distribution of the metal particles on the surface of the catalysts.
2. Monitoring the porosity of the catalyst support.
3. Interaction between supports/promoters and metal particles for favorable product distribution.

Several approaches such as wet impregnation, co-precipitation, colloids, micro-emulsions, and sol-gel methods have been reported in the literature for the preparation of catalysts for FT synthesis. Capillary impregnation or wet impregnation of metal salts are the most common methods where

TABLE 7.3
Significant Characteristics of Iron and Cobalt-Based Catalysts in the Fischer-Tropsch Synthesis

Characteristic	Iron-Based Catalysts	Cobalt-Based Catalysts
Feedstock	Flexible H_2/CO ratio in the range of 0.5–2.5 due to high water-gas shift activity	H_2/CO ratio in the range of 2–2.3 due to low water-gas shift activity
Operating conditions	Operates both at low temperatures (220°C–270°C) and high temperatures (300°C–350°C)	Operating mostly at low temperatures (190°C–240°C). At high temperatures, increased CH_4 selectivity leads to catalyst deactivation.
Catalytic activity	At higher space velocities, more active than Co catalyst	At lower space velocities, more active for higher CO conversion
Product distribution	*n*-Paraffins are the major products with a considerable number of α-olefins (α = 0.65–0.92). Lower paraffin/olefin ratio is obtained.	*n*-paraffins are the major products with a marginal number of α-olefins (α = 0.85–0.92). Higher paraffin/olefin ratio is obtained.
Suitable promoters	Ca, K, Li, Rb, and Na (alkali metals)	CeO_2, ZrO_2, and La_2O_3 (oxide promotors). Pd, Rh, Ru and Pt (noble metals)
Cost and lifetime of the catalyst	Less expensive and lower lifetime	More expensive and longer lifetime

porosity and the specific surface area of the supports are known. Co-precipitation has been modernized for iron-based catalyst than cobalt specifically for metallic precursors or supports such as Mn-Al, Zn-Al, or Zn oxides (Khasin et al. 2001). Co-precipitation is obtained when the solubility of the precursor changes by the blending of an alkali carbonate, hydroxide or ammonia. Nevertheless, to co-precipitate more than two components at the same time, all the components of the solubility need to be identical. The key factors for controlling the size of the metallic precursors for the formation of defined compounds between metallic oxides and supports are temperature, duration of precipitation, addition conditions, starting metallic salt, and nature of the precipitating agent. Sol-gel method has been technologically advanced to formulate metal/SiO_2 catalysts with tetraethoxysilane as the SiO_2 precursor. The other precipitation methods such as microemulsions and colloids have been reported elsewhere (Khodakov et al. 2007).

7.6 REACTORS USED IN THE FISCHER-TROPSCH PROCESS

Based on the operating temperatures, FT processes can be categorized as high-temperature Fischer-Tropsch (HTFT) and low-temperature Fischer-Tropsch (LTFT). Effective control over temperature and efficient recovery of waste heat is the most important design features of FT reactors for large-scale applications. Currently, there are four types of commercial FT reactors: multi-tubular fixed bed reactor, slurry reactor, entrained flow reactor, and circulating fluidized bed reactors. Circulating fluidized bed and entrained flow reactors are used for HTFT process, whereas multi-tubular and slurry reactors are used for the LTFT process.

The key difference between the LTFT and HTFT reactors is the absence of the liquid phase outside the catalyst particles in HTFT reactors. The presence of the liquid phase in the HTFT fluidized bed reactors leads to severe technical problems because of the loss of fluidization and particle agglomeration. The geometry of the reactor also influences the hydrocarbon chain growth, CO conversion, catalytic activity, and product selectivity (Steynberg et al. 1999; Davis 2005; Guettel et al. 2008). The main objective of all FT reactors is to keep a consistent temperature profile within the catalyst bed. Some limiting factors for the scalability of an efficient FT reactor are uniform heat distribution in the catalyst bed, pressure drop constraints, intra-particle diffusion limitations, risk of attrition due to small catalyst particle sizes, and catalyst separation from FT products.

Ceramic-structured monolithic catalysts are being developed for enhanced product distribution. In comparison to catalysts used in fixed bed reactors, diffusion limitation is an order of magnitude lesser for the monolithic catalyst. The capability to use thin catalyst layers with a changing thickness upon monolithic materials removes the diffusion limitations and provides ideal catalyst selectivity and activity. However, monoliths face internal diffusion limitations, fragility, and imperfect channel structures in large-scale applications (Hilmen et al. 2001).

7.7 CONCLUSIONS AND FUTURE PERSPECTIVES

The transformation of biomass into liquid transportation fuels through the FT process seems to be a promising technology. Besides, FT process is an established GTL technology that can be integrated to BTG technologies (gasification) to provide many technical and economic benefits. The prominent benefit of the FT process is the production of liquid transportation fuels with properties comparable to fossil-based diesel and gasoline. Therefore, the FT process could be a promising technology to supplement synthetic transportation fuels at an industrial scale that could potentially replace the fossil fuels without modifying the properties of the fuels and/or vehicular engine configurations.

Usually, biomass gasification with air produces syngas with less H_2 content, but the use of oxy-steam gasification or hydrothermal gasification increases the H_2 concentration to maintain a high H_2/CO ratio in the synthesis gas for the compatibility with the FT process. Depending on the nature of the desired FT product distribution, cobalt and iron-based catalysts are commercially used based on their many advantages. The choice of integration of gasifiers (BTG) with FT reactors (GTL) also necessitates a higher catalyst performance, stability, recovery, and reusability. Therefore, thorough understanding of the physical chemistry of the catalyst, reaction kinetics and catalytic mechanism along with optimal FT process conditions are recommended for desired product distribution. More studies on techno-economic analysis and life-cycle assessment will help determine the long-term economic feasibility of such BTG–GTL integrated systems.

REFERENCES

Ail, S. S., and S. Dasappa. 2016. Biomass to liquid transportation fuel via Fischer Tropsch synthesis—Technology review and current scenario. *Renewable and Sustainable Energy Reviews* 58:267–286.

BP Energy Outlook: 2018 edition. https://www.bp.com/content/dam/bp/en/corporate/pdf/energy-economics/energy-outlook/bp-energy-outlook-2018.pdf (accessed on April 21, 2018).

Conti, J., and P. Holtberg. 2011. World Energy Outlook 2011. Washington, DC: Department of Energy, U.S. Energy Information Administration. https://www.iea.org/publications/freepublications/publication/WEO2011_WEB.pdf (accessed on April 06, 2018).

Cristales, R. Z., J. Sessions, K. Boston, and G. Murphy. 2015. Economic optimization of forest biomass processing and transport in the Pacific Northwest USA. *Forest Science* 61:220–234.

Davis, H. B. 2005. Fischer–Tropsch synthesis: Overview of reactor development and future potentialities. *Topics in Catalysis* 32:143–168.

Dimitriou, I., H. Goldingay, and A. V. Bridgwater. 2018. Techno-economic and uncertainty analysis of Biomass to Liquid (BTL) systems for transport fuel production. *Renewable and Sustainable Energy Reviews* 88:160–175.

Drift, A. V. D., and H. Boerrigter. 2006. Synthesis gas from biomass for fuels and chemicals. *Energy Research Centre of the Netherlands*, Report ECN-C-06-001.

Drift, A. V. D., R. V. Ree, and H. J. Veringa. 2004. Entrained flow gasification of biomass. Ash behaviour, feeding issues, and system analyses. *Energy Research Centre of the Netherlands*, Report ECN-C-04-039.

Dry, M. 2004. Chemical concepts used for engineering purposes. *Studies in Surface Science and Catalysis* 152:196–257.

ETIP Bioenergy. 2018. Discontinued BtL Projects. http://www.etipbioenergy.eu/value-chains/conversion-technologies/advanced-technologies/biomass-to-liquids/discontinued-btl-projects (accessed on April 15, 2018).

Fischer, F., and H. Tropsch. 1923. Über die Synthese höherer Glieder der aliphatischen Reihe aus Kohlenoxyd. *European Journal of Inorganic Chemistry* 56:2428–2443.

Flory, P. J. 1936. Molecular size distribution in linear condensation polymers *Journal of American Chemical Society* 58:1877–1885.

Fontenelle, A., and F. A. Fernandes. 2011. Comprehensive polymerization model for Fischer–Tropsch synthesis. *Chemical Engineering Technology* 34:963–971.

Friedel, R. A., and R. B. Andreson. 1950. Composition of synthetic liquid fuels. I. Product distribution and analysis of C_5-C_8 paraffin isomers from cobalt catalyst. Contribution of the R&D branch from office of synthetic liquid fuels. *Bureau of Mines* 72:1212–1215.

Gong, M., S. Nanda, H. N. Hunter, W. Zhu, A. K. Dalai, and J. A. Kozinski. 2017a. Lewis acid catalyzed gasification of humic acid in supercritical water. *Catalysis Today* 291:13–23.

Gong, M., S. Nanda, M. J. Romero, W. Zhu, and J. A. Kozinski. 2017b. Subcritical and supercritical water gasification of humic acid as a model compound of humic substances in sewage sludge. *The Journal of Supercritical Fluids* 119:130–138.

Guettel, R., U. Kunz, and T. Turek. 2008. Reactors for Fischer-Tropsch synthesis. *Chemical Engineering Technology* 31:746–754.

Hamelinck, C. N., and A. P. C. Faaij. 2006. Outlook for advanced biofuels. *Energy Policy* 34:3268–3283.

Hao, X., M. E. Djatmiko, Y. Xu, Y. Wang, J. Chang, and Y. Li. 2008. Simulation Analysis of a Gas-to-Liquid Process Using Aspen Plus. *Chemical Engineering Technology* 31:188–196.

Hilmen, A. M., E. Bergene, O. A. Lindvåg, D. Schanke, S. Eri, and A. Holmen. 2001. Fischer–Tropsch synthesis on monolithic catalysts of different materials. *Catalysis Today* 69:227–232.

Khasin, A. A., T. M. Yurieve, V. V. Kaichev, V. I. Bukhtiyarov, A. A. Budneva, E. A. Paukshtis, and V. N. Parmon. 2001. Metal–support interactions in cobalt-aluminum co-precipitated catalysts: XPS and CO adsorption studies. *Journal of Molecular Catalysis A: Chemical* 175:189–204.

Khodakov, A. Y., W. Chu, and P. Fongarland. 2007. Advances in the development of novel cobalt Fischer–Tropsch catalysts for synthesis of long-chain hydrocarbons and clean fuels. *Chemical Reviews* 107:1692–1744.

Liu, K. 2009. Coal and syngas to liquids. In *Hydrogen and Syngas Production and Purification Technologies*, ed. K. Liu, C. Song, and V. Subramani, 486–521. Hoboken, NJ: John Wiley & Sons.

Nanda, S., A. K. Dalai, I. Gökalp, and J. A. Kozinski. 2016a. Valorization of horse manure through catalytic supercritical water gasification. *Waste Management* 52:147–158.

Nanda, S., J. Isen, A. K. Dalai, and J. A. Kozinski. 2016b. Gasification of fruit wastes and agro-food residues in supercritical water. *Energy Conversion and Management* 110:296–306.

Nanda, S., J. Mohammad, S. N. Reddy, J. A. Kozinski, and A. K. Dalai. 2014. Pathways of lignocellulosic biomass conversion to renewable fuels. *Biomass Conversion and Biorefinery* 4:157–191.

Nanda, S., M. Gong, H. N. Hunter, A. K. Dalai, I. Gökalp, and J. A. Kozinski. 2017. An assessment of pinecone gasification in subcritical, near-critical and supercritical water. *Fuel Processing Technology* 168:84–96.

Nanda, S., P. Mohanty, K. K. Pant, S. Naik, J. A. Kozinski, and A. K. Dalai. 2013. Characterization of North American lignocellulosic biomass and biochars in terms of their candidacy for alternate renewable fuels. *Bioenergy Research* 6:663–677.

Nanda, S., R. Azargohar, A. K. Dalai, and J. A. Kozinski. 2015a. An assessment on the sustainability of lignocellulosic biomass for biorefining. *Renewable and Sustainable Energy Reviews* 50:925–941.

Nanda, S., S. N. Reddy, H. N. Hunter, I. Butler, and J. A. Kozinski. 2015b. Supercritical water gasification of lactose as a model compound for valorization of dairy industry effluents. *Industrial and Engineering Chemistry Research* 54:9296–9306.

NREL. 2006. From biomass to biofuels. National Renewable Energy Laboratory. Report NREL/BR-510-39436. https://www.nrel.gov/docs/fy06osti/39436.pdf (accessed on June 04, 2018).

Paterson, J., M. Peacock, E. Ferguson, R. Purves, and M. Ojeda. 2017. In situ diffraction of Fischer–Tropsch catalysts: Cobalt reduction and carbide formation. *Chem Cat Chem* 9:3463–3469.

Phillips, S. D., J. K. Tarud, M. J. Biddy, and A. Dutta. 2011. Gasoline from woody biomass via thermochemical gasification, methanol synthesis, and methanol-to-gasoline technologies: A technoeconomic analysis. *Industrial and Engineering Chemistry Research* 50:11734–11745.

Rana, R., S. Nanda, J. A. Kozinski, and A. K. Dalai. 2017. Investigating the applicability of Athabasca bitumen as a feedstock for hydrogen production through catalytic supercritical water gasification. *Journal of Environmental Chemical Engineering* 6:182–189.

Reddy, S. N., S. Nanda, A. K. Dalai, and J. A. Kozinski. 2014. Supercritical water gasification of biomass for hydrogen production. *International Journal of Hydrogen Energy* 39:6912–6926.

Reddy, S. N., S. Nanda, and J. A. Kozinski. 2016. Supercritical water gasification of glycerol and methanol mixtures as model waste residues from biodiesel refinery. *Chemical Engineering Research and Design* 113:17–27.

Sandeep, K., and S. Dasappa. 2014. Oxy-steam gasification of biomass for hydrogen rich syngas production using down draft reactor configuration. *International Journal of Energy Research* 38:174–188.

Sara, H. J., B. Enrico, V. Mauro, D. C. Andrea, and N. Vincenzo. 2016. Techno-economic analysis of hydrogen production using biomass gasification-A small scale power plant study. *Energy Procedia* 101:806–813.

Steen, E. V., and H. Schulz. 1999. Polymerisation kinetics of the Fischer–Tropsch Co hydrogenation using iron and cobalt based catalysts. *Applied Catalysis A: General* 186:309–320.

Stelmachowski, M., and L. Nowicki. 2003. Fuel from synthesis gas-the role of process engineering. *Applied Energy* 74:85–93.

Steynberg, A. P., and Dry, M. E. (Eds.). 2004. *Fischer-Tropsch Technology*. Elsevier, Amsterdam, the Netherlands, Vol. 152, pp. 601–680.

Steynberg, A. P., M. E. Dry, B. H. Davis, B. B. Breman, and H. Schulz. 1999. Short history and present trends of Fischer–Tropsch synthesis. *Applied Catalysis A: General* 186:3–12.

Swanson, R. M., J. A. Satrio, and R. C. Brown. 2010. Techno-economic analysis of biofuels production based on gasification. *National Renewable Energy Laboratory*, Report NREL/TP-6A20-46587.

Veziroglu, T. N., and S. Sahin. 2008. 21st Century's energy: Hydrogen energy system. *Energy Conversion and Management* 49:1820–1831.

Yang, W., S. Rehman, X. Chu, Y. Hou, and S. Gao. 2015. Transition metal (Fe, Co and Ni) carbide and nitride nanomaterials: structure, chemical synthesis and applications. *Chemical Nano Materials* 1:376–398.

Zhang, W., H. Liu, I. U. Hai, Y. Neubauer, P. Schroder, H. Oldenburg, A. Seilkopf, and A. Kolling. 2012. Gas cleaning strategies for biomass gasification product gas. *International Journal of Low-Carbon Technologies* 7:69–74.

8 Production of Biolubricant Basestocks from Structurally Modified Plant Seed Oils and Their Derivatives

Venu Babu Borugadda, Vaibhav V. Goud, and Ajay K. Dalai

CONTENTS

8.1 Introduction 123
8.2 Chemical Composition, Structure, and Properties of Plant Seed Oils 124
8.3 Epoxidation 126
8.4 Ring Opening Reactions or Hydroxylation 128
8.5 Di-esters Formation through Esterification and Anhydrides Addition 129
8.6 Tri-ester and Tetra-esters Formation through Esterification and Anhydrides Addition 131
8.7 Current Scenario and Prospects 134
8.8 Conclusions 136
References 136

8.1 INTRODUCTION

A lubricant is a viscous substance used to improve the relative movement between two moving surfaces. Based on the origin, lubricant oils are broadly classified as mineral (fossil-derived), synthetic, and vegetable oils. In industry, they are categorized as engine and non-engine lubricants (Bart et al. 2013). Among these, mineral oils are the most widely used for material lubrication, and their consumption is around 95%. Conventional lubricants are typically the composites of basestock oils (70%–99%) blended with additives (1%–30%) to meet a desired requirement (Arbain and Salimon 2009). The additives are chemicals that are used to blend with basestocks to enhance the desirable characteristics that they lack or to improve the existing properties. The properties considered important for lubricant application are viscosity index, pour point, thermal and oxidative resistance (stability), flash point, acidity (neutralization number), friction properties, and biodegradability.

Mineral oils originate from crude oil, which contains many hydrocarbon molecules and traces of nitrogen, sulfur, oxygen, and metal salts. Synthetic oil is a blend of chemicals that are artificially made by a chemical modification of petroleum-derived products. They are more highly stable at high temperatures and oxidation atmospheres than many conventional lubricants. Conversely, synthetic and mineral oils contain harmful substances that pollute the natural ecosystem and are difficult to dispose of. According to Gawrilow (2004), annually, 10-15 MT of petroleum-based fuels and lubricants enter the ecosystem through accidental spills, total loss, and evaporation. The presence of these mineral-based lubricants in the environment pollutes the atmosphere, land, and groundwater posing many adverse risks to the health of humans, plants, and aquatic life. On the other hand, conventional oils are expensive, deplete at a faster rate, and are toxic to the environment because of their non-biodegradable nature. Therefore, due to these environmental threats, the mineral-based synthetic lubricants usually are not considered eco-friendly.

Nevertheless, replacing mineral-based and synthetic lubricants with eco-friendly lubricants is one of the promising ways to decrease the unfavorable properties (low biodegradability) on the environment caused by using traditional lubricants. The potential use of renewable raw materials in the production of lubricants is currently of major interest. As a result, vegetable oils, plant seed oils, and their esters are alternatives to conventional and synthetic lubricants because of their biodegradable nature, performance, and renewability. In comparison to conventional lubricants, plant seed oils and their derivatives have advantageous properties, such as higher viscosity, low friction coefficient, eco-friendly, renewable, biodegradable, non-water polluting, and improved fuel economy (Erhan and Asadauskas 2000; Mercurio et al. 2004). Even though vegetable oils contribute many advantages, they cannot be used directly in their natural (original) form as lubricant base oils and industrial fluids because of their lower oxidative stability and inadequate thermal stability (Becker and Knorr 1996; Nanda et al. 2019), poor cold flow behaviour, and other tribo-chemical degrading processes (i.e., oxidation and degradation) (Brophy and Zisman 1951).

Despite the instabilities of vegetable oils, there is a great necessity to use them to replace the conventional lubricant base oils. In view of this, structural or chemical modification, additive treatment, and genetic modification techniques could overcome the poor oxidative, hydrolytic stability, and high-temperature sensitivity of plant seed oils and their derivatives when they are applied as lubricant basestocks (Soni and Agarwal 2014). Among all the possible routes, structural modification of plant seed oils and their derivatives has been a promising technique to formulate biodegradable lubricant basestocks.

In formulating the lubricant base oil from vegetable oils, the chemical (fatty acid) composition of potential feedstock is a significant feature. Chemical modifications of plant seed oils and their derivatives can be categorized into two groups: reactions on the hydrocarbon chain (e.g., oxidative cleavage, metathesis, radical additions, the addition of carboxylic acids, dimerization, and epoxidation) and reaction on the carboxyl group (e.g., transesterification and hydrogenation). In general, more than 90% of structural modifications of plant seed oils and their derivatives occur at the carboxyl groups present in fatty acids and the remaining 10% at the fatty acid hydrocarbon chain (Bart et al. 2013). Among all the feasible chemical modifications, a reaction involved in the fatty acid hydrocarbon chain has much significance. From the detailed study of the chemical modification of vegetable oils, many feedstocks used were staple food crops in human nutrition (Soni and Agarwal 2014; Borugadda and Goud 2014). In addition, due to the high price of virgin oils, the end-product price also remains high. Therefore, to minimize the end-product cost, the non-edible and waste edible oils, which do have low commercial value in the market, can be used to prepare biolubricant basestocks. Hence, this chapter proposes the modification of fatty acid structures to formulate lubricant base oils with improved performance and biodegradable nature.

8.2 CHEMICAL COMPOSITION, STRUCTURE, AND PROPERTIES OF PLANT SEED OILS

Plant seed oils can be classified as edible and non-edible with chemical compounds known as fats or lipids. Currently, these materials are attracting more attention because they originate from renewable sources and have adaptability to the ecosystem and compatibility for use in diverse fields as feedstocks to synthesize value-added products. Chemically, plant seed oils and animal fats are triglycerides (tri-esters of glycerol with fatty acids) that are hydrophobic. Triglycerides are derived from different carboxylic acids and vary in the nature of the alkyl chain attached to glycerol, which constitutes the plant seed oil matrix. A generalized triglyceride structure is shown in Figure 8.1a as proposed by Boyde (2002).

The R_1, R_2, and R_3 presented in Figure 8.1a are different fatty acids attached to the glycerol molecule. Thus, the fatty acid composition mainly contributes to the physicochemical properties of plant seed oils, which is comprised of 90% fatty acids and 10% glycerol. In the last few years, many feedstocks have been identified for various practical applications, and the common feedstocks

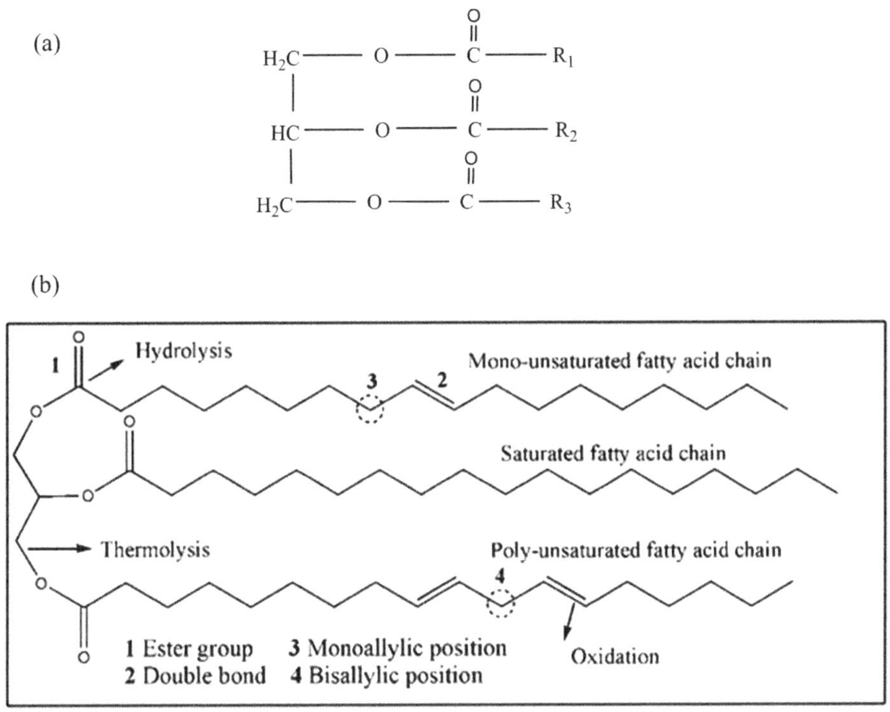

(a)

(b)

FIGURE 8.1 (a) Schematic structure of triglyceride molecule; (b) typical triglyceride structure presenting sites susceptible to chemical degradation.

available are edible oils, which play a major part in human nutrition. Therefore, attention needs to focus on feedstocks that are not edible, such as edible waste oils and non-edible oils.

A common demonstration of the triglyceride molecule along with vulnerable positions in its chemical structure is shown in Figure 8.1b. Mainly, in the fatty acid molecule at three positions a chemical or structural modification can be affected easily (i.e., ester moieties, double bonds along the fatty acid chains, and β-hydrogen) (Figure 8.1b). Alkenes (unsaturated fatty acids) comprise an electron-rich double bond that can be readily functionalized by reactions with electrophilic reagents. The glycerol portion of triglyceride is destructible and decomposes at higher temperatures through thermolysis and autoxidation. Further, vegetable oils undergo hydrolysis with water (Goyan et al. 1998).

Hwang and Erhan (2001) described that auto-oxidation of vegetable oils occurs by free radical formation, and peroxy radicals are formed by the reaction of free radicals with oxygen. The peroxy radical can further attack some lipid molecules to remove a hydrogen atom to form hydroperoxide and generate more free radicals to propagate the oxidation process. The complete mechanism for auto-oxidation of plant oils has been well reported by Moser and Erhan (2007). During oxidation, fatty acids in the plant seed oil can polymerize, thereby increasing the viscosity and acidity. Similarly, polyunsaturation in the fatty acids results in poor oxidation stability (Becker and Knorr 1996), while saturated fatty acids lead to poor performance in low-temperatures. This result limits the utilization of lipids for industrial applications in their original form. Therefore, to prepare the biolubricant basestocks from triglycerides, structural changes are necessary to enhance operational limitations of plant seed oils and their derivatives. Table 8.1 illustrates the influence of different chemical or structural modifications of fatty acids on certain lubricant basestocks.

Modification of fatty acids could compromise the following criteria: tribological performance, thermal and oxidative stability, low-temperature performance, and biodegradability. Consecutively, various modern technologies avoid or transform the unsaturation in the fatty acid chain to solve the issues regarding the practical application of vegetable oils as biolubricant base oils. An in-depth survey

TABLE 8.1
Effects of Modified Triglycerides Chemical Structure on Physical Properties of Plant Seed Oils and Their Derivatives

Structural Modification	Physical Properties
A higher degree of branching	(i) Excellent low-temperature performance
	(ii) Hydrophobic nature
	(iii) Lower viscosity index
Higher linearity	(i) Higher viscosity index
	(ii) Fairly low-temperature performance
Lower saturated fatty acids	(i) Increased low-temperature performance
	(ii) Limited stability for autoxidation
Higher saturated fatty acids	(i) Enhanced stability for auto-oxidation
	(ii) Inadequate low-temperature performance

revealed many practical solutions, such as genetic modification (Adhvaryu and Erhan 2002), additive treatment (Sharma et al. 2008b), selective hydrogenation of unsaturated sites (Wadumesthrige et al. 2009), transesterification (Bokade and Yadav 2007), and a series of structural or chemical modifications (Biswas et al. 2007; Campanella et al. 2008) to the fatty acids. Among these practical solutions, this chapter addresses the series of structural or chemical modifications to the fatty acids.

The following state-of-the-art methods reveal the processes that relate to the possible chemical modifications (as a single reaction or series of reactions) on various edible and non-edible oils along with the pros and cons to prepare biolubricant base stocks. Subsequently, the possible techniques to enhance the stability, low-temperature performance, biodegradability, and tribological properties have been discussed. In addition, this chapter also provides a substantial understanding on the perspective of biolubricant base stocks along with their formulation in the past and their future prospect.

8.3 EPOXIDATION

Epoxidation is one of the utmost significant double bond modification reactions. Epoxidation consists of peroxy acid formed by a reaction of formic or acetic acid and hydrogen peroxide (Saurabh et al. 2011). In the second stage, a reaction between peracid and a double bond produces epoxide. A simplified epoxidation reaction is summarized in Figure 8.2 as proposed by Saurabh et al. (2011).

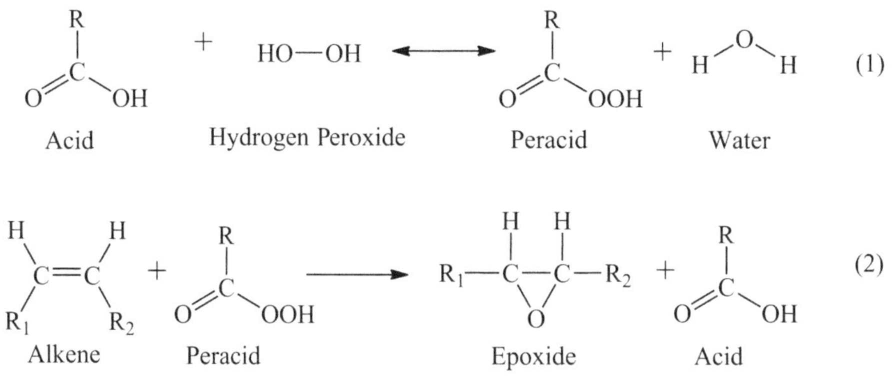

$$\underset{\text{Acid}}{R{-}C(=O){-}OH} + \underset{\text{Hydrogen Peroxide}}{HO{-}OH} \longleftrightarrow \underset{\text{Peracid}}{R{-}C(=O){-}OOH} + \underset{\text{Water}}{H{-}O{-}H} \quad (1)$$

$$\underset{\text{Alkene}}{R_1(H)C{=}C(H)R_2} + \underset{\text{Peracid}}{R{-}C(=O){-}OOH} \longrightarrow \underset{\text{Epoxide}}{R_1{-}\overset{H}{C}{-}\overset{H}{C}{-}R_2\ (\text{bridged by } O)} + \underset{\text{Acid}}{R{-}C(=O){-}OH} \quad (2)$$

FIGURE 8.2 Mechanism and reaction scheme for epoxide formation.

The epoxide is a cyclic ether consisting of three-membered rings. According to IUPAC norms, the epoxide is called an oxirane ring (Figure 8.2). Olefin epoxidation occurs through three reaction mechanisms, such as Sharpless asymmetric, Prilezhaev, Jacobsen-Katsuki, and enzymatic epoxidation. Every mechanism has its own advantages and disadvantages in terms of selectivity, catalyst life, and yield, as well as product separation and purification. The epoxidation by a Prilezhaev reaction has gained much significance since it leads to low oxirane degradation and higher selectivity depending on the catalyst and feedstock. The epoxidation process is categorized into four major types based on the catalyst used, such as metal-catalyzed ion exchange resin, sulfuric acid, and chemo-enzymatic epoxidation (Tan and Chow 2010). The ion exchange resin has been reported widely in the literature due to the ease of catalyst separation and minimal oxirane cleavage (Fiser et al. 2001).

Wu et al. (2000) studied rapeseed oil epoxidation and estimated significant physicochemical properties such as viscosity, pour point, biodegradability, oxidative stability, anti-oxidant study, and friction properties. The results of the study revealed improvement in the viscosity of epoxidized rapeseed oil (86.73 cSt) compared to rapeseed oil (34.75 cSt), whereas the pour point was –15°C and –12°C for rapeseed oil and epoxidized rapeseed oil, respectively. It also was reported that epoxidized rapeseed oil exhibits better friction properties and improved oxidative stability. The result of this study reveals increased pour point, which is contrary to the essential performance as a biolubricant.

A study on epoxidation of high oleic soybean oil through genetic modification was conducted by Adhvaryu and Erhan (2002). The study indicates that among the epoxidized soybean oil, high oleic soybean oil and epoxidized soybean oil showed maximum oxidative stability of 188.1°C. They also investigated the oxidative stability at various additive concentrations and concluded that 0.5 wt% additive concentration gave maximum oxidative stability. However, other important properties such as boundary lubrication and the deposit-forming tendency showed improved performance. Apart from that, the coefficient of friction was higher for epoxidized soybean oil than for soybean oil and high oleic soybean oil.

Salimon et al. (2012a) and Abdullah et al. (2014) investigated mono-epoxidation of linoleic acid using Novozyme 435 as a catalyst. They studied the influence of various process parameters on the epoxidation by D-optimal design and attained the optimum condition at H_2O_2 (15 μL), Novozyme-435 (120 mg), reaction time (7 hours) to yield 82.1% of mono-epoxidized oleic acid and 4.9 mass percent of oxirane oxygen content. In this work, the respective properties of linolenic acid and mono-epoxidized oleic acid were determined such as pour point (–2°C and –41°C), flash point (115°C and 128°C), viscosity index (224 and 130.8), and oxidative stability (189°C and 168°C). These studies concluded that mono-epoxidized oleic acid has a positive influence on pour point and flash point.

A recent study on epoxidation of passion fruit oil and moringa oil by Silva et al. (2015a, 2015b) revealed the synthesis of novel biolubricants with satisfactory properties. This study also tested the additive treatment of prepared epoxides. The outcomes of the study revealed that epoxides of passion fruit oil (EPFO) and moringa oil (EMO) had respective oxidation stability (16.9 and 204.7 minutes without additives; 24.6 and 311.8 minutes with additives), and viscosity (185.65 cSt and 80.37 cSt) was higher compared to passion fruit oil (7.5 minutes and 31.8 cSt) and moringa oil (28.3 minutes and 44.9 cSt). The oxirane oxygen content of the EPFO and EMO was 4.5 and 4.2 mass percent, respectively. Finally, the study concluded that epoxidized oil showed an improvement in additive solubility, fluid performance, exhibited superior lubrication performance properties, film formation, and lower friction co-efficient than the commercial mineral base fluids.

Although a vast pool of research material is available in the literature on epoxidation of edible and non-edible vegetable oils for diverse applications, there are few studies on understanding the effects of process parameters on epoxidation of oils and their methyl esters to determine the thermodynamic and kinematic parameters (Goud et al. 2006; Dinda et al. 2008; Mungroo et al. 2008; Derawi and Salimon 2010; Meshram et al. 2011; Sun et al. 2011; Fiser et al. 2012;

Monono et al. 2015; Borugadda and Goud 2013, 2014, 2015a, 2015b, 2016, 2018; Borugadda et al. 2017). In the early years of research on the synthesis of biolubricants, H_2SO_4 was used as a homogeneous acid catalyst because of its acid strength, selectivity, and short reaction time. The newer studies revealed the disadvantages of using H_2SO_4 as a catalyst because of the formation of unwanted by-products due to excess catalyst loading and difficulty in product purification. Therefore, acidic ion exchange resin as a heterogeneous catalyst has gained much significance for in-situ epoxidation because of its satisfactory outcomes in oxirane content, catalyst reusability, and feasibility of reaction conditions.

During the epoxidation, H_2O_2 is considered as an oxygen donor because of its high active oxygen content (47.1 wt%) and acetic acid as an oxygen carrier. By examining the physicochemical properties of epoxides from various feedstocks, it has been observed that epoxides have higher pour point, which is not suitable for low-temperature applications. However, all other properties exhibit satisfactory outcomes. Therefore, to enhance the pour point of epoxides, a series of structural modifications to the epoxide ring is neccessary. However, the opening of the epoxide ring (hydroxylation) creates a more favorable chemical site for reactions to take place (Borugadda and Goud 2012). Furthermore, functionalization of the epoxides enhances the pour point compared to raw oils, epoxides, and their respective esters. Hence, the following section discusses the possible series of structural modifications which can be carried out on epoxides.

8.4 RING OPENING REACTIONS OR HYDROXYLATION

The use of various catalysts during the ring opening is required for further functionalization of epoxides by ring opening to enhance the physicochemical properties of biolubricants. The process of introducing hydroxyl groups in the oxirane ring is known as hydroxylation or ring opening. The ring opening of epoxides occurs by cleavage of one of the carbon-oxygen bonds. Under acidic conditions, ring opening occurs in two key steps. First, the epoxide is protonated. Second, the nucleophile attacks at the most substituted position. The ring opening reaction can proceed by either S_N2 or S_N1 mechanisms in the presence of alcohols, acids, and water. Ring opening is a critical step in the production of polyol or hydroxylated products. A variety of factors that may affect the hydroxylation are clearly discussed for process optimization.

Sharma et al. (2006) prepared four hydroxy thio-ether derivatives using epoxidized soybean oil as a feedstock. The ring opening was carried out in the presence of a perchloric acid as catalyst using 1-butanethiol, 1-decanethiol, octadanethiol, and cyclohexyl mercaptan as alcohols. The resulting hydroxyl compounds showed increased oxidation stability, lubricity performance, better affinity for metal surfaces, improved friction, and wear behavior. The onset oxidation temperature of soybean oil, epoxidized soybean oil, and ring opening compound of 1-butanethiol derivative was 134°C, 195°C, and 181°C, respectively. Similarly, Sharma et al. (2008a) synthesized acyl derivatives of epoxidized soybean oil using several anhydrides (i.e., heptanoic, propionic, acetic, isobutyric, hexanoic, valeric, and butyric acid) in the presence of a BF_3 catalyst. Among all the anhydrides, hexanoic anhydride showed enhanced lubricity properties as an anti-wear and pour point depressant additive. The molecular weight of prepared epoxidized soybean oil and diesters was 845 Da (epoxidized soybean oil), 883 Da (acetic acid), 887 Da (propionic acid), 884 Da (iso-butyric acid), 890 Da (butyric acid), and 905 Da (hexanoic acid). Further, Campanella et al. (2010) reported the preparation of sunflower and high oleic sunflower oil epoxidation followed by ring opening with methanol and ethanol using sulfuric acid and fluoroboric acid as catalysts. The crystallization temperatures of soybean oil, sunflower oil, high oleic sunflower oil, and their epoxides were –65°C and –35°C, –68°C and –42°C, and –29°C and 4°C, respectively. The rheological study of the samples revealed Newtonian behavior.

Salimon et al. (2011) studied the optimization of epoxidized oleic acid ring opening by D-optimal design. The optimum condition of study was oleic acid/mono-epoxidized oleic acid (0.3:1 wt%), *p*-Toluenesulfonic acid/mono-epoxidized oleic acid (0.5:1 wt%), reaction temperature (110°C), reaction time (4.5 hours) with yield, and oxirane oxygen content of 84.6 and 0.1 mass percent, respectively. A study on Jatropha oil epoxidation and ring opening by Daniel et al. (2011) showed that for ring opened product the pour point was around –18°C and demonstrated improvement in the oxidative stability and pour point of epoxidized product.

Arumugam et al. (2012) studied the epoxidation and ring opening of sunflower, palm, rapeseed, and their methyl esters. The ring opening reaction was carried out in presence of methanol and water. Among all the samples, rapeseed oil showed favorable properties as a lubricant. The biodegradability of the epoxidized rapeseed oil and their ring opening products was greater than 95%. Similarly, other significant properties such as viscosity, viscosity index, and pour point was 90.1 and 35 cSt, 160 and 220, as well as –11°C and –15°C, respectively, for epoxidized rapeseed oil and their ring opening products.

Kulkarni et al. (2013) studied the epoxidation of mustard oil with further ring opening in presence of sulfuric, sulphamic, and methane sulphonic acids as catalysts using 2-ethyl hexanol. The results revealed that for ring opening products with an increase in catalyst loading and time, the viscosity index increases and pour point decreases. Among all the catalysts, sulphamic acid was the most suitable catalysts for the reaction. Madankar et al. (2013) epoxidized the canola oil and performed the ring opening in presence of *n*-butanol, amyl alcohol and, 2-ethyl hexanol using IR-15 as the catalyst. The respective kinematic viscosity, pour point, and onset temperatures were recorded as epoxide (151 cSt, 10°C, and 320°C), butylated (251.7 cSt, –5°C, and 355°C), amylated (190.5 cSt, –8°C, and 361°C), and 2-ethyl hexanol (85.5 cSt, –15°C, and 405°C). This study also suggested catalyst reusability up to four times. These investigations show that alcohols and anhydrides can be used to improve the pour point and lubricant properties. However, further structural modifications such as esterification and functionalization of various anhydrides can be carried out to develop tri-esters and tetra-esters for multi-grade and high-quality biolubricants.

8.5 DI-ESTERS FORMATION THROUGH ESTERIFICATION AND ANHYDRIDES ADDITION

The modification of a ring-opened product to di-ester, tri-ester, and tetra-esters must be carried out to improve low-temperature properties (pour point), stability, and viscosity of the products to obtain high-quality biolubricant basestocks. Epoxidized soybean oil was structurally modified via ring opening and esterification reactions. Hwang and Erhan (2001) esterified epoxidized soybean oil in the presence of methanol, 1-butanol, 2-butanol, 1-hexanol, cyclohexanol, 1-decanol, and 2,2-dimethyl di-propanol. Their study indicated that the esterification step required longer reaction time for higher molecular weight alcohols except for 2,2-dimethyl dipropanol. The esterified products of methanol, 1-butanol, and 1-hexanol exhibited pour points of –3°C, –6°C, and –9°C, respectively. Furthermore, esterification of ring opening product was carried out in the presence of acetic, butyric, 2-methyl propionic, and hexanoic anhydrides. The anhydride derivatives were further blended with a 1% pour point depressant, and the pour point was –39°C, –39°C, and –45°C, respectively, with acetic and hexanoic anhydrides as the esterification reagents. The proposed reaction scheme for the ring opening and esterification is shown in Figure 8.3a.

Hwang et al. (2003) during their study on esterification of epoxidized soybean oil with 2-ethyl hexanol observed the formation of transesterified products. The esterification with anhydrides (acetic and hexanoic) and addition of a 1% pour point depressant (PAO-4) (1.3 equiv. mol of 2-ethyl hexanol) showed a low-temperature stability of –36°C with a high viscosity index of 159. By blending with a 1% pour point depressant (3 equiv. mol of 2-ethyl hexanol), the pour point and onset temperature were –60°C and 165.6°C, respectively

Hwang and Erhan (2006) examined the ring opening reaction with 2-butyl octanol, decanol, 2-hexyl octanol, dodecanol, 2-octyl decanol, and acetylation. The study revealed the formation of

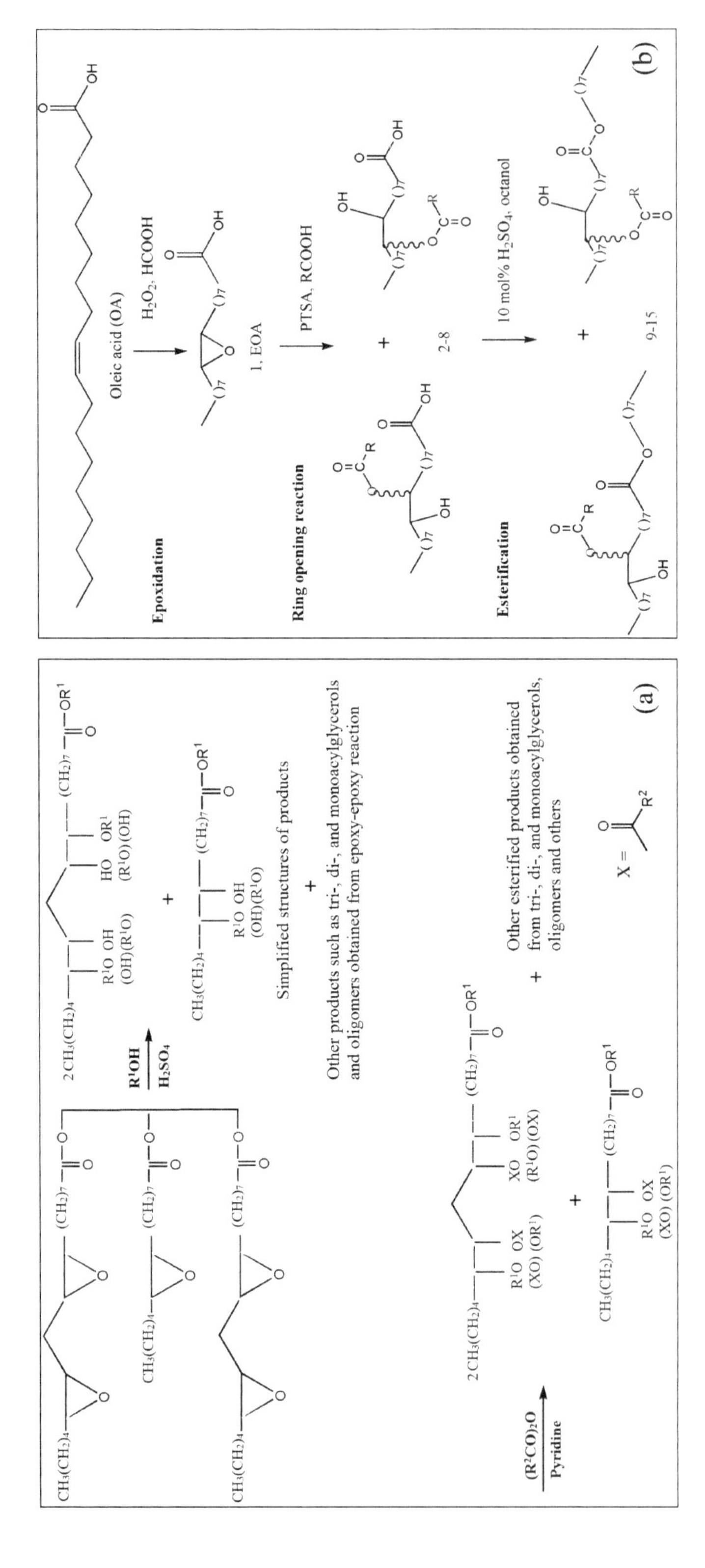

FIGURE 8.3 (a) Reaction scheme for ring opening and esterification of epoxidized soybean oil. (b) Epoxidation followed by ring opening using *p*-Toluenesulfonic acid as catalyst. RCOOH is variable to octanoic, nonanoic, lauric, myristic, palmetic, stearic, and behenic acid.

transesterified products varying from 0% to 100%, as well as the pour points of the ring opening products without additive and with a 1% additive. However, with a 1% additive, the pour point was in the range of –18°C to –36°C and –21°C to –46°C for transesterified products and ring opening products, respectively. Likewise, the pour point of acetylated products without additive and with a 1% additive was in the range of –27°C to –42°C and –30°C to –48°C, respectively. The viscosity of 0% and 100% transesterified product was 232.4 and 74.5 cSt, respectively. For acetylated products, it was around 103.4 and 41.5 cSt. From this study, it also was observed that ring opening products have lower viscosity index and acetylated products have higher viscosity index as well as oxidative stability. The properties obtained in this study were analogous with synthetic and mineral oil lubricants.

Sharma et al. (2006) studied the ring opening and acetylation of epoxidized soybean oil by using acetic, butyric, isobutyric, and hexanoic anhydride. The focus of the study was to enhance the thermo-oxidative stability of prepared anhydride derivatives. Maximum oxidative stability achieved with acetic anhydride with and without additives was 256°C and 174°C, respectively. However, Lathi and Mattiasson (2007) synthesized a biolubricant through a ring-opening reaction using an IR-15 catalyst with *n*-butanol, isoamyl alcohol, 2-ethyl hexanol, and further acetylation of epoxidized soybean oil by acetic anhydride. The results of this study revealed that the 3 mol of *n*-butanol in the esterification reaction gave a maximum viscosity of 312.9 cSt. The pour point of the acetylated products was greater than –5°C (*n*-butanol), –15°C to 0°C (isoamyl alcohol), and less than –10°C (2-ethyl hexanol). The rheology of the prepared products was examined and found to follow Newtonian fluid behavior. The results of the study revealed that the prepared products are biodegradable, and the catalyst can be reused for up to four times.

Salimon and Salih (2009a) studied oleic acid oxirane ring opening reaction in presence of various fatty acids using *p*-Toluenesulfonic acid (PTSA) as a catalyst (Figure 8.3b). The esterification with octanol revealed that improved low-temperature properties can be obtained with an increase in the fatty acid chain length. The product of the ring opening reaction with nonanoic acid derivative showed pour point and viscosity as –30°C and 67 cSt, respectively. Further esterification by behenic acid revealed pour point and viscosity as –50°C and 123 cSt, respectively (Figure 8.3b). On the other hand, a similar study by Salimon and Salih (2009b) with butanol as an esterification reagent showed utmost pour point and viscosity value as –56°C and 234 cP for the behenic fatty acid derivative.

Salimon and Salih (2010) also studied the ring opening of oleic acid epoxide with octanol and 2-ethyl hexanol (2-EH) followed by esterification of mono-esters with octanol and 2-ethyl hexanol. The unconverted free hydroxyl group was further reacted with oleic and stearic acid to give tri-esters. The prepared di-ester exhibited favorable cold flow properties and their pour point and viscosity were–23°C and 129 cP as well as –28°C and 149 cP, respectively. Similarly, the pour point and viscosity of the prepared tri-esters were –31°C and 159 cP as well as –35°C and 178 cP, respectively. From this study, it was concluded that the tri-esters yielded best performance properties.

8.6 TRI-ESTER AND TETRA-ESTERS FORMATION THROUGH ESTERIFICATION AND ANHYDRIDES ADDITION

Salimon et al. (2012b) reported a series of structural modifications (Figure 8.4a) such as epoxidation, ring opening, esterification, and acetylation to prepare tri-esters of oleic acid. The significant physicochemical properties of prepared tri-esters are pour point, flash point, viscosity index, and oxidation onset temperature. The lubricity properties were determined by the high-frequency reciprocating rig (HFRR). The outcomes of the study revealed that mid-chain esters exhibit positive impact on low-temperature performance, but shows a negative impact on the onset temperature. The results revealed that for mono-esters, di-esters, and tri-esters, maximum pour point and viscosity index were –43°C and 110 cP, –45°C and 128 cP, and –48°C and 145 cP, respectively, for behenic

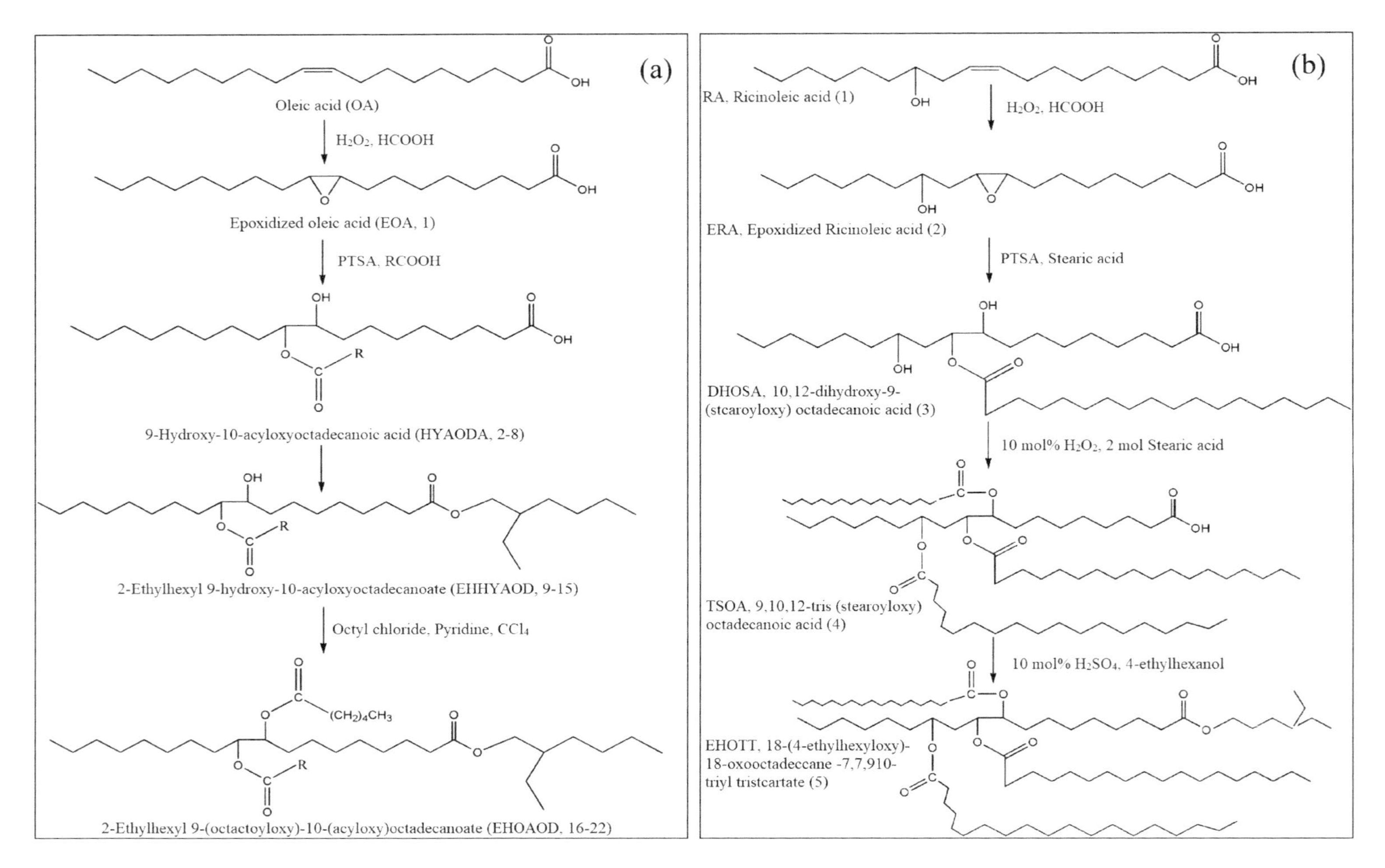

FIGURE 8.4 (a) Oleic acid-based tri-esters formation. (b) Typical structural modification scheme for tetra-esters.

acid derivatives. However, mono-, di-, and tri-esters of all octanoic acid derivatives showed maximum oxidative onset temperature as 113°C, 131°C, and 142°C, respectively. Acceptable friction properties also were observed for the prepared tri-esters. During the entire structural modifications, the yield was in the range of 55%–92%.

Salimon et al. (2011) reported a series of structural modifications (ring opening and esterification) of epoxidized ricinoleic acid to prepare biolubricant basestocks. Epoxidation and ring opening reactions were carried out according to their earlier reports on oleic acid using identical reagents and reaction conditions except for esterification step, which was carried out with butanol. The structurally modified epoxidized ricinoleic acid, ring opening, and esterified products showed maximum pour point and viscosity index of 9°C and 67, –23°C and 171, and –47°C and 243, respectively, for behenic acid derivatives. Salimon et al. (2012a) studied the esterification in presence of 2-ethyl hexanol and octanol. The measured properties such as pour point and viscosity index of 2-ethyl hexanol derivative were –42°C and 205, respectively. The pour point and dynamic viscosity for octanol-derived esters were –53°C and 283 mPa.s at 25°C. From these chemical alterations, it is evident that in the final product, there are two more single hydroxyl groups that can be functionalized further.

Salih et al. (2013) studied the functionalization of hydroxyl groups with 2 mol of stearic acid during esterification and 4-ethyl hexanol during acylation. The measured pour point, viscosity index, and oxidative onset temperature were –32°C, 193°C, and 245.5°C for tri-esters and –44°C, 257°C, and 282.1°C for tetra-esters, respectively. This study also determined the tribological properties of prepared esters. All the derivatives showed good anti-wear and friction-reducing properties. From this investigation, it could be concluded that ricinoleic acid di-esters with octanol as an esterification reagent exhibits lower pour point (–53°C) compared to tetra-esters of ricinoleic acid. The complete reaction pathway for the formation of tetra-esters is described by Salih et al. (2013) and presented in Figure 8.4b.

Salih et al. (2011a) prepared the oleic acid-based tri-esters as biolubricant. The oleic acid chemical structure was altered by a series of chemical modifications such as epoxidation, ring opening, esterification, and acetylation. The synthesis process adopted in this work was identical to their earlier studies except for the ring opening and acylated reaction. In this study, the esterification was carried out by *n*-pentanol (linear) and iso-pentanol (branched). Further acylation was carried out by oleic and stearic acid. The pour point and dynamic viscosity of di-esters (i.e., *n*-pentanol and iso-pentanol) were –20°C and 78 cP and –22°C and 88 cP, respectively. The pour point and dynamic viscosity for the tri-esters (i.e., oleic and stearic acid) were –24°C and 105 cP and –25°C and 113 cP, respectively. The enhancement in the pour point was due to the increased chain length of the branching agents.

In another study by Salih et al. (2011b) using octanol as an esterification reagent and lauroyl chloride and carbon tetrachloride (CCl_4) as acylation reagent, hydrogen bonding was a vital parameter impacting the performance of formulated basestocks. A further increase in the polar functionalities of fatty acid structures showed a positive impact on lubricity properties (i.e., wear protection and strong absorption on metal surfaces). Among all the derivatized products, maximum pour point was –45°C for the behenic acid derivative. Salimon et al. (2012b) carried out ring opening with 2-ethyl hexanol and acylation by caproic, octanoic, capric, lauric, and myristic fatty acids using H_2SO_4 as the catalyst. The increase in the branching and chain length resulted in an increase in the pour points of esterified (–24.33°C) and acylated (–47.19°C) products of myristic acid derivatives. The oxidative stability of the epoxidized, ring-opened, esterified, and acylated products (myristic acid derivative) were 75.1°C, 112.7°C, 116.3°C, and 80.5°C, respectively.

A similar kind of study was carried out by Mahajan et al. (2013) on epoxidation of mustard oil and its further esterification with sec-amyl alcohol and stearic acid in presence of H_2SO_4 as a catalyst. The study claimed that with an increase in the catalyst loading and reaction time, specific gravity and viscosity index of ring opening products increases. The pour point was favorable at lower catalysts loading compared to higher catalyst loading. Considerable lubricant properties were determined for the prepared esters. The pour point of the ring opening product was –30°C and –26°C, respectively at 1 wt% and 2 wt% of catalyst loading.

Sharma and Dalai (2013) described a one-pot process for the formulation of biolubricant basestocks from epoxidized canola oil. One-pot synthesis of epoxy ring opening followed by esterification was carried out with acetic anhydride using a sulfated Ti-SBA-15(10) catalyst. The optimal reaction conditions were 130°C reaction temperature, 4.5 grams of acetic anhydride, 10 wt% sulfated Ti-SBA-15(10) catalyst loading for 5 hours of reaction time, and stirring speed of 1,000 rpm. Furthermore, the effects of various catalysts on the conversion of epoxy canola oil to the esterified product were examined. The sulfated Ti-SBA-15 demonstrated 100% conversion of esterified product. The reaction kinetics, catalyst reusability, and lubricity properties were also determined. Finally, the study concluded that the oxidative property of biolubricant was outstanding (56.1 hours) with pour point of –9°C and a kinematic viscosity of 670 cSt at 100°C. The prepared esters also demonstrated excellent lubricity properties with a wear scar of 130 μm.

Li and Wang (2015) derived biolubricants with improved auto-oxidation stability and low-temperature performance through epoxidation of used cooking oil esterification by methanol and transesterification by iso-octanol, iso-tridecanol, and iso-octadecanol. Among all the transesterified products, iso-octadecanol derivative showed interesting results in kinematic viscosity (43.4 cSt), pour point (–24°C), and oxidative onset temperature (194.5°C). The investigation of the tribological properties suggested that transesterified products showed more improved performance than waste cooking oil.

8.7 CURRENT SCENARIO AND PROSPECTS

During the last few decades, there has been important progress towards the use of biodegradable lubricants from renewable resources. Currently, the growth in the lubricant market is supported by the expansion of automobile industry, transportation sector, manufacturing, and other industrialized activities. From the previous information from various lubricant industries across the globe, it was found that during the past decade the global lubricant market has undergone remarkable changes. International demand for lubricants has continued at around 630 kilotons in 2014 with a 6.9% compounding annual growth rate (CAGR) from 2016 to 2024 (Chemical World 2018).

Presently, the global market for biolubricants is comprised of automotive and industrial applications at 1.3 million pounds (U.S. $2.1 billion) in 2014 and is expected to reach 1.9 million pounds (U.S. $3 billion) by 2020 with a CAGR of 5.5% since 2010 (Bio-lubricants 2014). Among all the biolubricant sectors, the automotive sector boosts the bio-lubricants application because of higher inherent biodegradation rate, low toxic nature to aquatic organisms, and low level of bioaccumulation. Similarly, owing to fuel efficiency and maintenance in light-duty and heavy-duty vehicles, biolubricants also occupy a large share, which is estimated to grow at 7% CAGR from 2016 to 2024 (Petrochemicals 2015).

Strict regulatory actions from the U.S. Government include a minimum specified level of renewable content for a wide variety of fuels and lubricants. In addition, the U.S. Air Force is encouraging biodegradable lubricants as a strategic and essential pitch to national security that is escalating the biolubricants industry. The chemical registration policies in Europe and China have been playing a vital role in incentivizing larger adoption of biolubricants resulting in further expansion in their production. The demand for high oxidation resistance plant seed oils and their derivatives to enhance the functionality and to improve the performance of biolubricants drives research and development. Figure 8.5 presents the worldwide lubricant market partitioning by region and application type. Table 8.2 shows the present biolubricant market segmentation according to the source, application, end use, geography, and vendors (marketsandmarkets.com 2018). Table 8.3 shows some key biolubricants manufacturers in the world (Biolubricants 2014).

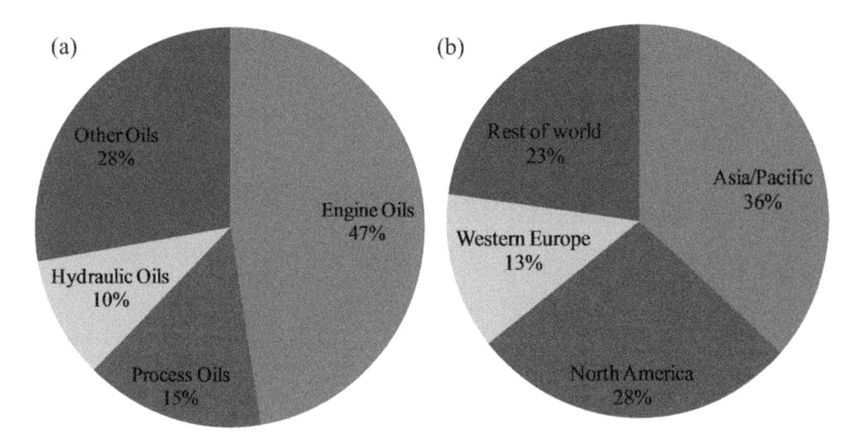

FIGURE 8.5 Current worldwide lubricant market segmented by (a) application area and (b) geography.

TABLE 8.2
Biolubricant Market Segmentation According to the Source, Application, End Use, Origin, and Vendors

Segment	
Source	(i) Animal fats
	(ii) Edible and non-edible oils
	(iii) Others
Application	(i) Automotive
	(ii) Chainsaw oils
	(iii) Engine oils
	(iv) Greases
	(v) Industrial
	(vi) Metal works
	(vii) Oilfield
	(viii) Process oils
	(ix) Transportation
	(x) Others
End use	(i) Automotive
	(ii) Commercial
	(iii) Construction
	(iv) Marine
	(v) Oil and gas
	(vi) Others
Vendors	(i) British Petroleum
	(ii) Cargill
	(iii) China National Petroleum Corporation
	(iv) Exxon Mobile
	(v) Shell

TABLE 8.3
Global Key Manufactures of Biolubricants

Manufacturers	Country
Albemarle Corporation	USA
Binol Biolubricants	Sweden
British Petroleum	UK
Chevron Corporation	USA
China National Petroleum Corporation	China
Emery Oleochemicals	Malaysia
Exxon Mobile	USA
Fuchs Petrolub	Germany
JXTG Nippon Oil & Energy	Japan
Klüber Lubrication München KG	Germany
Lukoil	Russia
Panolin AG	Switzerland
PetroChina Company Limited	China
Royal Dutch Shell plc	Netherlands
Total S.A.	France

8.8 CONCLUSIONS

The formulation of biolubricant basestocks by structural or chemical modification of plant seed oils and their derivatives is a favorable technology. Biolubricants are highly in demand because they are renewable, biodegradable, eco-friendly, and relatively non-toxic. Plant seed oils and their derivatives are promising substitutes for mineral oils for producing biolubricants and biodegradable greases. Increased research can result in better oxidation resistance of plant seed oils and their derivatives with improved functionality, which is expected to drive the biolubricants segment globally. Furthermore, the implementation of environmental regulations can regulate the wide-scale applications for biolubricants for transportation, aviation, and marine sectors. Significant research and development initiatives with technological innovations can further increase the biolubricant market on a worldwide scale.

REFERENCES

Abdullah, B. M., N. Salih, and J. Slimon. 2014. Optimization of the chemo-enzymatic mono-epoxidation of linoleic acid using D-optimal design. *Journal of Saudi Chemical Society* 18:276–287.

Adhvaryu, A., and S. Z. Erhan. 2002. Epoxidized soybean oil as a potential source of high temperature lubricants. *Industrial Crops and Products* 15:247–254.

Arbain, H. N., and J. Salimon. 2009. Synthesis and characterization of ester trimethylolpropane (TMP) based *Jatropha curcas* oil as biolubricant base stock. *Journal of Science and Technology* 2:47–58.

Arumugam, S., G. Sriram, and L. Subadhra. 2012. Synthesis, chemical modification and tribological evaluation of plant oil as biodegradable low temperature lubricant. *Procedia Engineering* 38:1508–1517.

Bart, J. C. J., E. Gucciardi, and S. Cavallaro. 2013. *Biolubricants: Science and Technology*. Oxford, UK: Woodhead Publishing Limited.

Becker, R., and A. Knorr. 1996. An evolution of antioxidants for vegetable oils at elevated temperatures. *Lubrication Science* 8:95–117.

Biolubricants. 2014. Report on biolubricants—A Global Market Overview. Report ID:2112733. https://www.reportbuyer.com/product/2112733/biolubricants-a-global-market-overview.html (accessed on June 4, 2018).

Biswas, A., A. Adhvaryu, D. G. Stevenson, B. K. Sharma, J. L. Willet, and S. Z. Erhan. 2007. Microwave irradiation effects on the structure, viscosity, thermal properties and lubricity of soybean oil. *Industrial Crops and Products* 25:1–7.

Bokade, V. V., and G. D. Yadav. 2007. Synthesis of bio-diesel and bio-lubricant by transesterification of vegetable oil with lower and higher alcohols over heteropolyacids supported by clay (K-10). *Process Safety and Environmental Protection* 85:372–377.

Borugadda, V. B., A. K. R. Somidi, and A. K. Dalai. 2017. Chemical/structural modification of canola oil and canola biodiesel: Kinetic studies and biodegradability of the alkoxides. *Lubricants* 5:1–11.

Borugadda, V. B., and V. V. Goud. 2012. Biodiesel production from renewable feed stocks: Status and opportunities. *Renewable and Sustainable Energy Reviews* 16:4763–4784.

Borugadda, V. B., and V. V. Goud. 2013. Comparative studies of thermal, oxidative and low temperature properties of waste cooking oil and castor oil. *Journal of Renewable and Sustainable Energy* 063104:1–14.

Borugadda, V. B., and V. V. Goud. 2014. Synthesis of waste cooking oil epoxide as a bio-lubricant base stock: Characterization and optimization study. *Journal of Bioprocess Engineering and Biorefinery* 3:57–72.

Borugadda, V. B., and V. V. Goud. 2015a. Response surface methodology for optimization of bio-lubricant basestock synthesis from high free fatty acids castor oil. *Energy Science and Engineering* 3:371–383.

Borugadda, V. B., and V. V. Goud. 2015b. Improved low-temperature properties of chemically modified high free fatty acid castor oil–methyl esters: Blending and optimization study. *Journal of Energy Engineering* 142:040150201-10.

Borugadda, V. B., and V. V. Goud. 2016. Improved thermo-oxidative stability of structurally modified waste cooking oil methyl esters for bio-lubricant application. *Journal of Cleaner Production* 112:4515–4524.

Borugadda, V. B., and V. V. Goud. 2018. Long-term storage stability of epoxides derived from vegetable oils and their methyl esters. *Energy & Fuels* 32:3428–3435.

Boyde, S. 2002. Green lubricants: Environmental benefits and impacts of lubrication. *Green Chemistry* 4:293–307.

Brophy, J. E., and W. A. Zisman. 1951. Surface chemical phenomena in lubrication. *Annals of the New York Academy of Sciences* 53:836–861.

Campanella, A., C. Fontanini, and M. A. Baltanás. 2008. High yield epoxidation of fatty acid methyl esters with performic acid generated in-situ. *Chemical Engineering Journal* 144:466–475.

Campanella, A., E. Rustoy, A. Baldessari, and M. A. Baltanas. 2010. Lubricants from chemically modified vegetable oils. *Bioresource Technology* 101:245–254.

Chemical World. 2018. Report on trend towards sustainable development to drive biolubricants industry demand.

Daniel, L., A. R. Ardiyanti, B. Schuur, R. Manurung, A. A. Broekhuis, and H. J. Heeres. 2011. Synthesis and properties of highly branched jatropha curcas L. oil derivatives. *European Journal of Lipid Science and Technology* 113:18–30.

Derawi, D., and J. Salimon. 2010. Optimization on epoxidation of palm olein by using performic acid. *European Journal of Chemistry* 7:1440–1448.

Dinda, S., A. V. Patwardhan, V. V. Goud, and N. C. Pradhan. 2008. Epoxidation of cottonseed oil by aqueous hydrogen peroxide catalysed by liquid inorganic acids. *Bioresource Technology* 99:3737–3744.

Erhan, S. Z., and S. Asadauskas. 2000. Lubricant base stocks from vegetable oils. *Industrial Crops and Products* 11:277–282.

Fiser, S. S., M. Jankovic, and O. Borota. 2012. Epoxidation of castor oil with peracetic acid formed in situ in the presence of an ion exchange resin. *Chemical Engineering and Processing: Process Intensification* 62:106–113.

Fiser, S. S., M. Jankovic, and Z. S. Petrovic. 2001. Kinetics of in situ epoxidation of soybean oil in bulk catalyzed by ion exchange resin. *Journal of American Oil Chemists Society* 78:725–731.

Gawrilow, I. 2004. Vegetable oil usage in lubricants. *Oleochemicals Inform* 15:702–705.

Goud, V. V., A. V. Patwardhan, and N. C. Pradhan. 2006. Studies on the epoxidation of mahua oil (madhumica indica) by hydrogen peroxide. *Bioresourc Technology* 97:1365–1371.

Goyan, R. L., R. E. Melley, P. A. Wissner, and W. C. Ong. 1998. Biodegradable lubricants. *Lubricating Engineering* 54:10–17.

Grandviewresearch.com. 2018. Lubricants Market Size, Share & Trends Analysis Report By Application (Industrial, Automotive, Marine, Aerospace), By Region (North America, Europe, APAC, CSA, MEA), Competitive Landscape, And Segment Forecast, 2018 – 2025. Report ID: 978-1-68038-123-8.

https://chemicalworldnet.wordpress.com/2018/03/29/trend-towards-sustainable-development-to-drive-biolubricants-industry-demand/ (accessed on June 4, 2018).

Hwang, H. S., A. Adhyaryu, and S. Z. Erhan. 2003. Preparation and properties of lubricant base stocks from oxidised soybean oil and 2-ethylhexanol. *Journal of American Oil Chemists Society* 80:811–815.

Hwang, H. S., and Z. S. Erhan. 2001. Modification of epoxidized soybean oil for lubricant formulations with improved oxidative stability and low pour point. *Journal of American Oil Chemists Society* 78:1179–1184.

Hwang, H. S., and Z. S. Erhan. 2006. Synthetic lubricant basestocks from epoxidised soybean oil and Guerbet alcohols. *Journal of American Oil Chemists Society* 23:311–317.

Kulkarni, R. D., P. S. Deshpande, S. U. Mahajan, and P. P. Mahulikar. 2013. Epoxidation of mustard oil and ring opening with 2-ethylhexanol for biolubricants with enhanced thermo-oxidative and cold flow characteristics. *Industrial Crops and Products* 49:586–592.

Lathi, P. S., and B. Mattiasson. 2007. Green approach for the preparation of biodegradable lubricant base stock from epoxidised vegetable oil. *Applied Catalysis B: Environmental* 69:207–212.

Li, W., and X. Wang. 2015. Biolubricants derived from waste cooking oil with improved oxidation stability and low temperature properties. *Journal of Oleo Science* 64:367–374.

Madankar, C. S., A. K. Dalai, and S. N. Naik. 2013. Green synthesis of biolubricant basestocks from canola oil. *Industrial Crops and Products* 44:139–144.

Mahajan, S. U., P. R. Kulkarni, R. D. Kulkarni, and P. P. Mahulikar. 2013. Synthesis and characterization of chemically modified epoxidised mustard oil for biolubricant properties. *International Journal Applied Engineering Research* 17:2023–2030.

Mercurio, P., K. A. Burns, and A. Negri. 2004. Testing the ecotoxicology of vegetable versus mineral based lubricating oils: 1. degradation rates using tropical marine microbes. *Environmental Pollution* 129:165–173.

Meshram, P. D., R. G. Puri, and H. V. Patil. 2011. Epoxidation of wild safflower (Carthamus Oxyacantha) oil with peroxy acid in presence of strongly acidic cation exchange resin IR- 122 as catalyst. *International Journal of ChemTech Research* 3:1152–1163.

Monono, E. M., D. M. Haagenson, and D. P. Wiesenborn. 2015. Characterizing the epoxidation process conditions of canola oil for reactor scale-up. *Industrial Crops and Products* 67:364–372.

Moser, B. R., and S. Z. Erhan. 2007. Preparation and evaluation of a series of α-hydroxy ethers from 9, 10-epoxystreates. *European Journal of Lipid Science and Technology* 109:206–213.

Mungroo, R., N. C. Pradhan, V. V. Goud, and A. K. Dalai. 2008. Epoxidation of canola oil with hydrogen peroxide catalysed by acidic ion exchange resin. *Journal of American Oil Chemists Society* 85:887–896.

Nanda, S., R. Rana, H. N. Hunter, Z. Fang, A. K. Dalai, and J. A. Kozinski. 2019. Hydrothermal catalytic processing of waste cooking oil for hydrogen-rich syngas production. *Chemical Engineering Science* 195:935–945.

Petrochemicals. 2015. Report on synthetic Lubricants Market Analysis, Market Size, Application Analysis, Regional Outlook, Competitive Strategies and Forecast, 2016 To 2024. https://www.hexaresearch.com/research-report/synthetic-lubricants-industry (Accessed on June 4, 2018).

Salih, N., J. Salimon, and E. Yousif. 2011a. The physicochemical and tribological properties of oleic acid based triester biolubricants. *Industrial Crops and Products* 34:1089–1096.

Salih, N., J. Salimon, and E. Yousif. 2011b. Synthesis of oleic acid based esters as potential basestock for biolubricant production. *Turkish Journal of Engineering and Environmental Science* 35:115–123.

Salih, N., J. Salimon, E. Yousif, and M. Abdullah. 2013. Biolubricant basestocks from chemically modified plant oils: Ricinoleic acid based –tetraesters. *Chemistry Central Journal* 7:128–141.

Salimon, J., and N. Salih. 2009a. Oleic acid diesters: Synthesis, characterization and low temperature properties. *European Journal of Scientific Research* 32:216–222.

Salimon, J., and N. Salih. 2009b. Substituted esters of octadecanoic acid as a potential biolubricants. *European Journal of Scientific Research* 31:273–279.

Salimon, J., and N. Salih. 2010. Chemical modification of oleic acid oil for biolubricant industrial applications. *Australian Journal of Basic and Applied Sciences* 4:1999–2003.

Salimon, J., B. M. Abdullah, and N. Salih. 2011. Optimization of the oxirane ring opening reaction in biolubricant base oil production. *Arabian Journal of Chemistry* 9:S1053–S1058.

Salimon, J., N. Salih, and B. M. Abdullah. 2012a. Production of chemo-enzymatic catalyzed mono-epoxide biolubricant: Optimization and physico-chemical characteristics. *Journal of Biomedicine and Biotechnology* 2012:1–11.

Salimon, J., N. Salih, and E. Yousif. 2012b. Biolubricant basestocks from chemically modified ricinoleic acid. *Journal of King Saud University* 24:11–17.

Saurabh, T., M. Patnaik, S. L. Bhagt, and V. C. Renge. 2011. Epoxidation of vegetable oils: A review. *International Journal of Advances in Engineering and Technology* 2:491–501.

Sharma, B. K., A. Adhvaryu, and S. Z. Erhan. 2006. Synthesis of hydroxy thio-ether derivatives of vegetable oil. *Journal of Agricultural and Food Chemistry* 54:9866–9872.

Sharma, B. K., K. M. Doll, and S. Z. Erhan. 2008b. Ester hydroxy derivatives of methyl oleate: Tribological, oxidation and low temperature properties. *Bioresource Technology* 99:7333–7340.

Sharma, R. V., and A. K. Dalai. 2013. Synthesis of biolubricant from epoxy canola oil using sulphated Ti-SBA-15 catalyst. *Applied Catalysis B: Environmental* 142–143:604–614.

Sharma, Y. C., B. Singh, and S. N. Upadhyay. 2008a. Advancements in development and characterization of biodiesel: A review. *Fuel* 87:2355–2373.

Silva, M. S., E. L. Foletto, S. M. Alves, T. N. C. Dantas, and A. A. D. Neto. 2015b. New hydraulic biolubricants based on passion fruit and moringa oils and their epoxy. *Industrial Crops and Products* 69:362–370.

Silva, M. S., H. J. Arimateja, G. F. Silva, N. A. A. Dantas, and D. T. N. Castro. 2015a. New formulations for hydraulic biolubricants based on epoxidised vegetable oils: Passion fruit (*Passiflora edulis* Sims f. *flavicarpa* Degener) and moringa (*Moringa oleifera Lamarck*). *Brazilian Journal of Petroleum Gas* 9:27–36.

Soni, S., and M. Agarwal. 2014. Lubricants from renewable energy resources-a review. *Green Chemistry Letters and Reviews* 7:359–382.

Sun, S., X. Ke, L. Cui, G. Yang, Y. Bi, F. Song, and X. Xu. 2011. Enzymatic epoxidation of *Sapindus mukorossi* seed oil by perstearic acid optimized using response surface methodology. *Industrial Crops and Products* 33:676–682.

Tan, S. G., and W. S. Chow. 2010. Biobased epoxidized vegetable oils and its greener epoxy blends: A review. *Polymer-Plastics Technology and Engineering* 49:1581–1590.

Technical Report by marketsandmarkets.com. 2018. Bio-Lubricant Market by Base Oil Type (Vegetable Oil, Animal Fat), Application (Hydraulic Fluids, Metalworking Fluids, Chainsaw Oils, Mold Release Agents), End Use (Industrial, Commercial Transport, Consumer Automobile)—Global Forecast to 2022. Report Code: CH 3544.

Wadumesthrige, K., S. O. Salley, and K. Y. N. Simong. 2009. Effects of partial hydrogenation, Epoxidation and hydroxylation on the fuel properties of fatty acid methyl esters. *Fuel Processing Technology* 90:1292–1299.

Wu, X., Z. Xingang, Y. Shengring, C. Haigang, and W. Dapu. 2000. The study of epoxidised rapeseed oil used as potential biodegradable lubricant. *Journal of American Oil Chemists Society* 77:561–563.

9 Recent Advances in Consolidated Bioprocessing for Microbe-Assisted Biofuel Production

Prakash Kumar Sarangi and Sonil Nanda

CONTENTS

9.1 Introduction ... 141
9.2 Lignocellulosic Biomass Structure ... 143
9.3 Progress in Consolidated Bioprocessing ... 144
9.4 Bacterial Agents in Consolidated Bioprocessing ... 144
9.5 Filamentous Fungi in Consolidated Bioprocessing ... 147
9.6 Yeasts in Consolidated Bioprocessing ... 147
9.7 Native and Artificial Cellulosomes ... 149
9.8 Consolidated Bioprocessing for Biobutanol Production ... 149
9.9 Conclusions ... 151
References ... 152

9.1 INTRODUCTION

The global interest in the production of renewable transportation fuels has been increasing due to the rising fuel prices, huge demand for present fossil-based fuels, greenhouse gas (GHG) emissions, and global warming. The rapid escalation for energy demands across the globe and the diminishing fossil fuels have motivated efforts to find out new energy sources that are cleaner, renewable, and environmentally sustainable (Chu and Majumdar 2012; Nanda et al. 2017a). Renewable energy systems, especially those biomass-based, tend to meet the energy supply requirements and increase the food security and the economic growth of a country (Yilman and Selim 2013; Nanda et al. 2015). Biomass from plant resources, particularly lignocellulosic biomass, is an abundantly available renewable resource worldwide at a production of nearly $150–170 \times 10^9$ tons annually (Pauly and Keegstra 2008). Lignocellulosic biomasses are the most abundant, renewable, and economical carbon sources that can substitute the food-based biomass (starchy feedstocks) for large amounts of fuels and chemical production (Wen et al. 2009; Nanda et al., 2016). Lignocellulosic biofuels are gaining attention as the future transportation fuels, and significant research in this area is being focused on developing eco-friendly and economical technologies for the conversion of lignocellulosic biomasses into biofuels (Perlack et al. 2005).

Lignocellulose biomasses mostly include agricultural crop residues such us cereals, straw, sugarcane bagasse, coconut and banana residues, rice husks, corn cobs, cotton stalks and so on including forest residues and energy crops (Naik et al. 2010; Parisutham et al. 2014; Sarangi et al. 2016; Nanda et al., 2018). The fuels derived from lignocellulosic biomasses tend to reduce GHG emissions because fresh plants utilize the CO_2 generated from their combustion during photosynthesis (Nanda et al. 2014b). Lignocellulosic biomass are second-generation feedstocks because they do not compete with food crops or arable agricultural lands. Due to their abundant availability, cost effectiveness, renewable nature, and potential for global distribution, lignocellulosic biomasses have attracted widespread attention towards producing alternative liquid fuels. Lignocellulosic biomass is composed of 60%–80% complex carbohydrates of C_5 and C_6 sugar units. The major challenge for the extraction of these fermentable sugars from lignocellulosic biomass and the production of bioalcohols is the presence of lignin. To overcome such barriers, efficient pretreatment methods are needed that can help to break down the structure making this biomass more susceptible to hydrolytic enzymes and fermenting microorganisms.

Complete degradation of cellulose into simple sugars requires a complex consisting of three enzymes. The first enzyme, cellobiohydrolase or CBH (EC 3.2.1.91), an exo-1, 4-β-D-glucanase, cleaves cellobiose units at the ends of cellulose chains. The second enzyme, endo-1,4-β-D-glucanases or EG (EC 3.2.1.4), breaks the internal β-1,4-glucosidic bonds inside the cellulose chain. The final process of hydrolysis hydrolyzes cellobiose to glucose, and the subsequent separation of glucose components from different soluble oligosaccharides is done by 1,4-β-D-glucosidases or BG (EC 3.2.1.21). Thus, a synergistic action of these three cellulolytic enzymes is required for the complete degradation of cellulose.

The bioconversion of lignocellulosic biomass into alcohol-based biofuels requires a multistep process such as pretreatment, enzymatic hydrolysis, and fermentation (Fougere et al. 2016). Therefore, it is a great challenge for the valorization of lignocellulosic biomass for formation of bioalcohols simultaneously reducing the production cost. Different fermentation processes are used for biomass conversion such as simultaneous saccharification and fermentation (SSF) and simultaneous saccharification and co-fermentation (SSCF) to produce bioalcohols. However, consolidated bioprocessing (CBP) is emerging as a single-step conversion process of lignocellulosic biomass to alcohol-based fuels through a combination of all the unit operations in a single reactor using a single microorganism or microbial consortium to pretreat, hydrolyze, and ferment the biomass (Figure 9.1). An ideal microorganism for CBP must possess all the desired features of enzyme production, cellulose saccharification, and ethanol fermentation. In CBP, the processing cost of biomass is lower compared to other fermentation processes, and due to process integration, it also avoids the cost of external enzymes. In CBP, typically four biologically mediated events occur such as (Lynd et al. 2002):

1. The production of depolymerizing enzymes like cellulases and hemicellulases
2. Hydrolysis of the polysaccharide part of pretreated biomass
3. Fermentation of the available hexose and pentose sugars
4. Fermentation of sugars to ethanol or butanol

This chapter describes the recent progresses in the CBP systems focusing on different types of microorganism. This chapter also discusses the production of bioethanol and biobutanol using CBP and assesses the process integrations making it economically feasible. Also highlighted in this chapter is the association of the microbial consortium system, recombinant microorganisms, and metabolic engineering with the CBP system to produce biofuel from lignocellulosic biomass.

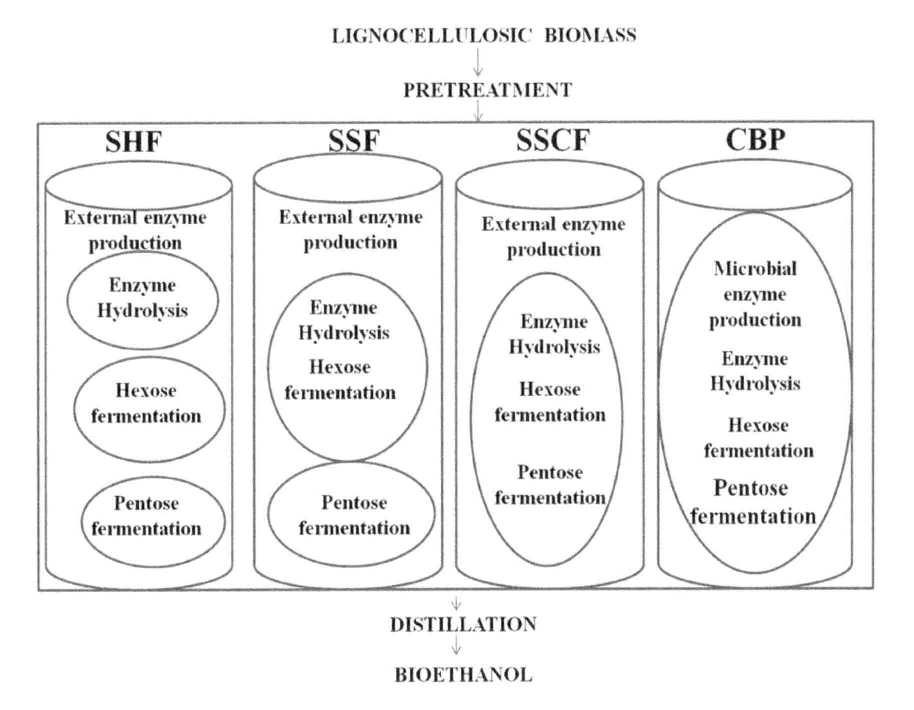

FIGURE 9.1 Different bioprocessing strategies for production of bioethanol from lignocellulosic biomass. Abbreviations: separate hydrolysis and fermentation (SHF), simultaneous saccharification and fermentation (SSF), simultaneous saccharification and co-fermentation (SSCF), and consolidated bioprocessing (CBP).

9.2 LIGNOCELLULOSIC BIOMASS STRUCTURE

The carbohydrate composition in lignocellulosic materials has a considerable effect on the alcohol yields and varies significantly with various factors like geographical location, growth conditions, and crop maturity (Nanda et al. 2013). The lignocellulosic component largely provides the structural integrity of the plant and usually is present in roots, stalks, and leaves. The three major polymers such as cellulose, hemicelluloses, and lignin along with pectin, extractives, and ash are in the lignocellulosic biomass depending on their types, species, and sources (Agbor et al. 2011). Extractives are the non-structural biomass components soluble in neutral organic solvents or water that consist of various biopolymers like terpenoids, fats, lipids, steroids, waxes, resin acids, and other phenolic components. Different types of bonding such as intermolecular bridges, covalent bonding, and van der Waals forces render lignocellulose an intricate structure strong enough to resist an enzymatic attack. In general, the composition of a typical lignocellulosic biomass is cellulose (30%–60%), hemicellulose (20%–40%) and lignin (15%–25%) on dry basis (Nanda et al. 2013).

Cellulose is the most plentiful polysaccharide consisting of polymeric chain of D-glucose subunits joined together by β (1$\rightarrow$4) linkages. The repeating cellulose chains in the range of 2,000–20,000 sub-units form cellobiose and later forms microfibrils combining by different bonding such as van der Waals forces, covalent bonding, and hydrogen bonding (Agbor et al. 2011). Cellulose occurs in both amorphous and crystalline forms.

Hemicellulose is the second most abundant polymer in nature. It is primarily composed of xylan, which is a branched polymer of a pentose sugar (i.e., xylose). The degree of polymerization of hemicellulose is 50–200, which is shorter than cellulose. The hydrolysis of hemicellulose $(C_6H_{10}O_5)_n$

generally produces xylose ($C_5H_{10}O_5$), which can be further converted to furfural or ethanol. Unlike cellulose, hemicellulose has a lack of homogeneity in structure and its short lateral side chains. Hemicellulose also has a lower molecular weight than cellulose.

In lignocellulosic biomass, lignin binds with cellulose and hemicellulose acting as a major barrier for its conversion towards cellulosic bioethanol. Being a polymer, different branched phenolic compounds and various side chain groups restrict its solubility in most solvents. To ferment the sugars (cellulose and hemicellulose) present in lignocellulosic material, the lignin matrix must be broken down. An efficient pretreatment technology can alter the network of cellulose, hemicellulose, and lignin and reduce the recalcitrance of lignocellulosic biomass prior to its bioconversion.

9.3 PROGRESS IN CONSOLIDATED BIOPROCESSING

Various microorganisms have the capability to secrete the enzymes required to hydrolyze the polysaccharides in lignocelluloses. However, the efficient simultaneous hydrolysis of polysaccharides and their fermentation to a desired alcohol product by a wild type microorganism is still in progress (Hahn-Hägerdal et al. 2006). Hence, strain development is essential for the conversion of lignocellulosic materials to biofuels in the context of CBP. Many wild type microorganisms having cellulolytic abilities on wide-ranging substrates generally undergo reduced growth characteristics or deprived product formation. On the other hand, some organisms have high product formation ability but lack the hydrolytic activities on cellulose and hemicellulose. Table 9.1 lists a few wild type microorganisms with cellulolytic abilities.

The ideal microorganism for CBP of lignocellulosic biomass must possess a few characteristics like inhibitor tolerance, low nutrient and stable pH requirements, as well as high temperature tolerance, thereby having generally regarded as safe (GRAS) characteristics. Therefore, a CBP microorganism acting as biocatalyst for the direct conversion of pretreated lignocellulosic biomass into a specific metabolic product in a bioreactor is still a big challenge. Two major efforts can be adopted to resolve the issues in CBP such as:

1. Metabolic engineering of naturally occurring cellulolytic microorganisms to improve their fermentative capabilities and efficiencies
2. Genetic engineering of efficient fermentative microorganisms to express a heterologous cellulolytic system and facilitate the pre-fermentation stage in vivo

9.4 BACTERIAL AGENTS IN CONSOLIDATED BIOPROCESSING

Among bacteria, the thermophilic bacteria group shows great prospective as CBP agents, which have abilities of cellulose degradation and subsequent ethanol production under thermophilic conditions. There are great advantages to high temperature hydrolysis and fermentation since it saves energy because the reactor does not need cooling before inoculation and further heating

TABLE 9.1
Wild Type Cellulolytic Microorganisms

Organisms	Carbon Source	Fuel Product (g/L)	Reference
Clostridium acetobutylicum	Grass	0.6 (butanol)	Berezina et al. (2008)
Clostridium cellulolyticum	Cellulose	0.5 (ethanol)	Guedon et al. (2002)
Clostridium phytofermentans	Corn stover	2.8 (ethanol)	Tolonen et al. (2011)
Clostridium thermocellum	Cellulose	1.4 (ethanol)	Argyros et al. (2011)
Other *Clostridia* strains	Cellulose	<0.04 (butanol)	Virunanon et al. (2008)

towards distillation. In addition, the reaction rate becomes double with the increase in temperature by approximately 10°C, thereby decreasing the enzymatic load (Ibrahim and El-diwan 2007). Generally, in CBP, high temperatures favor the hydrolysis process because cellulose hydrolysis and sugar release are the rate-limiting steps. Due to their thermostable nature, these enzymes can resist higher temperatures, thereby having longer half-lives. In addition, the chance of bacterial contamination decreases at high fermentation temperatures (≥60°C). A major challenge for microorganisms used in CBP is their reduced resistance to temperature and other by-product inhibitors.

One of the most accepted microorganisms used in the CBP systems is *Clostridium thermocellum*, which produces an extracellular multi-enzyme complex bearing different glycoside hydrolase like cellulase, hemicellulose, and carbohydrate esterase on the surface of cell membranes. This bacterium has a high capability for hydrolysis of different cellulosic materials such as crystalline cellulose (Shao et al. 2011), poplar, wheat straw (Hörmeyer et al. 1988), and switch grass (Yee et al. 2012). *C. thermocellum*, an anaerobic, acetogenic, cellulolytic, and thermophilic bacterium has been extensively used for CBP. Adapting to unique culture conditions like high temperatures, anaerobic conditions, and cellulose conversion rates of nearly 2.5 gram/liter/hour (g/L/h), this thermostable bacterium reduces the operational costs for ethanol recovery. *C. thermocellum* has attained great interest in CBP development due to its novel characteristics such as temperature-resistant enzymes for biomass conversion and the tendency to synthesize different value-added products (Deng et al. 2013). Wild type *C. thermocellum*, *Caldicellulosiruptor bescii,* and *Caldicellulosiruptor obsidiansis* were used for CBP ethanol production from dilute acid-pretreated transgenic and wild type switch grass (Yee et al. 2012).

The hydrolytic pathways of *C. thermocellum* from cellulosic substrates into ethanol, acetate, and lactate are different from other microorganisms. With the help of an ATP-binding cassette, *C. thermocellum* performs the transportation of cellobiose instead of a phosphotransferase system (Nataf et al. 2009). Contrasting with pentose-phosphate pathway (PPP) or the Entner-Doudoroff pathway, cellobiose and glucose require phosphorylation to glucose 1-phosphate before glycolysis. After the conversion of pyruvate into acetyl-CoA, the reduced FdH2 and CO_2 facilitate conversion to lactic acid by lactate dehydrogenase. The formation of acetyl-CoA during this process makes the production of acetaldehyde easy and subsequently ethanol through NAD^+ linked dehydrogenase.

Clostridium phytofermentans ATCC 700394 was used as a CBP microorganism to produce ethanol from ammonia fiber expansion (AFEX)-pretreated corn stover (Jin et al. 2011). However, compared to *C. thermocellum, C phytofermentans* can degrade the available sugars including xylose to ethanol and acetate. This strain could hydrolyze 76% and 88.6% of glucan and xylan, respectively. When AFEX-treated corn stover was used as the sole carbon source with nutrients supplementation, glucan and xylan conversions were recorded at 48.9% and 77.9%, respectively, and ethanol concentration was 7.0 grams/liter (g/L) after 264 hours (Jin et al. 2012a).

A single-step conversion over wide ranges of cellulosic and hemicellulosic substrates to produce ethanol, acetate, and lactate was detected by a cellulose-degrading thermophilic anaerobic *Clostridium* sp. DBT-IOC-C19 strain. In this case, ethanol was the important fermentation product (Singh et al. 2017). The bacterium group *Geobacillus*, thermophilic bacilli, can ferment various types of sugar substrates such as glucose, xylose, and arabinose at a wide range of temperatures from 55°C to 70°C producing different fermentation products such as lactic acid, formic acid, acetic acid, and ethanol (Barnard et al. 2010). The bioconversion of lignocelluloses by enzymes like cellulases, xylanases, and lignases also are attributed by certain species of *Geobacillus.*

Another thermophilic anaerobe of the *Thermoanaerobacterium* group shows the xylanolytic activities and the potential to ferment xylose, mannose, galactose, and glucose. In contrast to *Clostridium* strains, which are cellulolytic in nature, *Thermoanaerobacterium* exhibit the hemicellulolytic activities (Schuster and Chinn 2013). The growth of such a bacterial group is well visualized at temperatures between 4°C and 65°C and pH between 4.0 and 6.5 exhibiting endoxylanase, β-xylosidase, and other xylanolytic enzyme activities (Shaw et al. 2012). The experiments on bioethanol and biohydrogen production (CHE) production were conducted by using the *Themoanaerobacterium aciditolerans* strain AK54 (Sigurbjornsdottir and Orlygsson 2011).

Another thermophilic anaerobe, *Themoanaerobacterium calidifontis* nov. strain (RX1), shows bioconversion ability on xylan, starch, glucose, and xylose into ethanol (Shang et al. 2013). Experiments using an extreme thermophilic bacterium, *Caldicellulosiruptor saccharolyticus* DSM 8903, have been conducted for a one-step bioconversion of switch grass, microcrystalline cellulose, and glucose as the substrates (Talluri et al. 2013). Different *Clostridium* strains used for CBP system are summarized in Table 9.2.

TABLE 9.2
Comparison of *Clostridium* Strains Used for Consolidated Bioprocessing

Strain	Substrate	Ethanol (mM)	Acetic Acid (mM)	Lactic Acid (mM)	References
27405	Milled filter Paper	17.4	9.1	–	Lv et al. (2012)
27405	Pretreated switchgrass	4.3	8.5	–	Wilson et al. (2013)
27405	Pretreated poplar	18.0	13.5	–	Wilson et al. (2013)
27405	Cellobiose	27	51	7	Ellis et al. (2012)
27405	Avicel	37	45	6	Ellis et al. (2012)
27405	Microcrystalline cellulose	23.7	25.2	27	Tachaapaikoon et al. (2012)
27405	α-cellulose	75	105	150	Levin et al. (2006)
AS-39	Cellobiose	47	23	–	Lamed et al. (1988)
AS-39	Cellulose	52	23	–	Lamed et al. (1988)
AS-39	Avicel 10.3	37	46	6	Tyurin et al. (2004)
ATCC 27405	Avicel 5	18	13.8	–	Deng et al. (2013)
ATCC 35609	MN300 8	31.2	22.5	–	Wilson et al. (2013)
BC1	Whatman No. 1 filter paper	10	2	–	Galbe et al. (2007)
CS7	Milled filter paper	17.1	8.3	0.8	Saddler and Chan (1984)
CS8	Milled filter paper	17.1	5.4	–	Saddler and Chan (1984)
DBT-IOC-C19	Avicel 10	32.6	18.7	5.1	Singh et al. (2017)
DSM 1237	α-cellulose	70	35	8	Chinn et al. (2007)
DSM1237	Whatman paper	18.5	15.2	–	Argyros et al. (2011)
DSM1313	Avicel	28.7	46.4	27.6	Geng at al. (2010)
JN4	Avicel 10	6.2	7.6	13	Izquierdo et al. (2014)
JW20	Avicel 10	25.4	13	9.4	Rani et al. (1997)
LQRI	Cellobiose	26	24	–	Lamed et al. (1988)
LQRI	Cellulose	28	27	–	Lamed et al. (1988)
LQRI	Solka floc	30.8	27.3	–	Ng et al. (1981)
LQRI	SO_2-treated wood	16.9	13.6	–	Ng et al. (1981)
LQRI	Steam-exploded wood	14.9	15.2	–	Ng et al. (1981)
LQRI	Untreated wood	2.2	6.4	–	Ng et al. (1981)
LQRI	MN300	31.2	22.5	–	Ng et al. (1981)
M1570	Avicel	121.8	2.7	1.2	Argyros et al. (2011)
S14	Microcrystalline cellulose	41.2	63	8.2	Bayer et al. (2004)
SS19	Filter paper	296.5	–	–	Balusu et al. (2005)
SS21	Filter paper	311.3	–	–	Rani et al. (1998)
SS22	Filter paper	289.6	–	–	Rani et al. (1998)
YM 4	Avicel 10	44.7	26.5	20.4	Dumitrache et al. (2016)
YS	Avicel 20	36	12	–	Sato et al. (1993)

9.5 FILAMENTOUS FUNGI IN CONSOLIDATED BIOPROCESSING

Though the global research in CBP has emphasized the emergence of novel, cost-effective, and sustainable technologies, major bottlenecks also have been identified for complete saccharification of pretreated biomass because of long conversion times and the need for large quantities of cellulolytic enzymes. Hence, there is a need for a microbial source with a high capacity of cellulolytic enzymes production to use large amounts of lignocellulosic biomass on industrial scales. Fungi have been considered "microbial biofactories" to resolve these problems because of the potential to produce adequate amount of cellulases. As far as the production of cellulases is concerned by fungi, *Trichoderma reesei* produces more than 100 g/L of cellulase enzyme compared to only a few grams of cellulase production by bacterium (Xu et al. 2009; Vitikainen et al. 2010). The presence of endoplasmic reticulum and the Golgi complex inside the cytoplasm of fungi helps to implement genetic engineering for an efficient CBP as compared to bacterial cell (Xu et al. 2009). Furthermore, a wide range of sugar substrates can be assimilated by filamentous fungi to produce ethanol that have more resistance to temperature than bacteria (Taherzadeh and Karimi 2007).

Some native *Mucor circinelloides* strains (NBRC 6746 and NBRC4572) have been isolated to produce chitinolytic ability on N-acetylglucosamine and chitin substrates to produce bioethanol (Inokuma et al. 2013). However, due to the low enzymatic activities, the standardization of culture conditions and genetic engineering of these strains are required for further utilization into CBP systems. Among fungi, *Fusarium oxysporum* has the ability for ethanol production through CBP system using various straw biomasses (Ali et al. 2012). This fungus can digest the biomass into sugars and subsequently into ethanol by fermenting both pentose and hexose molecules (Anasontzis and Christakopoulos 2014) with a production level of about 0.35 g/g ethanol. The major advantages of using fungi over bacteria in CBP is for the simultaneous hydrolyzing and fermenting capabilities of lignocellulosic biomass (Hennessy et al. 2013).

Among the filamentous fungi, *T. reesei* is the major agent to satisfy many challenges relating to stability after genetic modification (Kuck and Hoff 2010) and efficient cellulolytic machinery (Silva-Rocha et al. 2014). Among other filamentous fungi, other possible CBP agents for ethanol production are *Fusarium verticillioides* (de Almeida et al. 2013), *Aspergillus oryzae* (Hossain 2013), *Paecilomyces variotii* (Zerva et al. 2014), *Flammulina velutipes* (Mizuno et al. 2009a), and *Phlebia* sp. (Kamei et al. 2012a). Among white-rot basidiomycetes, *Peniophora cinerea* and *Trametes suaveolens* have the potential to convert hexose to ethanol (Okamoto et al. 2010).

9.6 YEASTS IN CONSOLIDATED BIOPROCESSING

Saccharomyces cerevisiae is a well-established industrial microorganism for ethanol production. However, it lacks the hydrolytic activities for the conversion of pretreated cellulosic component of biomass into simple sugar. Hence, genetic engineering has allowed the expression of celluloses and hemicelluloses into yeast strains so that yeasts can be used for saccharolytic and ethanologenic activities. There is need for multiple enzymes for cellulose hydrolysis that includes endoglucanase (EG), exoglucanase (CBH), and β-glucosidase (BGL). The major role of endoglucanases is to break internal bonds at the amorphous sites of cellulose chain, whereas exoglucanases cleaves from the end, generating cellobiose, and β-glucosidases cleave the cellobiose into glucose.

Due to ethanologenic nature, yeast can be used as a CBP agent if the cellulases from different cellulose hydrolytic microorganisms such as *C. thermocellum, C. cellulovorans,* and *C. cellulolyticum,* and aerobic cellulolytic fungi are expressed heterologously. Most ethanol-fermenting microorganisms prefer mesophilic conditions (28°C–37°C) for ethanol production. On the other hand, the maximum activity of cellulases occurs at higher temperatures (50°C). As a result, significant decrease in ethanol production efficiency is detected when mesophilic ethanol-fermenting microorganisms are used. Thus, the implementation of thermo-stable microorganisms for high fermentation capacity at elevated temperatures results in a significant increase in ethanol production efficiency.

Khuyveromyces marxianusis, a thermotolerant yeast, is one of the CBP agents that can produce ethanol (Flores et al. 2013). Growing at temperatures up to 52°C, this microorganism has a short generation time (Rajoka et al. 2003), which can convert xylose to ethanol. The high potential of *K. marxianus* for CBP ethanol production using different feedstocks at high temperatures was well studied by many researchers (Yuan et al. 2012; Flores et al. 2013). Flores et al. (2013) screened a few *K. marxianus* strains for their fructanases activity and ethanol production from *Agave tequilana fructans* (ATF) as the substrate. They found a few strains with fructanases activity and CBP ethanol production capability (20 g/L ethanol).

Another yeast strain, *Clavispora,* has an ability to use cellobiose as the substrate to produce enough amounts of β-glucosidase towards cellulosic ethanol accumulation (Liu et al. 2012). Having tolerance to some major inhibitors during pretreatment process, the strain produces 23 g/L of ethanol from 25% xylose-extracted corncob residue at 37°C in a simultaneous saccharification and fermentation system without the addition of any exogenous β-glucosidases. Cryophilic yeasts *Mrakia* spp. and *Mrakiella* spp. are yeasts predominantly found in Arctic and Antarctic regions. Tsuji et al. (2013, 2014) isolated and characterized *Mrakia blollopis* SK-4 from Nagaike Lake in Skarvsnes ice-free area (East Antarctica) with the ability to ferment various sugars substrates like glucose, sucrose, maltose, raffinose, and fructose at low temperatures.

Cellulolytic yeasts efficiently express multiple enzymes that act synergistically to hydrolyze cellulosic polymer into simple monomers. A yeast strain was developed that expresses three types of cellulolytic enzymes to achieve a one-step conversion of cellulose to ethanol. Although their effort at combining cellulose hydrolysis with ethanol-producing ability advances the goal of achieving a cost-efficient CBP, the efficiency of cellulose hydrolysis still lags and must be enhanced for commercial scale biofuel production. Table 9.3 describes the cellulase engineering in *S. cerevisiae*.

TABLE 9.3
Cellulase Engineering in *Saccharomyces cerevisiae*

Cellulase Expression	Cellulase Formation	Ethanol	Carbon Sources	Reference
CipA3 (*C. thermocellum*); EG (*T. reesei*); CBH (*T. reesei*); BGL (*A. aculeatus*)	Display of trifunctional mini-cellulosome	1.8 g/L	Phosphoric acid-swollen cellulose	Wen et al. (2010)
EG (*T. reesei*); BGL (*Saccharomycopsis fibuligera*)	Free enzymes	1.0 g/L	Phosphoric acid-swollen cellulose	Den Haan et al. (2007)
EG (*T. reesei*); CBH (*T. reesei*); BGL (*A. aculeatus*)	Co-display on cell surface	7.5 g/L	Phosphoric acid-swollen cellulose and pretreated rice straw	Yamada et al. (2011)
EG (*T. reesei*); CBH (*T. reesei*); BGL (*A. aculeatus*)	Co-display on cell surface	3.0 g/L	Phosphoric acid-swollen cellulose	Fujita et al. (2004)
EG (*Thermoascus aurantiacus*); CBH (*T. reesei*); BGL (*A. aculeatus*)	Display of single cellulase on cell surface followed by co-culture	2.1 g/L	Phosphoric acid-swollen cellulose	Baek et al. (2012)
Scaf (*C. thermocellum*, *C. cellulolyticum* and *Ruminococcus flavefaciens*); EG CelA (*C. thermocellum*); CBH CelE (*C. cellulolyticum*); BGL (*T. aurantiacus*)	Display of Scaf on cell surface, or secretion of either EG, CBH, or BGL, followed by co-culture	1.87 g/L	Phosphoric acid-swollen cellulose	Tsai et al. (2010)

9.7 NATIVE AND ARTIFICIAL CELLULOSOMES

The attachment of proteins to the bacterial cell surface by the function of cell wall binding domains (CWBDs) has been extensively reported. Surface-layer homology (SLH) domains regarded as the major examples of CWBDs are observed on the amino- or carboxyl-terminal of many structural proteins and enzymes including glycosyl hydrolases. Found in double-repeats or triple-repeats, these domains non-covalently attach to either teichoic acid or teichuronic acid in the bacterial cell wall. The anaerobic gram-positive thermophilic *C. thermocellum* is one of the best examples of a SLH-displaying cellulolytic bacterium containing a well-characterized multi-enzyme complex termed a cellulosome. Anaerobic microorganisms have developed an elaborately structured multi-enzyme complex called a cellulosome to maximize their biocatalytic efficiency (Bayer et al. 2004).

Cellulosomes are extracellular self-assembled multi-enzyme complexes produced and displayed on the cellulolytic anaerobic microorganisms having cellulose degradation capabilities. These are composed of a large non-catalytic scaffolding protein that facilitates the binding of a multitude of different glycosyl hydrolases and other enzymes that contain corresponding dockerin modules to intermittent cohesion domains along the scaffolding structure. This macromolecular system delivers numerous enzymes with proximity for the substrate by providing better catalytic efficiency than soluble enzymes. As the cellulosome system exhibited much greater degradation potential than the non-complexed cellulase system, it may be regarded as a "quantum leap" in the progress of biomass conversion to future biofuel technology. The attachment of a multitude of cellulases near one another increases the synergistic cellulolytic activity between them that results in enhancing the potential of *C. thermocellum* to utilize various cellulosic substrates like Avicel, filter paper, and pretreated mixed hardwood.

The major differences between the cellulosome and free enzyme systems are the scaffolding protein bearing several cohesin domains and enzymatic subunits having dockerin domains. Apart from the multiple cohesins, scaffolding also contains at least one cellulose-binding domain (CBD), which acts as a targeting agent to help the binding of the substrate to the catalytic domains. Cohesin dockerin interaction is highly species-specific, which provides highly ordered and position-specific assembly of catalytic domains. This protein multiplex is attached to the cell membrane indirectly by anchoring protein. This anchoring protein is attached to the scaffolding protein by Type II cohesion-dockerin interactions and is strongly attached to the cell surface using SLH domains. The attachment of this large multi-protein complex near the cell ensures that most of the liberated sugars from cellulose are absorbed and used by the cellulosome-expressing bacterium. Significant progress has been made in the heterologous expression and display of an attached cellulolytic system in non-cellulolytic microorganisms. One example is a genetically engineered *Escherichia coli* LY01 strain that expresses and displays an endoglucanase, exoglucanase and β-glucosidase from *Clostridium cellulolyticum* on its cell surface.

9.8 CONSOLIDATED BIOPROCESSING FOR BIOBUTANOL PRODUCTION

Butanol is an essential fuel or chemical having vital roles in the manufacture of solvents, butyl amines, plasticizers, amino resins, and butyl acetates (Nanda et al. 2017b). Having energy content comparable to gasoline, butanol can achieve the same energy output occupying less volume. Butanol retains a lower vapor pressure compared to ethanol and can be used in existing vehicle engines without engine modification (Qureshi et al. 2013; Wen et al. 2014a; Jouzani and Taherzadeh 2015).

The acetone-butanol-ethanol (ABE) fermentation is carried out using anaerobic bacterium such as *Clostridium acetobutylicum* or *Clostridium beijerinckii* in a biphasic process involving acidogenesis and solventogenesis. Although the acidogenic phase involves the production of acids (e.g., acetic acid and butyric acid), the solventogenic phase is related to the accumulation of solvents (e.g., acetone, butanol, and ethanol). The ABE-producing bacteria can utilize both starchy and lignocellulosic substrates. However, the later must be hydrolyzed using a suitable pretreatment method (i.e., dilute acid and enzymatic hydrolysis). Different biomass such as wheat straw (Nanda et al. 2014a), rice straw (Gottumukkala et al. 2014), barley straw (Qureshi et al. 2010a), corn stover (Parekh et al. 1988; Qureshi et al. 2010b), corncobs and fibers (Guo et al. 2013), palm kernel cake (Shukor et al. 2014), cassava starch (Li et al. 2014a, 2014b), pinewood (Nanda et al. 2014a), timothy grass (Nanda et al. 2014a), switch grass (Qureshi et al. 2010b), and sago pith (Linggang et al. 2013) have been used as substrates for ABE fermentation.

The major drawback in butanol production from lignocellulosic biomass is the microbial intolerance towards butanol toxicity, which results in low final butanol titer levels (Qureshi et al. 2008; Ezeji et al. 2012). However, the cost of exogenous cellulase use has made these processes economically uncompetitive because solventogenic *Clostridia* are unable to utilize lignocellulose as a raw material directly (Bellido et al. 2014; Wen et al. 2014a, 2014b). Although various attempts have been made to resolve these problems in butanol production, CBP is regarded as a competent and cost-effective method for butanol production from low-cost renewable feedstock. To realize the true potentials of CBP, a single wild type microorganism, microbial co-culture or consortium system, or genetically engineered single microorganisms must be supplemented and adapted to use lignocellulose to produce butanol (Wen et al. 2014a, 2014b). Table 9.4 summarizes some recent reports on CBP strategies for butanol production using lignocellulosic biomass.

TABLE 9.4
List of Microorganisms and Consolidated Bioprocessing Systems for Butanol Production from Lignocellulosic Biomass

Biomass	Microorganism	CBP System	Butanol Concentration/ Yield	References
Birch wood xylan	*Clostridium* strain BOH3 and *Kluyvera* strain OM3	Co-culture	1.2 g/L	Xin and He (2013)
Cassava starch	*Clostridium beijerinckii* and *Clostridium tyrobutyricum*	Continuous co culture in fibrous bed reactor	6.66 g/L (yield: 0.18 g/g)	Li et al. (2013)
Cellobiose	*Clostridium thermosaccharolyticum*	Recombinant DNA technology: overexpression of *bcs* operon	5.1 mM	Bhandiwad et al. (2013)
Cellulose	*Clostridium acetobutylicum* and *Clostridium cellulolyticum*	Co-culture	350 mg/L	Salimi and Mahadevan (2013)
Cellulose	*Clostridium cellulolyticum*	Recombinant DNA technology: expressing enzymes that convert pyruvate to isobutanol	660 mg/L	Higashide et al. (2011)
Cellulose	*Clostridium thermocellum* and *Clostridium saccharoperbutylacetonicum*	Co-culture	7.9 g/L	Nakayama et al. (2011)
Cellulose and lichenan	*Clostridium beijerinckii* NCIMB 8052	Recombinant DNA technology: transferring genes encoding glycoside hydrolases (*celA* and *celD*)	4.9 g/L	Lopez-Contreras et al. (2001)

(*Continued*)

TABLE 9.4 (*Continued*)
List of Microorganisms and Consolidated Bioprocessing Systems for Butanol Production from Lignocellulosic Biomass

Biomass	Microorganism	CBP System	Butanol Concentration/ Yield	References
Corn cobs	*Clostridium cellulovorans* strain 743B and *Clostridium beijerinckii* strain NCIMB 8052	Co-culture	8.30 g/L	Wen et al. (2014a)
Corn stover	*Clostridium cellulolyticum*	Metabolic engineering: sporulation abolishment and carbon overload alleviation	0.42 g/L	Li et al. (2014c)
Corn stover	*Trichoderma reesei* and *Escherichia coli*	Co-culture	1.88 g/L and 62% theoretical	Minty et al. (2013)
Sugarcane bagasse	*Klebsiella pneumoniae* CGMCC1.9131	Response surface methodology	26.2 g/L	Song et al. (2012)
Switch grass	*Escherichia coli*	Recombinant DNA technology: expression of hydrolase and butanol pathway genes	28 mg/L	Bokinsky et al. (2011)
Xylan and xylose	*Clostridium* strain BOH3	Single native strain	16 g/L	Rajagopalan et al. (2014)

9.9 CONCLUSIONS

Due to huge availability, low-cost, and less environmental impacts, biofuel production from lignocellulosic biomass has been a sustainable option to substitute for fossil fuels. However, biofuel yield from renewable sources have a lot challenges during their bioproduction. In the context of global use and effects on food security, the production level of biofuels may be enhanced by the application low-cost technologies. Bioconversion technologies must develop frameworks for cost benefit analysis with high product yields. By exploring innovative methods for process integration by minimizing total cost investment and maximizing energy potential, improving overall process economics could address many major drawbacks of second-generation biofuel production. CBP can consolidate new, efficient, and economic pretreatment technologies, fermentation processes, and product recovery techniques to provide solutions for the future energy crisis. CBP can support the lowest cost route for bioconversion of lignocellulosic biomass to fuels through hydrolysis and fermentation in a single bioreactor. CBP also can improve bioconversion technologies by improved hydrolysis and fermentation processes without supplementing exogenous saccharolytic enzymes and eliminating inhibition effects.

Substantial work has been carried out by adopting native and genetically engineered microorganisms in various combinations of CBP for biofuel production. Native single strains with both cellulolytic and ethanologenic activities can be used as CBP along with microbial co-cultures and genetically engineered microorganisms. However, the application of genetic and metabolic engineering on microorganism to retain CBP behavior can solve some problems for high biofuel production. The major concern of the improvement of fermentation capacity is developing tolerance by the microorganisms to toxic by-products and main products accumulated during biomass pretreatment and fermentation. The implementation of metabolic engineering and research on cell surface engineering can potentially address such challenges.

REFERENCES

Agbor, V., N. Cicek, R. Sparling, A. Berlin, and D. Levin. 2011. Biomass pretreatment: Fundamentals toward applications. *Biotechnology Advances* 29:675–685.

Ali, S. S., M. Khan, B. Fagan, E. Mullins, and F. M. Doohan. 2012. Exploiting the inter-strain divergence of *Fusarium oxysporum* for microbial bioprocessing of lignocellulose to bioethanol. *AMB Express* 2:1–9.

Anasontzis, G. E., and P. Christakopoulos. 2014. Challenges in ethanol production with *Fusarium oxysporum* through consolidated bioprocessing. *Bioengineered* 5:393–395.

Argyros, D., S. Tripathi, and T. Barrett. 2011. High ethanol titers from cellulose by using metabolically engineered thermophilic, anaerobic microbes. *Applied and Environmental Microbiology* 77:8288–8294.

Baek, S. H., S. Kim, K. Lee, J. K. Lee, and J. S. Hahn. 2012. Cellulosic ethanol production by combination of cellulase-displaying yeast cells. *Enzyme Microbiology Technology* 51:366–372.

Balusu, R., R. Paduru, S. Kuravi, G. Seenaya, and G. Reddy. 2005. Optimization of critical medium components using response surface methodology for ethanol production from cellulosic biomass by *Clostridium thermocellum* SS19. *Process Biochemistry* 40:3025–4030.

Barnard, D., A. Casanueva, M. Tuffin, and D. Cowan. 2010. Extremophiles in biofuel synthesis. *Environmental Technology* 31:871–888.

Bayer, E. A., J. P. Belaich, Y. Shoham, and R. Lamed. 2004. The cellulosomes: Multi enzyme machines for degradation of plant cell wall polysaccharides. *Annual Review in Microbiology* 58: 521–554.

Bellido, C., M. L. Pinto, M. Coca, G. González-Benito, and M. T. García-Cubero. 2014. Acetone–butanol–ethanol (ABE) production by *Clostridium beijerinckii* from wheat straw hydrolysates: Efficient use of penta and hexa carbohydrates. *Bioresource Technology* 167:198–205.

Berezina, O. V, S. P. Sineokiĭ, G. A. Velikodvorskaia, W. Schwarz, and V. V. Zverlov. 2008. Extracellular glycosyl hydrolase activity of the clostridia producing acetone, butanol, and ethanol. *Prikl Biokhim Mikrobiol* 44:49–55.

Bhandiwad, A., A. Guseva, and L. Lynd. 2013. Metabolic engineering of *Thermoanaerobacterium thermosaccharolyticum* for increased *n*-butanol production. *Advanced Microbiology* 3:46–51.

Bokinsky, G., P. P. Peralta-Yahya, A. George, B. M. Holmes, E. J. Steen, J. Dietrich, T. S. Lee et al. 2011. Synthesis of three advanced biofuels from ionic liquid-pretreated switchgrass using engineered *Escherichia coli*. *Proceedings of the National Academy of Sciences* 108:19949–19954.

Chinn, M., S. Nokes, and H. Strobel. 2007 Influence of process conditions on end product formation from *Clostridium thermocellum* 27405 in solid substrate cultivation on paper pulp sludge. *Bioresource Technology* 98:2184–2193.

Chu, S., and A. Majumdar. 2012. Opportunities and challenges for a sustainable energy future. *Nature* 488:294–303.

de Almeida, M. N., V. M. Guimarães, D. L. Falkoski, E. M. Visser, G. A. Siqueira, A. M. Milagres, and S. T. de Rezende. 2013. Direct ethanol production from glucose, xylose and sugarcane bagasse by the corn endophytic fungi *Fusarium verticillioides* and *Acremonium zeae*. *Journal of Biotechnology* 168:71–77.

Den Haan, R., S. H. Rose, L. R. Lynd, and W. H. van Zyl. 2007. Hydrolysis and fermentation of amorphous cellulose by recombinant *Saccharomyces cerevisiae*. *Metabolic Engineering* 9:87–94.

Deng, Y., D. Olson, J. Zhou, C. Herring, A. Shaw, and L. Lynd. 2013. Redirecting carbon flux through exogenous pyruvate kinase to achieve high ethanol yields in *Clostridium thermocellum*. *Metabolic Engineering* 15:151–158.

Dumitrache, A., H. Akinosho, M. Rodriguez, X. Meng, C. G. Yoo, J. Natzke, N. L. Engle et al. 2016. Consolidated bioprocessing of Populus using *Clostridium* (*Ruminiclostridium*) *thermocellum*: A case study on the impact of lignin composition and structure. *Biotechnology for Biofuels* 9:31.

Ellis, L., E. Holwerda, D. Hogsett, S. Rogers, X. Shao, T. Tschaplinski, P. Thorne, and L. R. Lynd. 2012. Closing the carbon balance for fermentation by *Clostridium thermocellum* (ATCC 27405). *Bioresource Technology* 103:293–299.

Ezeji, T. C., N. Qureshi, and H. P. Blaschek. 2012. Microbial production of a biofuel (acetone-butanol-ethanol) in a continuous bioreactor: Impact of bleed and simultaneous product recovery. *Bioprocess Biosystems Engineering* 3:109–116.

Flores, J. A., A. Gschaedler, L. Amaya-Delgado, E. J. Herrera-López, M. Arellano, and J. Arrizon. 2013. Simultaneous saccharification and fermentation of *Agave tequilana fructans* by *Kluyveromyces marxianus* yeasts for bioethanol and tequila production. *Bioresource Technology* 146:267–273.

Fougere, D., S. Nanda, K. Clarke, J. A. Kozinski, and K. Li. 2016. Effect of acidic pretreatment on the chemistry and distribution of lignin in aspen wood and wheat straw substrates. *Biomass and Bioenergy* 91:56–68.

Fujita, Y., J. Ito, M. Ueda, H. Fukuda, and A. Kondo. 2004. Synergistic saccharification, and direct fermentation to ethanol, of amorphous cellulose by use of an engineered yeast strain codisplaying three types of cellulolytic enzyme. *Applied and Environmental Microbiology* 70:1207–1212.

Geng, A., Y. He, C. Qian, X. Yan, and Z. Zhou. 2010. Effects of key factors on hydrogen production from cellulose in a co-culture of *Clostridium thermocellum* and *Clostridium thermopalmarium*. *Bioresource Technology* 101:4029–4033.

Gottumukkala, L. D., B. Parameswaran, S. K. Valappil, A. Pandey, and R. K. Sukumaran. 2014. Growth and butanol production by *Clostridium sporogenes* BE01 in rice straw hydrolysate: Kinetics of inhibition by organic acids and the strategies for their removal. *Biomass Conversion Biorefinery* 4:227–283.

Guedon, E., M. Desvaux, and H. Petitdemange. 2002. Improvement of cellulolytic properties of *Clostridium cellulolyticum* by metabolic engineering. *Applied and Environmental Microbiology* 68:53–58.

Guo, T., A. Y. He, T. F. Du, D. W. Zhu, D. F. Liang, M. Jiang, and P. K. Ouyang. 2013. Butanol production from hemicellulosic hydrolysate of corn fiber by a *Clostridium beijerinckii* mutant with high inhibitor-tolerance. *Bioresource Technology* 135:379–385.

Hahn-Hägerdal, B., M. Galbe, M. F. Gorwa-Grauslund, G. Lidén, and G. Zacchi. 2006. Bioethanol—The fuel of tomorrow from the residues of today. *Trends in Biotechnology* 24:549–556.

Hennessy, R. C., F. Doohan, and E. Mullins. 2013. Generating phenotypic diversity in a fungal biocatalyst to investigate alcohol stress tolerance encountered during microbial cellulosic biofuel production. *PloS ONE* 8:e77501.

Higashide, W., Y. Li, Y. Yang, and J. C. Liao. 2011. Metabolic engineering of *Clostridium cellulolyticum* for production of isobutanol from cellulose. *Applied and Environmental Microbiology* 77:2727–2733.

Hörmeyer, H. F., P. Tailliez, J. Millet, H. Girard, G. Bonn, O. Bobleter, and J.P. Aubert. 1988. Ethanol production by *Clostridium thermocellum* grown on hydrothermally and organosolv-pretreated lignocellulosic materials. *Applied Microbiology and Biotechnology* 29:528–535.

Hossain, S. M. 2013. Bioethanol fermentation from non-treated and pretreated corn stover using *Aspergillus oryzae*. *Chemical Engineering Research Bulletin* 16:33–44.

Ibrahim, A. S. S., and A. El-diwany. 2007. Isolation and identification of new cellulases producing thermophilic bacteria from an Egyptian hot spring and some properties of the crude enzyme. *Australian Journal of Basic and Applied Sciences* 1:473–478.

Inokuma, K., M. Takano, and K. Hoshino. 2013. Direct ethanol production from N-acetylglucosamine and chitin substrates by *Mucor* species. *Biochemical Engineering Journal* 72:24–32.

Izquierdo, J. A., S. Pattathil, A. Guseva, M. G. Hahn, and L. R. Lynd. 2014. Comparative analysis of the ability of *Clostridium clariflavum* strains and *Clostridium thermocellum* to utilize hemicellulose and unpretreated plant material. *Biotechnology for Biofuels* 7:136.

Jin, M., C. Gunawan, V. Balan, and B. E. Dale. 2012a. Consolidated bioprocessing (CBP) of AFEX™-pretreated corn stover for ethanol production using *Clostridium phytofermentans* at a high solids loading. *Biotechnology and Bioengineering* 109:1929–1936.

Jin, M., V. Balan, C. Gunawan, and B. E. Dale. 2011. Consolidated bioprocessing (CBP) performance of Clostridium phytofermentans on AFEX-treated corn stover for ethanol production. *Biotechnology and Bioengineering* 108:1290–1297.

Jouzani, G. S., and M. J. Taherzadeh. 2015. Advances in consolidated bioprocessing systems for bioethanol and butanol production from biomass: A comprehensive review. *Biofuel Research Journal* 5:152–195.

Kamei, I., Y. Hirota, T. Mori, H. Hirai, S. Meguro, and R. Kondo. 2012a. Direct ethanol production from cellulosic materials by the hypersaline-tolerant white-rot fungus *Phlebia sp.* MG-60. *Bioresource Technology* 112:137–142.

Kuck, U., and B. Hoff. 2010. New tools for the genetic manipulation of flamentous fungi. *Applied Microbiology and Biotechnology* 86:51–62.

Lamed, R., J. Lobos, and T. Su. 1988. Effects of stirring and hydrogen on fermentation products of *Clostridium thermocellum*. *Applied and Environmental Microbiology* 54:1216–1221.

Levin, D., R. Islam, N. Cicek, and R. Sparling. 2006. Hydrogen production by *Clostridium thermocellum* 27405 from cellulosic biomass substrates. *International Journal of Hydrogen Energy* 31:1496–1503.

Li, H. G., F. K. Ofosu, K. T. Li, Q. Y. Gu, Q. Wang, and X. B. Yu. 2014b. Acetone, butanol, and ethanol production from gelatinized cassava flour by a new isolate with high butanol tolerance. *Bioresource Technology* 172:276–282.

Li, H. G., W. Luo, Q. Wang, and X. B. Yu. 2014a. Direct fermentation of gelatinized cassava starch to acetone, butanol, and ethanol using *Clostridium acetobutylicum* mutant obtained by atmospheric and room temperature plasma. *Applied. Biochemistry. Biotechnology* 172:3330–3341.

Li, J., N. R. Baral, and A. K. Jha. 2014c. Acetone–butanol–ethanol fermentation of corn stover by *Clostridium* species: Present status and future perspectives. *World Journal of Microbiology and Biotechnology* 30:1145–1157.

Li, L., H. Ai, S. Zhang, S. Li, Z. Liang, Z. Q. Wu, and J. F. Wang. 2013. Enhanced butanol production by coculture of *Clostridium beijerinckii* and *Clostridium tyrobutyricum*. *Bioresource Technology* 143:397–404.

Linggang, S., L. Y. Phang, H. Wasoh, and S. Abd-Aziz. 2013. Acetone–butanol–ethanol production by *Clostridium acetobutylicum* ATCC 824 using sago pith residues hydrolysate. *Bioenergy Research* 6:321–328.

Liu, Z. L., S. A. Weber, M. A. Cotta, and S. Z. Li. 2012. A new β-glucosidase producing yeast for lower-cost cellulosic ethanol production from xylose-extracted corncob residues by simultaneous saccharification and fermentation. *Bioresource Technology* 104:410–416.

López-Contreras, A. M., H. Smidt, J. van der Oost, P. A. Claassen, H. Mooibroek, and W. M. de Vos. 2001. *Clostridium beijerinckii* cells expressing *Neocallimastix patriciarum* glycoside hydrolases show enhanced lichenan utilization and solvent production. *Applied and Environmental Microbiology* 67:5127–5133.

Lynd, L. R., P. J. Weimer, W. H. Van Zyl, and I. S. Pretorius. 2002. Microbial cellulose utilization: Fundamentals and biotechnology. *Microbiology and Molecular Biology Reviews* 66:506–577.

Minty, J. J., M. E. Singer, S. A. Scholz, C. H. Bae, J. H, Ahn, C. E. Foster, and X. N. Lin. 2013. Design and characterization of synthetic fungal-bacterial consortia for direct production of isobutanol from cellulosic biomass. *Proceeding of National Academy of Science* 110:14592–14597.

Mizuno, R., H. Ichinose, T. Maehara, K. Takabatake, and S. Kaneko. 2009a. Properties of ethanol fermentation by *Flammulina velutipes*. *Bioscience Biotechnology and Biochemistry* 73:2240–2245.

Naik, S., V. V. Goud, P. K. Rout, K. Jacobson, and A. K. Dalai. 2010. Characterization of Canadian biomass for alternative renewable biofuel. *Renewable Energy* 35:1624–1631.

Nakayama, S., K. Kiyoshi, T. Kadokura, and A. Nakazato. 2011. Butanol production from crystalline cellulose by cocultured *Clostridium thermocellum* and *Clostridium saccharoperbutylacetonicum* N1-4. *Applied and Environmental Microbiology* 77:6470–6475.

Nanda, S., A. K. Dalai, and J. A. Kozinski. 2014a. Butanol and ethanol production from lignocellulosic feedstock: Biomass pretreatment and bioconversion. *Energy Science Engineering* 2:138–148.

Nanda, S., A. K. Dalai, and J. A. Kozinski. 2016. Supercritical water gasification of timothy grass as an energy crop in the presence of alkali carbonate and hydroxide catalysts. *Biomass and Bioenergy* 95:378–387.

Nanda, S., D. Golemi-Kotra, J. C. McDermott, A. K. Dalai, I. Gökalp, and J. A. Kozinski. 2017b. Fermentative production of butanol: Perspectives on synthetic biology. *New Biotechnology* 37:210–221.

Nanda, S., J. Mohammad, S. N. Reddy, J. A. Kozinski, and A. K. Dalai. 2014b. Pathways of lignocellulosic biomass conversion to renewable fuels. *Biomass Conversion Biorefinery* 4:157–191.

Nanda, S., P. Mohanty, K. K. Pant, S. Naik, J. A. Kozinski, and A. K. Dalai. 2013. Characterization of North American lignocellulosic biomass and biochars in terms of their candidacy for alternate renewable fuels. *Bioenergy Research* 6:663–677.

Nanda, S., R. Azargohar, A. K. Dalai, and J. A. Kozinski. 2015. An assessment on the sustainability of lignocellulosic biomass for biorefining. *Renewable and Sustainable Energy Reviews* 50:925–941.

Nanda, S., R. Rana, Y. Zheng, J. A. Kozinski, and A. K. Dalai. 2017a. Insights on pathways for hydrogen generation from ethanol. *Sustainable Energy & Fuels* 1:1232–1245.

Nanda, S., S. N. Reddy, D. V. N. Vo, B. N. Sahoo, and J. A. Kozinski. 2018. Catalytic gasification of wheat straw in hot compressed (subcritical and supercritical) water for hydrogen production. *Energy Science and Engineering* 6:448–459.

Nataf, Y., S. Yaron, F. Stahl, R. Lamed, E. A. Bayer, T. H. Scheper, A. L. Sonenshein, and Y. Shoham. 2009. Cellodextrin and laminaribiose ABC transporters in *Clostridium thermocellum*. *Journal of Bacteriology* 191:203–209.

Ng, T., A. Ben-Bassat, and J. Zeikus. 1981. Ethanol production by thermophilic bacteria: fermentation of cellulosic substrates by cocultures of *Clostridium thermocellum* and *Clostridium thermohydrosulfuricum*. *Applied and Environmental Microbiology* 41:1337–1343.

Okamoto, K., K. Imashiro, Y. Akizawa, A. Onimura, M. Yoneda, Y. Nitta, N. Maekawa, and H. Yanase. 2010. Production of ethanol by the white-rot basidiomycetes *Peniophora cinerea* and *Trametes suaveolens*. *Biotechnology Letters* 32:909–913.

Parekh, S. R., R. S. Parekh, and M. Wayman. 1988. Ethanol and butanol production by fermentation of enzymatically saccharified SO_2 pretreated lignocellulosics. *Enzyme Microbial Technology* 10:660–668.

Parisutham, V., T. H. Kim, and S. K. Lee. 2014. Feasibilities of consolidated bioprocessing microbes: From pretreatment to biofuel production. *Bioresource Technology* 161:431–440.

Pauly, M., and K. Keegstra. 2008. Cell wall carbohydrates and their modification as a resource for biofuels. *The Plant Journal* 54:559–568.

Perlack, R. D., L. L. Wright, A. F. Turhollow, R. L. Graham, B. J. Stokes, and D. C. Erbach. 2005. *Biomass as a Feedstock for a Bioenergy and Bioproducts Industry: The Technical Feasibility of a Billion-ton Annual Supply*. Oak Ridge, TN: USDOE and USDA.

Qureshi, N., B. C. Saha, R. E. Hector, B. Dien, S. Hughes, S. Liu, L. Iten, M. J. Bowman, G. Sarath, and M. A. Cotta. 2010b. Production of butanol (a biofuel) from agricultural residues: Part II—Use of corn stover and switchgrass hydrolysate. *Biomass and Bioenergy* 34:566–571.

Qureshi, N., B. Saha, B. Dien, R. Hector, and M. Cotta. 2010a. Production of butanol (a biofuel) from agricultural residues: Part I - Use of barley straw hydrolysate. *Biomass and Bioenergy* 34:559–565.

Qureshi, N., S. Liu, and T. C. Ezeji. 2013. Cellulosic butanol production from agricultural biomass and residues: Recent advances in technology. In *Advanced Biofuels and Bioproducts*, ed. J. W. Lee, Vol. 15, 247–265. New York: Springer.

Qureshi, N., T. Ezeji, J. Ebener, B. Dien, M. Cotta, and H. Blaschek. 2008. Butanol production by *Clostridium beijerinckii*. Part I: Use of acid and enzyme hydrolyzed corn fiber. *Bioresource Technology* 99:5915–5922.

Rajagopalan, G., J. He, and K. L. Yang. 2014. Direct fermentation of xylan by *Clostridium* strain BOH3 for the production of butanol and hydrogen using optimized culture medium. *Bioresource Technology* 154:38–43.

Rajoka, M. I., S. Khan, and R. Shahid. 2003. Kinetics and regulation studies of the production of b-galactosidase from *Kluyveromyces marxianus* grown on different substrates. *Food Technology Biotechnology* 41:315–320.

Rani K. S., M. V. Swamy, and G. Seenayya. 1997. Increased ethanol production by metabolic modulation of cellulose fermentation in *Clostridium thermocellum*. *Biotechnology Letters* 19:819–823.

Rani, K., M. Swamy, and G. Seenaya. 1998. Production of ethanol from various pure and natural cellulosic biomass by *Clostridium thermocellum* strains SS21 and SS22. *Process Biochemistry* 33:435–440.

Saddler, J. N, and M. K. H. Chan. 1984. Conversion of pretreated lignocellulosic substrate to ethanol by *Clostridium thermocellum* in mono-culture and co-culture with *Clostridium thermosaccharolyticum* and *Clostridium thermohydrosulphuricum*. *Canadian Journal of Microbiology* 130:212–220.

Salimi, F., and R. Mahadevan. 2013. Characterizing metabolic interactions in a clostridial co-culture for consolidated bioprocessing. *BMC Biotechnology* 13:95.

Sarangi. P. K., N. J. Singh, and T. A. Singh. 2016. Agricultural crop residues: Unutilized biomass having huge energy potential. In *Contemporary Renewable Energy Technologies for Sustainable Agriculture*, ed. M. K. Ghosal. Narosa Publishing House. New Delhi, India, 47–61.

Sato, K., M. Tomita, S. Yonemura, S. Goto, K. Sekine, E. Okuma, Y. Takagi, K. Hon-Nami, and T. Saikit. 1993. Characterization of and ethanol hyper-production by *Clostridium thermocellum* I-1-B. *Bioscience, Biotechnology, and Biochemistry* 57:2116–2121.

Schuster, B. G., and M. S. Chinn. 2013. Consolidated bioprocessing of lignocellulosic feedstocks for ethanol fuel production. *Bioenergy Research* 6:416–435.

Shang, S. M., L. Qian, X. Zhang, K. Z. Li, and I. Chagan. 2013. *Themoanaerobacterium calidifontis sp.* nov., a novel anaerobic, thermophilic, ethanol-producing bacterium from hot springs in China. *Archives of Microbiology* 195:439–445.

Shao, X., M. Jin, A. Guseva, C. Liu, V. Balan, D. Hogsett, B. Dale, and L. Lynd. 2011. Conversion for Avicel and AFEX pretreated corn stover by *Clostridium thermocellum* and simultaneous saccharification and fermentation: Insights into microbial conversion of pretreated cellulosic biomass. *Bioresource Technology* 102:8040–8045.

Shaw, A. J., S. F. Covalla, B. B. Miller, B. T. Firliet, D. A. Hogsett, and C. D. Herring. 2012. Urease expression in a *Thermoanaerobacterium saccharolyticum* ethanologen allows high titer ethanol production. *Metabolic Engineering* 14:528–532.

Shukor, H., N. K. N. Al-Shorgani, P. Abdeshahian, A. A. Hamid, N. Anuar, N. A. Rahman, and M. S. Kalil. 2014. Production of butanol by *Clostridium saccharoperbutylacetonicum* N1-4 from palm kernel cake in acetone–butanol–ethanol fermentation using an empirical model. *Bioresource Technology* 170:565–573.

Sigurbjornsdottir, M. A., and J. Orlygsson. 2012. Combined hydrogen and ethanol production from sugars and lignocellulosic biomass by *Thermoanaerobacterium* AK 54, isolated from hot spring. *Applied Energy* 97:785–791.

Silva-Rocha, R., L. D. S. Castro, A. C. C. Antoniêto, M. E. Guazzaroni, G. F. Persinoti, and R. N. Silva. 2014. Deciphering the cis-regulatory elements for XYR1 and CRE1 regulators in *Trichoderma reesei*. *PLoS ONE* 9:e99366.

Singh, N., A. S. Mathur, D. K. Tuli, R. P. Gupta, C. J. Barrow, and M. Puri. 2017. Cellulosic ethanol production via consolidated bioprocessing by a novel thermophilic anaerobic bacterium isolated from a Himalayan hot spring. *Biotechnology for Biofuels* 10:73.

Song, Y., Q. Li, X. Zhao, Y. Sun, and D. Liu. 2012. Production of 2,3-Butanediol by *Klebsiella pneumonia* from enzymatic hydrolyzate of Sugarcane bagasse. *BioResources* 7:4517–4530.

Tachaapaikoon, C., A. Kosugi, and P. Pason. 2012 Isolation and characterization of a new cellulosome-producing *Clostridium thermocellum* strain. *Biodegradation* 23:57–68.

Taherzadeh, M. J., and K. Karimi. 2007. Acid-based hydrolysis processes for ethanol from lignocellulosic materials: A review. *Bioresources* 2:472–499.

Talluri, S., S. M. Raj, and L. P. Christopher. 2013. Consolidated bioprocessing of untreated switchgrass to hydrogen by the extreme thermophile *Caldicellulosiruptor saccharolyticus* DSM 8903. *Bioresource Technology* 139:272–279.

Tolonen, A. C., W. Haas, A. C. Chilaka, J. Aach, S. P. Gygi, and G. M. Church. 2011. Proteome-wide systems analysis of a cellulosic biofuel-producing microbe. *Molecular Systems Biology* 7:461.

Tsai, S. L., G. Goyal, and W. Chen. 2010. Surface display of a functional minicellulosome by intracellular complementation using a synthetic yeast consortium and its application to cellulose hydrolysis and ethanol production. *Applied and Environmental Microbiology* 76:7514–7520.

Tsuji, M., T. Goshima, A. Matsushika, S. Kudoh, and T. Hoshino. 2013. Direct ethanol fermentation from lignocellulosic biomass by Antarctic basidiomycetous yeast *Mrakia blollopis* under a low temperature condition. *Cryobiology* 67:241–243.

Tsuji, M., Y. Yokota, S. Kudoh, and T. Hoshino. 2014. Improvement of direct ethanol fermentation from woody biomasses by the Antarctic basidiomycetous yeast, *Mrakia blollopis,* under a low temperature condition. *Cryobiology* 68:303–305.

Tyurin, M. V., S. G. Desai, and L. R. Lynd. 2004. Electro transformation of *Clostridium thermocellum*. *Applied Environmental Microbiology* 70:883–890.

Virunanon, C., S. Chantaroopamai, J. Denduangbaripant, and W. Chulalaksananukul. 2008. Solventogenic-cellulolytic *clostridia* from 4-step-screening process in agricultural waste and cow intestinal tract. *Anaerobe* 14:109–117.

Vitikainen, M., M. Arvas, T. Pakula, M. Oja, M. Penttila, and M. Saloheimo. 2010. Array comparative genomic hybridization analysis of *Trichoderma reesei* strains with enhanced cellulase production properties. *BMC Genomics* 11:441.

Wen, F., J. Sun, and H. Zhao. 2010. Yeast surface display of trifunctional minicellulosomes for simultaneous saccharification and fermentation of cellulose to ethanol. *Applied Environmental Microbiology* 76:1251–1260.

Wen, F., N. U. Nair, and H. Zhao. 2009. Protein engineering in designing tailored enzymes and 18 micro-organisms for biofuels production. *Current Opinion in Biotechnology* 20:412–419.

Wen, Z., M. Wu, Y. Lin, L. Yang, J. Lin, and P. Cen. 2014a. Artificial symbiosis for acetone-butanol-ethanol (ABE) fermentation from alkali extracted deshelled corn cobs by co-culture of *Clostridium beijerinckii* and *Clostridium cellulovorans*. *Microbial Cell Factories* 13:92.

Wen, Z., M. Wu, Y. Lin, L. Yang, J. Lin, and P. Cen. 2014b. A novel strategy for sequential co-culture of *Clostridium thermocellum* and *Clostridium beijerinckii* to produce solvents from alkali extracted corn cobs. *Process Biochemistry* 49:1941–1949.

Wilson, C. M., M. Rodriguez, Jr., and C. M. Johnson. 2013. Global transcriptome analysis of *Clostridium thermocellum* ATCC 27405 during growth on dilute acid pretreated Populus and switchgrass. *Biotechnology for Biofuels* 6:179–198.

Xin, F., and J. He. 2013. Characterization of a thermostable xylanase from a newly isolated *Kluyvera* species and its application for biobutanol production. *Bioresource Technology* 135:309–315.

Xu, Q., A. Singh, and M. E. Himmel. 2009. Perspectives and new directions for the production of bioethanol using consolidated bioprocessing of lignocellulose. *Current Opinion in Biotechnology* 20:364–371.

Yamada, R., N. Taniguchi, T. Tanaka, C. Ogino, H. Fukuda, and A. Kondo. 2011. Direct ethanol production from cellulosic materials using a diploid strain of *Saccharomyces cerevisiae* with optimized cellulase expression. *Biotechnology for Biofuels* 4:8.

Yee, K. L., M. Rodriguez Jr., T. J. Tschaplinski, N. L. Engle, M. Z. Martin, C. Fu, and J. R. Mielenz. 2012. Evaluation of the bioconversion of genetically modified switchgrass using simultaneous saccharification and fermentation and a consolidated bioprocessing approach. *Biotechnology for Biofuels* 5:81.

Yilman, S., and H. Selim. 2013. A review on the methods for biomass to energy conversion systems design. *Renewable and Sustainable Energy Review* 25:420–430.

Yuan, W. J., B. L. Chang, J. G. Ren, J. P. Liu, F. W. Bai, and Y. Y. Li. 2012. Consolidated bioprocessing strategy for ethanol production from Jerusalem artichoke tubers by *Kluyveromyces marxianusunder* high gravity conditions. *Journal of Applied Microbiology* 112:38–44.

Zerva, A., A. L. Savvides, E. A. Katsifas, A. D. Karagouni, and D. G. Hatzinikolaou. 2014. Evaluation of *Paecilomyces variotii* potential in bioethanol production from lignocellulose through consolidated bioprocessing. *Bioresource Technology* 62:294–299.

10 Cultivation and Conversion of Algae for Wastewater Treatment and Biofuel Production

Priyanka Yadav, Sivamohan N. Reddy, and Sonil Nanda

CONTENTS

10.1 Introduction 159
10.2 Algal Metabolism 161
10.3 Role of Algae in Carbon Sequestration and Wastewater Treatment 162
10.4 Technologies for the Cultivation of Algae 164
10.5 Technologies for Harvesting of Algae 166
10.5.1 Bulk Harvesting Step 166
10.5.2 Dewatering Step 167
10.5.3 Methods for Algae Harvesting 167
10.6 Conversion of Algae to Biofuels 168
10.7 Hydrothermal Liquefaction of Algae 170
10.8 Conclusions 172
References 173

10.1 INTRODUCTION

The exploding population growth around the world with communities raising their standard of living has resulted in high-energy demands that have an intense effect on the current energy supplies. Fossil fuels, especially coal, natural gas, and petroleum contribute nearly 80% of the total energy supply (Milano et al. 2016). The transportation, industry, and power generation sectors majorly contribute to the greenhouse gas emissions (e.g., CO_2, CO, NO_x, and SO_x) leading to global warming (Hosseini et al. 2013). The high rate of consumption of fossil fuels poses major challenges related to reserves and CO_2 emissions. Therefore, there is an immediate need for finding alternatives fuels that are cleaner, safer, non-toxic, renewable, and economical. The renewable resources such as hydro, solar, geothermal, and biomass can potentially supplement future energy demands.

To reduce the burden of fossil fuels and considering environmental safety, carbon-neutral renewable sources, particularly from biomass, have received attention as potential energy sources. Waste renewable resources include lignocellulosic biomass including agricultural and forestry biomass (Nanda et al. 2016e, 2017a, 2018), energy crops (Nanda et al. 2016a), cattle manure (Nanda et al. 2016c), food waste (Nanda et al. 2016d), and sewage sludge (Gong et al. 2017a, 2017b). In addition, industrial effluents from dairy industries (Nanda et al. 2015b), biodiesel industries (Reddy et al. 2016), waste cooking oil (Nanda et al. 2019) and fossil fuel residues (e.g., bitumen, asphaltene and petroleum coke) (Rana et al. 2017, 2018) have also been investigated for biofuel production through thermochemical biomass-to-gas conversion pathways. Through specific thermochemical technologies (e.g., pyrolysis, gasification, and liquefaction) and biochemical technologies (e.g., fermentation and anaerobic digestion), waste biomass and organics can be converted to a variety of biofuels such as bio-oil, char, synthesis gas, hydrogen, methane, bioethanol, and biobutanol (Nanda et al. 2014, 2017b).

Biofuels derived from biomass can be an alternative to fossil fuels, and their economic production can meet current energy demands. Biofuels are derived from the biomass and are cleaner fuel because their combustion produces CO_2, which plants use during their growth (Nanda et al. 2013). Therefore, the use of biofuels is considered carbon-neutral since there is no net increase in the emissions of CO_2 to the atmosphere. However, the biorefinery process also can be made carbon-negative by carbon capture and sequestration techniques (Nanda et al. 2016f). These techniques are achieved by the application of biochar or use of algae. Biochar is a product of thermochemical conversion of biomass, particularly pyrolysis and gasification. It is rich in a stable form of carbon that is not converted to either bio-oil or gases. Hence, using the biochar as a soil amendment agent can not only enhance the soil fertility but also sequester the carbon in the soil for decades or centuries (Nanda et al. 2016b).

Microalgae, also considered as the third-generation biofuel feedstock, not only capture CO_2 from the atmosphere for photosynthesis but also grow on polluted wastewater to absorb the environmental contaminants (Rawat et al. 2011). Algae naturally fix the CO_2 as lipids and polysaccharides within their cells, which can also be transformed to bio-oils, fuel gases, and other value-added industrial products (Nanda et al. 2016f). CO_2 released into the atmosphere was taken up by the plants in the presence of sunlight through the photosynthesis route to yield the plant biomass. For an annual production of 100 gigatons of biomass, nearly 77 gigatons of CO_2 is consumed by the plants (Gupta and Tuohy 2013). Other advantages of biofuels are related to energy security and socioeconomic benefits, which help to boost the rural economy by creating employment opportunities and using the degraded or barren lands for the cultivation of third-generation energy crops (Nanda et al. 2015a).

Algae are one of the most promising biofuel feedstocks because of their higher demand, photosynthetic efficiency, and marginal use of land or water (Renuka et al. 2013). Algae are emerging as a third-generation biomass because of their rapid growth rates, their ability to grow by using nutrients present in the wastewater, and their use of the CO_2 from the atmosphere or flue gas. In addition, algae uptakes nearly 183 gigatons of CO_2 for 100 gigatons of algal biomass at higher photosynthetic efficiency than conventional plants (Najafi et al. 2011). The algal oil contains approximately 35,800 kilojoules per kilogram (kJ/kg) of energy, which is equivalent to about 80% of the energy in petroleum (Iqbal et al. 2017). Several environmental applications of algae are illustrated in Figure 10.1.

Microalgae have the tendency to accumulate increasing amounts of lipids compared to other oil seed crops such as rapeseed and sunflower. Microalgae also have the advantage of metabolic flexibility, which means that the biochemical composition of the biomass can be varied by changing the growth conditions. Many researchers have been focusing on the metabolic engineering of microalgae for the development of high lipid content within their cells and production of bio-oils in large-scale cultivation by using open pond systems and capturing CO_2 from the power plants. In general, algae typically contain 80%–85% water with photosynthesis leading to the dissociation of one mole of water per mole of CO_2 (Murphy and Allen 2011).

Lipids can be polar and neutral in nature and their content in microalgae can range up to 1.5–75 wt% (Alaswad et al. 2015). They are water-insoluble but dissolve in most organic solvents. Some examples of polar lipids are glycolipids and phospholipids, whereas neutral lipids can be acylglycerides and free fatty acids (FFA). Neutral lipids are used as an energy source in algae while polar lipids are used up in cell membranes. Different quantities of lipids can be produced in algae depending on their culture medium and growth conditions.

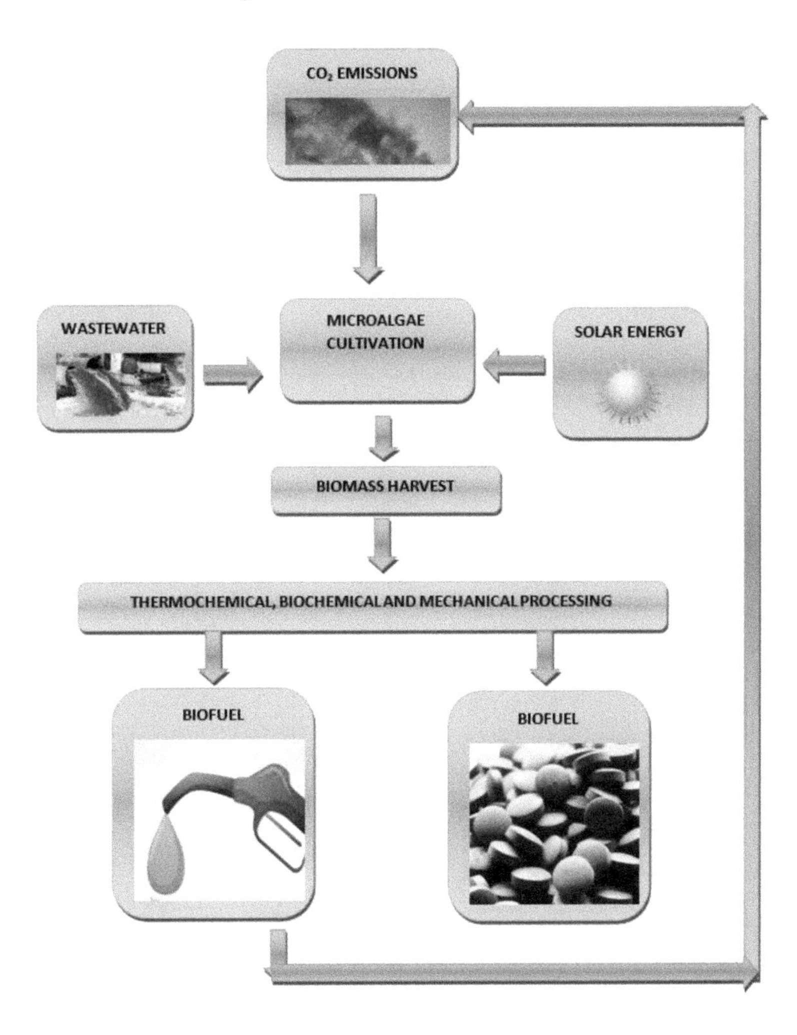

FIGURE 10.1 Various uses of algal biomass.

10.2 ALGAL METABOLISM

Algae are a diverse group of photosynthetic aquatic organisms and the branch of life science that deals with their study is called phycology. The algal metabolic process involves biochemical and transport mechanisms in which the nutrients present in water substrate are converted into the basic organic elements required to cultivate algae. Most of the metabolic activities that take place in algae are also common in other plants. Metabolic processes occurring in algae mainly include the processing of light harvesting, carbon acquisition, assimilation of nitrogen and sulfur, and the formation of specific supporting metabolites. The consequences of growth in severe conditions such as nutrient limitation and exposure to lights (e.g., visible and ultraviolet light) for algal metabolism are also considered. The growth of algae requires nutrients such as nitrogen, phosphorous, and carbon. Wastewater from a wide variety of sources, which include domestic, agricultural, refinery, and industry, can act as the source of these nutrients for algal growth.

Photosynthesis in algae consists of two main steps—light-controlled reactions and the Calvin cycle—which occur in the chloroplast. In light-controlled reactions, the light energy is converted to adenosine triphosphate (ATP) and nicotineamide adenine dinucleotide phosphate (NADPH) where water splits into electrons, hydrogen, and oxygen discharged into the atmosphere as the by-products. The next step is the Calvin cycle in which the energy and electrons from ATP and NADPH, respectively, are used to convert CO_2 into carbohydrates, lipids, and other by-products. The chloroplast is a good source to produce protein and hydrogen. Unicellular algae can produce hydrogen gas upon illumination with preceding anaerobic incubation of the cells in the absence of light. In chloroplast, the synthesis of protein and fatty acids takes place.

Controlling the metabolic reaction pathways can alter many cellular functions in the synthesis of selective products and enhance the processing of microalgae. One way is to manipulate the culture medium conditions in which nutrient regimens direct the algal metabolic pathways. The alternative way is metabolic engineering in which the organism's cellular machinery can be controlled directly through mutagenesis or by insertion of transgenes. Naturally, microalgae can store energy obtained from carbon sources in the form of lipids. When algae are exposed to stress factors or nutrient-limiting conditions such as unavailability of nitrogen, the oil accumulates in the starved cells. Under such conditions, polyunsaturated fatty acids (PUFA) can be obtained from microalgae that can be transformed to biodiesel through transesterification and other chemical reactions. In the genetic transformation approach, the engineered DNA fragment is incorporated into the host algae nucleus or chloroplastic genome. The chloroplast and nucleus both contain organelles, which consist of individual genomes that are capable of incorporation of the transgene. With the help of genetic transformation, recombinant microalgae are developed for different purposes such as biosynthesis of the augmented lipids, trophic conversion, engineering light-harvesting antennae, and mass production of recombinant proteins. Therefore, by changing the metabolism pathways of algae, different valuable products can be obtained (Rosenberg et al. 2008).

10.3 ROLE OF ALGAE IN CARBON SEQUESTRATION AND WASTEWATER TREATMENT

There are several stages in wastewater treatment as illustrated in Figure 10.2. The different stages in the treatment of wastewaters and industrial effluents are primary, secondary, tertiary, and quaternary treatments. The primary treatment involves the removal of suspended solids through several physical processes such as screening, filtration, sedimentation, flocculation, and flow equalization. The secondary treatment involves chemical processes such as adsorption, disinfection, and dechlorination to create ambient conditions for algal growth. The tertiary treatment involves the removal of dissolved organic matter through a combination of physical and chemical processes such as anaerobic digestion, aerated lagoons, trickling filters, and so on. The final stage of wastewater remediation is quaternary treatment that is mostly through biological means to remove dissolved nitrogen and phosphorous using algae.

Microalgae have higher CO_2 fixation ability as compared to other plants with a potential to produce value-added products. The cultivation of microalgae can be performed using the wastewater effluents from a variety of industries. Algal cultivation either takes place in open raceway ponds or closed photo-bioreactors. The raceway ponds consist of closed spiral coil and elliptically shaped recirculation channels, whereas photo-bioreactors consist of an arrangement of glass and tubular array to harness sunlight or visible light, absorb pollutants, and enhance constant cell growth. An airlift-driven raceway reactor for microalgae cultivation has the highest CO_2 consumption rate. By increasing the retention time of CO_2 in the photo-bioreactor, significant enhancement of CO_2 fixation efficiency can be achieved. Hence, algal-based CO_2 sequestration is one of the promising technologies along with biofuel production (Farrelly et al. 2013).

Microalgae seem to be one of the most proficient biomass sources used to produce biofuels due to their high lipid content, rapid growth rate, less maintenance, lower chances of inhibition by other microorganisms, and less cost investment. For an industrial microalgal production system, the site specifications can be varied. Integrated CO_2 bio-fixation, production of biofuel, and value-added

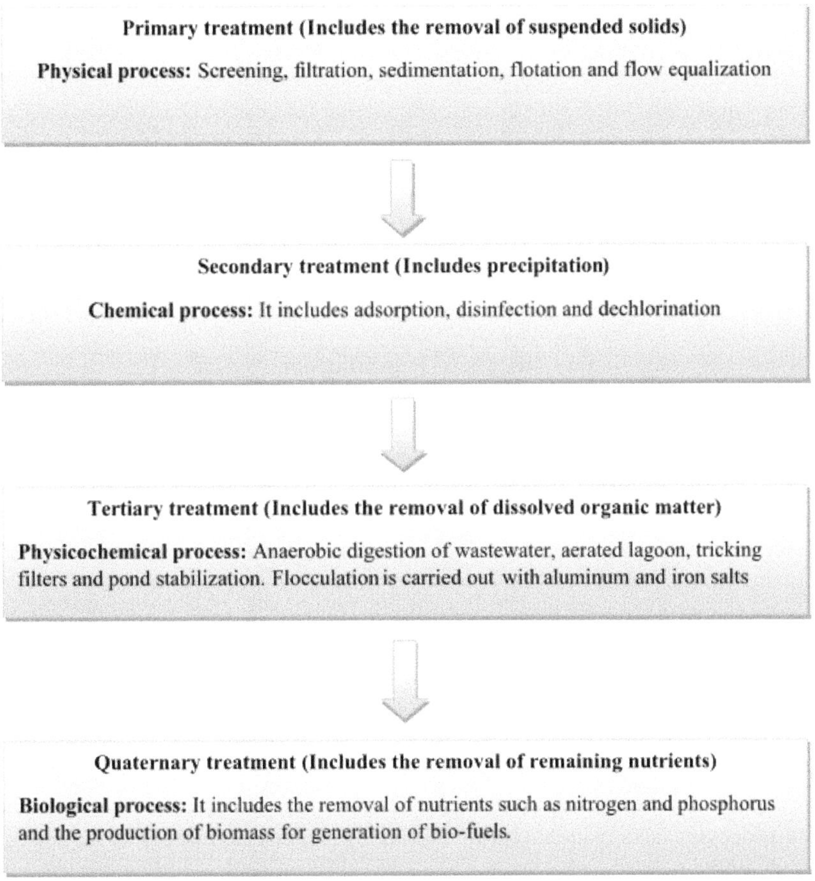

Primary treatment (Includes the removal of suspended solids)

Physical process: Screening, filtration, sedimentation, flotation and flow equalization

↓

Secondary treatment (Includes precipitation)

Chemical process: It includes adsorption, disinfection and dechlorination

↓

Tertiary treatment (Includes the removal of dissolved organic matter)

Physicochemical process: Anaerobic digestion of wastewater, aerated lagoon, tricking filters and pond stabilization. Flocculation is carried out with aluminum and iron salts

↓

Quaternary treatment (Includes the removal of remaining nutrients)

Biological process: It includes the removal of nutrients such as nitrogen and phosphorus and the production of biomass for generation of bio-fuels.

FIGURE 10.2 Different stages of wastewater treatment.

algal biomasses as a substitute to current CO_2 mitigation strategies are required for a circular bioeconomy. Generally, microalgae are used for an economical combination of CO_2 bio-fixation, wastewater treatment, and synthesis of lipids for biofuel production. Several studies have demonstrated the CO_2 capture technologies using algae can reduce emissions of greenhouse gases and accept the prospects and challenges based on cost and engineering of CO_2 sequestration (Mohan et al. 2016; Singh et al. 2017).

The cultivation of microalgae for wastewater treatment can be carried out in two systems such as a suspended-cell cultivation system or immobilized-cell cultivation system. In a suspended-cell cultivation system, huge quantities of wastewater can be treated and processed. The limitations of such systems are the harvesting of microalgae and disposal of residual wastewaters. In immobilized-cells cultivation systems, the harvesting of microalgae is easier compared to the disposal problem of the treated wastewater. The immobilization matrix cells have higher resistance to harsh operating conditions such as metal toxicity and pH. The limitations of such processes are the requirement of a high surface area and high capital cost because of the expensive polymeric matrix.

Understanding of mechanisms involved in the removal of nutrients by microalgae is very important to the advancement of wastewater remediation processes. Carbon molecules of microalgae can fix CO_2 from the atmosphere by photosynthesis. HCO_3^- is then converted to CO_2 inside the cells by ribulose biphosphate carboxylase oxygenase enzyme, which produces phosphoglycerate. The common inorganic forms of nitrogen are nitrate, nitrite, nitric acid, molecular nitrogen, nitrogen dioxide, ammonia, ammonium, nitrous oxide, and nitric oxide. Microalgae fix the molecular nitrogen from the atmosphere and then convert it into the ammonia-nitrogen (Barsanti and Gualtieri 2006). Several essential reactions involved in algae-based wastewater remediation are represented in the following equations.

$$N_2 + 16\ ATP + 8H^+ + 8e^- \rightarrow 2NH_3 + 16\ ADP + 16P_i + H_2 \tag{10.1}$$

$$NO_{3^-} + 2e^- + 2H^+ \rightarrow H_2O + NO_{2^-} \tag{10.2}$$

$$NO_{2^-} + 2e^- + 8H^+ \rightarrow 2H_2O + NH_{4^+} \tag{10.3}$$

$$\text{Glutamine} + NH_{4^+} + ATP \rightarrow \text{Glutamine} + ADP + P_i \tag{10.4}$$

$$ADP + P_i \rightarrow ATP \tag{10.5}$$

Microalgae assimilated the fixed N_2 as ammonium-nitrogen, nitrate-nitrogen, and nitrite-nitrogen. The assimilation of nitrogen requires its reduction to ammonium-nitrogen in a two-step process in the presence of nitrate and nitrite reductases enzymes. In the initial step, the nitrate-nitrogen reduces to nitrite-nitrogen by the nitrate reductase enzyme in the presence of NADPH as a reducing agent. Furthermore, nitrite-nitrogen is reduced to ammonium-nitrogen by the nitrite reductase enzyme, which further uses ferredoxin to catalyze the electron transfer reactions. Ammonium-nitrogen formed by the reduction of nitrate-nitrogen and nitrite-nitrogen is further converted into amino acids by the glutamine synthetase-glutamate synthase pathway in the presence of the glutamine synthase enzyme. Phosphorus enters microalgae cells through the plasma membrane in the form of HPO_4^{2-} and $H_2PO_4^-$. Further, phosphate-phosphorus is converted into organic compounds by processes such as phosphorylation, oxidative phosphorylation, and photophosphorylation. In these processes, adenosine diphosphate (ADP) is converted into ATP by an energy input (Martinez et al. 1999).

10.4 TECHNOLOGIES FOR THE CULTIVATION OF ALGAE

Several stages in the typical lifecycle of algae for biofuel production and environmental remediation from "cradle to grave" are cultivation, harvesting and dewatering, biomass processing, oil extraction, biofuel production (biodiesel, bio-oil, and so on), analysis, and distribution (Figure 10.3). Several technologies for cultivation of algae include wastewater stabilization ponds, algal high-rate ponds, open ponds, photo-bioreactors, and hybrid technologies. Wastewater stabilization pond is a biological treatment method in which the microalgae stabilizes the wastewater and decreases the pathogenic microorganisms. The growth limiting parameters for algae are retention time, temperature, sunlight, and loading rate of biochemical oxygen demand. Algal high-rate ponds consist of a shallow pond (0.3–0.6 m) operating at a lower hydraulic retention time. A large paddle wheel vane pump is used for generating the speed of about 10–30 centimeters per second (cm/s) for agitation. The open ponds for microalgae cultivation also use the paddle wheels for mixing and aeration purposes.

The photo-bioreactor systems consist of both indoor and outdoor enclosures system, which use sunlight or artificial light. Mechanical pumping is used for the mixing purposes. Several types of photo-bioreactors include vertical columns, flat-plates, tubular, and internally illuminated types. Vertical photo-bioreactors are easy to operate, compact in size, and comparatively cheaper. It was found that the bubble-column, airlift, and narrow tubular photo-bioreactors have the comparable values of biomass concentration and the specific growth rate. The flat plate photo-bioreactors have the advantage of their large illumination surface area. These types of photo-bioreactors are made

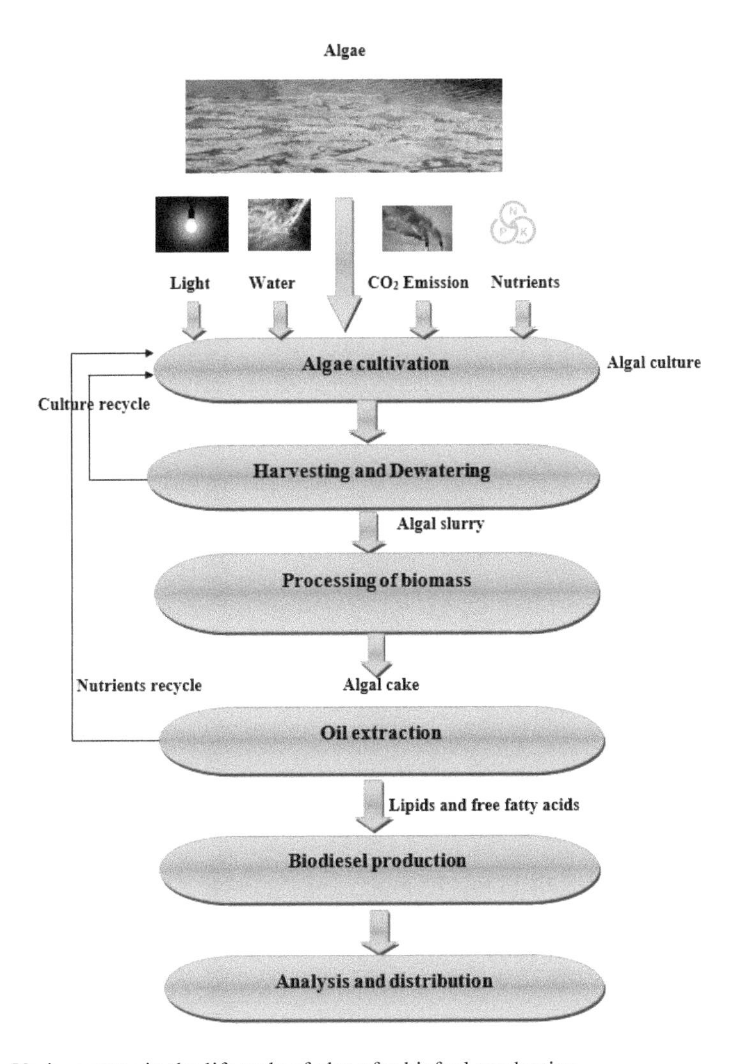

FIGURE 10.3 Various steps in the lifecycle of algae for biofuel production.

from translucent materials so that the maximal solar energy can be utilized. The tubular photo-bioreactors are most appropriate for the outdoor cultures. These types of photo-bioreactors allow the mass cultures to be circulated within the system using a pump. The tubular photo-bioreactor exists in the various forms such as horizontal, vertical, near horizontal, conical, and inclined. The internally illuminated photo-bioreactor employs fluorescent lamps for illumination. Algal cultures are mixed by the impellers, which are equipped with the photo-bioreactor. The major advantage of internally illuminated photo-bioreactor is that the contamination can be reduced because it can be sterilized at high pressures. An algal biofilm photo-bioreactor system comprises a growth surface for adherence of biofilm as well as systems for illumination and recirculation of nutrient medium. Peristaltic pumps can be used for adding medium over the surface. Fluorescent lamps are used for illumination. The surface is then inoculated with algal cultures and the process continues for several days (Ozkan et al. 2012).

Hybrid systems are a combination of photo-bioreactors and open-pond systems. In such systems, the contamination by inhibiting species is minimized by controlling the operating parameters, thus subsequently enhancing the division of algal cells. Furthermore, the cells are exposed to nutrients, which increase the algal lipid synthesis. *Haematococcus pluvialis* culture was used to produce

TABLE 10.1
Advantages and Disadvantages of Algal Production Systems

Production System	Advantages	Disadvantages
Open pond	• Relatively cheaper • Low energy requirement • Easy maintenance	• Low productivity • Large space requirement • Contamination of cultures
Tubular photo-bioreactor	• Relatively cheaper • High productivity • High illumination surface area	• Fouling • Large space requirement
Flat-plate photo-bioreactor	• High productivity • Easily sterilized • High illumination surface area	• Hydrodynamic stress • Scale-up issues • Temperature control is required
Column photo-bioreactor	• Compact design • Easily sterilized • Low stress • Low energy requirement	• Smaller illumination surface area

biofuel and it was reported that average biofuel production rate was greater than 10 toe/ha/annum. It was also found that the rate could be increased up to 76 toe/ha/annum using different species with higher oil and photosynthetic efficiency (Huntley and Redalje 2007). Table 10.1 summarizes the advantages and limitations of several algal production systems.

10.5 TECHNOLOGIES FOR HARVESTING OF ALGAE

The harvesting of algae is classified into two steps: bulk harvesting step and dewatering step.

10.5.1 Bulk Harvesting Step

1. Coagulation/flocculation: In coagulation, aggregates of microalgal cells are formed depending on their size and density. The flocks are extracted through the sedimentation process with the most commonly used coagulants such as ferric iron chloride, aluminum sulfate, and certain polymeric materials.
2. Electro-coagulation: In electro-coagulation, the positive ions are produced by the reactive metallic electrodes, which cause the coagulation of algal cells. The aggregates of microalgae cells are formed and settled at the bottom of the surface by the sedimentation process. The three methods available are electrolytic flocculation, electrolytic flotation, and electrolytic coagulation.
3. Flotation: Flotation combines the flocculents to form aggregates of algal cells by bubbling the air. The aggregates of cells settled at the bottom are further separated either by using the traditional skimming techniques or through the dissolved-air flotation method.
4. Sedimentation: The sedimentation process uses the gravitational force to remove the suspended particles present in the wastewater stream.
5. Immobilization: The immobilization technique uses the polymer encapsulation of algae. Suitable polymeric materials can be added at the time between the algal cultivation processes. The most widely used immobilizing agent is alginate.

10.5.2 Dewatering Step

1. Filtration: Filtration is a mechanical method that uses a membrane so that only the smaller particles can pass through, but the retentate contain oversized particles. For larger algae (>70 µm), the vacuum filter is used and for the smaller sized cells microfiltration or ultrafiltration membranes are used.
2. Belt press: In the belt press, the solution is placed between the two belts with the application of pressure. The belts then pass through the decreasing diameter rolls to dewater and separate the algal cells.
3. Centrifugation: The centrifugation process uses an apparatus that rotates at high speeds and separates the particles of different densities.

10.5.3 Methods for Algae Harvesting

1. Chemical-based methods: In the chemical-based harvesting operation, chemical flocculation is performed due to the smaller size of algal cell. This process increases the overall biomass size by the formation of aggregates of algal cells before harvesting. Electrolytes are used for the coagulation, whereas synthetic polymers are used for the flocculation of cells.
2. Mechanical-based methods: Mechanical-based methods can be further classified into centrifugation, filtration, sedimentation, and dissolved air flotation. Centrifugation is the most reliable method for the removal of suspended cells of algae. In the centrifugation operation, the centrifugal force separates the suspended cells of algae by density difference. The filtration method is used for harvesting the filamentous algal strains. Tangential flow filtration is used for the smaller suspended cells of algae. Sedimentation is used for harvesting the algal cells to give a solid concentration of nearly 15%. Sedimentation is relatively low at settling rates of around 0.1–2.6 centimeters per hour (cm/h). Dissolved air flotation is used for the removal of the wastewater-treated sludge. Dissolved air flotation is preferred more than sedimentation for the algae-rich waters. Table 10.2 makes a comparison of different mechanical-based harvesting methods available for algae.
3. Electrical-based methods: Electrophoresis of the algae cells is classified as the electrical-based separation of suspended algal cells. This process usually does not require the involvement of chemicals, hence eliminating the need for chemical separations at disposal stages.
4. Biological-based methods: Biological-based methods consist of bio-flocculation in which flocculation is done by the secreted biopolymers. In microbial flocculation, flocculating microorganisms are added to the algae culture. Better efficiencies can be achieved from use of the flocculates from soil microorganisms rather than from use of aluminum sulfate or poly-acryl amide for harvesting algae.

TABLE 10.2
Comparison of Different Mechanical-Based Methods of Harvesting Algae

Methods	Solids (wt%)	Advantages	Disadvantages
Centrifugation	12–22	Higher solid concentration	Energy intensive and higher cost
Tangential filtration	5–27	Higher solid concentration	Membrane fouling and higher cost
Gravity sedimentation	0.5–3	Lower costs	Higher costs
Dissolved air flotation	3–6	Large-scale applications	Flocculants are required

10.6 CONVERSION OF ALGAE TO BIOFUELS

Algal biomass is an organic material that stores and converts solar energy in the form of chemical energy during photosynthesis. With the use of suitable conversion technologies (pyrolysis, combustion, gasification, liquefaction, fermentation, anaerobic digestion, and so on), the biomass can be transformed to release heat, electricity, chemicals, and fuels (Nanda et al. 2014). Different technologies available classified under physicochemical, thermochemical, and biochemical methods for biomass (algae) conversion into energy and fuels are illustrated in Figure 10.4. The physiochemical conversion includes the mechanical extraction of oil from algae through pressing and then conversion of the pressed oil to biodiesel by esterification. Biochemical conversion includes fermentation and anaerobic degradation. The pressed algal cakes contain the residual polysaccharides, which can be pretreatment using acids, alkalis, and ionic liquids followed by enzymatic hydrolysis and fermentation using suitable fungi and bacteria to produce alcohol-based biofuels such as ethanol and butanol. Anaerobic degradation of the pressed algal cakes using methanogenic bacteria produces biogas (or methane). The thermochemical conversion includes combustion, gasification, pyrolysis, and liquefaction.

Combustion is an exothermic process in which different organic components react with oxygen to generate heat energy and steam that can be used to turn turbines to generate electricity. Gasification is thermal degradation of organic matter with a limited supply of oxygen to produce combustible synthesis gas (a mixture of CO and H_2). Pyrolysis is the thermal degradation of organic substances in an inert atmosphere to produce bio-oil, biochar, and gases. Liquefaction is a biomass-to-liquid conversion technology that mostly produces bio-oil and traces of tar and char. The bio-oil obtained from liquefaction contains less oxygen and moisture compared to pyrolysis-derived bio-oil, which results in its high-energy value (Nanda et al. 2014). Hydrothermal liquefaction is basically a hydrothermal conversion technique that uses hot-pressurized water acting as the reaction medium to solubilize the biomass directly to bio-oil. In the hydrothermal gasification process, the reaction temperature is greater than 350°C in the absence of an oxidizing agent, which generates a gas phase containing H_2, CO, CO_2, CH_4, and C_{2+} components. Hydrothermal gasification also uses subcritical and supercritical water to hydrothermally decompose algae to produce H_2-rich syngas. Hydrothermal carbonization transforms biomass into hydrochar at a comparatively lower temperature in the range of 180°C–250°C and pressures of 2–10 MPa.

Algal biomass can be converted into several types of biofuels depending upon the different techniques used and the part of biomass cell being used. The lipid content of the algal biomass is extracted and converted into biodiesel or green petroleum-based fuels. Biofuels can be used either in their pure form or blended with petroleum or diesel at different concentrations. The biocrude oil obtained from algal biomass can be used as an energy-efficient alternative for the fossil fuels. However, crude oils from algae have a high content of nitrogen-rich aromatic heteroatoms, which is a major hurdle in their upgrading to synthetic drop-in fuels using the existing biorefinery infrastructures. BG100 or 100% bio-gasoline can be directly used as a replacement for gasoline in any conventional gasoline-fueled motor engine. Bio-gasoline can be used in the same fuelling infrastructures because their fuel properties are comparable to fossil-based gasoline (Wang et al. 2014).

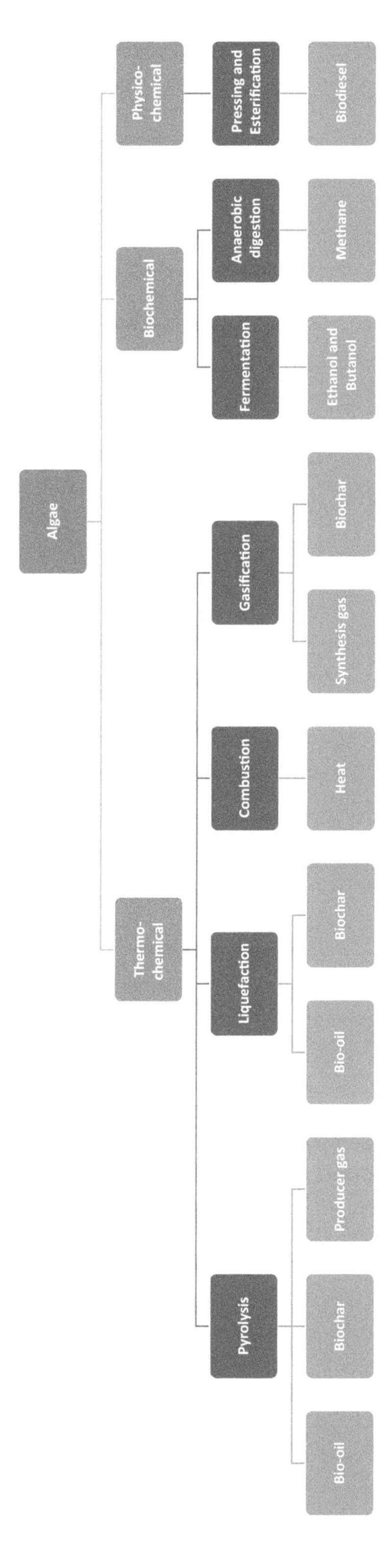

FIGURE 10.4 Conversion of algae to different biofuels through thermochemical, biochemical, and physicochemical pathways.

10.7 HYDROTHERMAL LIQUEFACTION OF ALGAE

Hydrothermal liquefaction can directly convert the wet algal biomass to energy dense bio-oil (Guo et al. 2015). It is a hydrothermal biomass-to-liquid conversion technique that uses hot compressed water as the reaction medium to decompose the biomass directly into bio-oil. Using algae, seaweed, and other aquatic plants such as water hyacinth for bio-oil production through hydrothermal liquefaction is advantageous and cost-effective because it does not require the addition of a pre-drying process because the reaction medium is water (Sasaki et al. 2004; Liu et al. 2006). The typical operating conditions required for hydrothermal liquefaction are 250°C–375°C and 10–20 MPa pressure with microalgae biomass fractions in the range of 5%–20% in the feed slurry (Liu et al. 2006).

The properties of water at normal and critical phases are mentioned in Table 10.3. Water occurs in three phases (liquid water, subcritical water, and supercritical water) depending on its critical temperatures and pressures. Liquid water occurs at ambient room temperature and atmospheric pressures, whereas subcritical water occurs at temperatures ($T_C \leq 374$°C) and pressures ($P_C \leq 22.1$ MPa) lower than the critical points of water. In contrast, supercritical water occurs at temperatures ($T_C \geq 374$°C) and pressures ($P_C \geq 22.1$ MPa) higher than the critical points of water. The density of liquid water at ambient conditions is 1 g/cm^3, whereas that of subcritical water is 0.6–0.8, and supercritical water is 0.17–0.58 (Toor et al. 2011). Similarly, the kinematic viscosity of water also decreases with the transition of liquid water (0.89 mPa s) to its subcritical (0.06–0.11 mPa s) and supercritical (0.03–0.07 mPa s) states. The heat capacity of water increases as the critical temperature and pressure increase. For example, the heat capacity of liquid water is 4.2 kJ/kg/K while that of subcritical water and supercritical water is 4.9–10.1 and 6.8–13 kJ/kg/K, respectively (Toor et al. 2011).

The thermophysical properties of water change at higher pressure and temperature, which results in the deterioration of the macromolecules present in biomass and the polymerization of smaller molecules into the larger molecules to form biofuels (Barreiro et al. 2013). At high pressure and temperature, the hydrogen bond of water molecules is much lower than at room temperature, which results in a low dielectric constant (Reddy et al. 2014). The dielectric constant of water decreases as the temperature and pressure of the hydrothermal reaction medium increases. For example, the dielectric constant of liquid water is 78.5 F/m, while that of subcritical water and supercritical water is 14.1–27.1 and 5.9–10.5, respectively (Toor et al. 2011). A smaller value of dielectric constant enhances the solubility of less polar organic solvents present in the water and also increases the ionic product constant (K_W) that advances the acid- and base-catalyzed hydrolysis reactions. Owing to the high ionic product of water at elevated temperature, [H^+] and [OH^-] ions are large which results in the heterolytic fission of aromatics compounds and catalyzes acid and base reactions.

Under subcritical conditions during hydrothermal liquefaction, the microcrystalline cellulose in biomass is hydrolyzed to amorphous cellulose and polymers where saccharides are released. In contrast, under supercritical conditions, cellulose appears to swell and the saccharides are released later. In general, cellulose hydrolyzes to glucose and fructose, which further decompose to smaller intermediates or furfural with the elimination of water. Furfural is converted to phenol, which may further degrade to smaller intermediates (Nanda et al. 2015b). Ultimately, smaller intermediates are converted to H_2, CH_4, CO_2, and CO using free-radical reactions. The proteins are hydrolyzed to amino acids (Muller-Feuga 2000). During hydrothermal degradation, lignin also is converted to phenols through hydrolysis of ether-bonds. These products are further degraded by hydrolysis of methoxy groups, but the benzene ring remains in a stable form (Kumar et al. 2018). During hydrothermal carbonization, cellulose is hydrolyzed to glucose and lignin is converted to phenol. Different free-radical reactions take place to form CH_4 and CO_2.

Hydrothermal liquefaction technique converts wet algal biomass into solid biochar, liquid bio-oil, and gas phase (H_2, CO, CO_2, and CH_4). This method also results in by-products that can be recovered

TABLE 10.3
Bio-oil Yields from Liquefaction of Algae at Different Operating Conditions

	Operating Conditions							
Feedstock	Temperature (°C)	Pressure (MPa)	Residence Time (min)	Reactor Type	Catalyst	Concentration	Oil Yield	Reference
Nannochloropsis	240–360	6–8	30	Batch high-pressure stainless steel reactor of 1.8 L capacity	Ni/TiO_2	120 g algae	48.2% at 300°C with catalyst	Wang et al. (2018)
Nannochloropsis salina	310–350	–	5–60	–	–	Algae slurry of 5 and 10 wt%.	High lipid algae- 54.3% for light oil, low lipid algae-27.5% for light oil	Cheng et al. (2017)
Cyanophyta	260–420	–	5–75	2.7 mL micro batch reactor	–	0.02–0.3 algae/water by weight	29.24% at 350°C, 60 min	Song et al. (2017)
Nannochloropsis, Neochloris oleoabundans, Chlorella vulgaris and *Botryococcus braunii*	600 (fast) and 350 (isothermal)	–	10 and 60	Batch reactor (stainless steel SS316)	–	15 wt% algae slurry with 11 vol% water	67 wt% dry basis for fast pyrolysis	Faeth and Savage (2016)
Nannochloropsis	210, 230 and 250	–	60	500 mL batch reactor	Nano-Ni/SiO_2 synthesized zeolite Na_2CO_3	10 g algae, 150 mL water and 0.5 g catalyst	30% at 250°C for nano-Ni/SiO_2	Saber et al. (2016)
Cyanobacteria and *Bacillariophyta*	250–350	–	15–90	300 mL batch reactor	–	10 g feed and 100 mL water	21.1 wt% dry basis	Huang et al. (2016)

to produce industrially relevant products such as fine chemicals. Hydrothermal liquefaction converts the protein, lipids, and carbohydrates present in algal biomass to bio-oil with a higher efficiency than other processes. Hydrothermal liquefaction also enables the separation and recycling of various nutrients such as calcium, iron, magnesium, nitrogen, potassium, and other minerals (Guo et al. 2015).

Hydrothermal liquefaction consists of three steps: depolymerization, decomposition, and recombination or re-polymerization. In the depolymerization phase, larger biopolymer chains are converted into shorter dimers or monomers. In the decomposition phase, water at high pressure breaks cellulose and other polymers into monomers with the removal of molecules like CO_2, water, and so on. The products formed in this phase are highly soluble in water. In the recombination or re-polymerization phase, the reactive fragments that were formed polymerize to generate biocrude oil, tar, gas and char.

The high-pressure and high-temperature requirements make the process expensive with concerns for the long-term durability of the reactor material. Although the pressure can be maintained under optimal and process safety ranges by lowering the operating temperatures, it also should be noted that the bio-oil yield is compromised at lower temperature (Saber et al. 2016). Therefore, a catalyst can be used to avoid the lower temperature concern and it results in a reduction in the formation of char and tar. The use of a catalyst also improves the efficiency of the process, which may lead to an increase in the bio-oil yield.

The catalysts usually used in hydrothermal liquefaction and gasification can be grouped as homogeneous and heterogeneous catalysts. The homogeneous catalyst includes NaOH, KOH, Na_2CO_3, K_2CO_3, and other alkali and carbonate salts. Alkali salts lower the chances of char and tar generation with the improvement of bio-oil yields through enhanced water-gas shift reaction. However, the recovery of the homogeneous catalysts is expensive because of the cost-intensive and energy-consuming separation process (Kumar and Gupta 2009). Heterogeneous catalysts include nickel, platinum, palladium, ruthenium, metal oxides, transition metal chlorides, carbon nanotubes, and so on. The separation and reuse of the heterogeneous catalysts are feasible, but their initial cost is relatively higher because of the high cost of the metals. Last, but not the least, several reactors found to be beneficial for hydrothermal processing of biomass are batch reactors, continuous stirred tank reactors (CSTR), autoclave reactors, quartz tube reactors, microreactors, tubular reactors, and fixed bed reactors (Savage et al. 2010). Table 10.3 summarizes some selected works on liquefaction of algae to produce bio-oil.

10.8 CONCLUSIONS

Most of the research related to the cultivation and harvesting of algae is limited to laboratory-scale and small-scale studies. The major challenges for large-scale studies are nutrient supply, availability of water, gas exchange, environmental control, and culture stability. This chapter provided insights on different microalgae used in the various wastewater treatment processes together with biofuel production. There is a crucial need to integrate CO_2 bio-fixation, biofuel production, and value-added product recovery from algal biomass as a substitute to the current CO_2 mitigation strategies. Commonly used processes for algal biomass harvesting are the combination of thickening and dewatering processes. The hydrothermal liquefaction is a promising hydrothermal technology for algal biomass conversion to bio-oils. It usually operates at relatively lower temperatures than the conventional pyrolysis and gasification processes, which leads to low-cost equipment requirement and less energy input. Bio-oils resulting from the liquefaction of microalgae can be upgraded to synthetic fuels or blended with the diesel or gasoline for drop-in fuels. The bio-oil produced from liquefaction has lower oxygen content with high energy density. By using the new and innovative technologies for cultivation, harvesting, and processing of algae, biodiesel can potentially compete with petroleum and capacities for scalability.

REFERENCES

Alaswad, A., M. Dassisti, T. Prescott, and A. G. Olabi. 2015. Technologies and developments of third generation biofuel production. *Renewable and Sustainable Energy Reviews* 51:1446–1460.

Barreiro, D. L., W. Prins, F. Ronsse, and W. Brilman. 2013. Hydrothermal liquefaction (HTL) of microalgae for biofuel production: State of the art review and future prospects. *Biomass and Bioenergy* 53:113–127.

Barsanti, L., and P. Gualtieri. 2006. *Algae-anatomy, Biochemistry and Biotechnology*, 2nd ed. Boca Raton, FL: CRC Press/Taylor & Francis Group.

Cheng, F., Z. Cui, L. Chen, J. Jarvis, N. Paz, T. Schaub, N. Nirmalakhandan, and C. E. Brewer. 2017. Hydrothermal liquefaction of high- and low-lipid algae: Bio-crude oil chemistry. *Applied Energy* 206:278–292.

Faeth, J. L., and P. E. Savage. 2016. Effects of processing conditions on biocrude yields from fast hydrothermal liquefaction of microalgae. *Bioresource Technology* 206:290–293.

Farrelly, D. J., C. D. Everard, C. C. Fagan, and K. P. McDonnell. 2013. Carbon sequestration and the role of biological carbon mitigation: A review. *Renewable and Sustainable Energy Reviews* 21:712–727.

Gong, M., S. Nanda, H. N. Hunter, W. Zhu, A. K. Dalai, and J. A. Kozinski. 2017a. Lewis acid catalyzed gasification of humic acid in supercritical water. *Catalysis Today* 291:13–23.

Gong, M., S. Nanda, M. J. Romero, W. Zhu, and J. A. Kozinski. 2017b. Subcritical and supercritical water gasification of humic acid as a model compound of humic substances in sewage sludge. *The Journal of Supercritical Fluids* 119:130–138.

Guo, Y., T. Yeh, W. Song, D. Xu and S. Wang. 2015. A review of bio-oil production from hydrothermal liquefaction of algae. *Renewable and Sustainable Energy Reviews* 48:776–790.

Gupta, V. K., and M. G. Tuohy. 2013. *Biofuel Technologies: Recent Developments*. New Delhi, India: Springer.

Hosseini, S. E., M. A. Wahid, and N. Aghili. 2013. The scenario of greenhouse gases reduction in Malaysia. *Renewable and Sustainable Energy Reviews* 28:400–409.

Huang, Y., Y. Chen, J. Xie, H. Liu, X. Yin, and C. Wu. 2016. Bio-oil production from hydrothermal liquefaction of high-protein high-ash microalgae including wild *Cyanobacteria* sp. and cultivated *Bacillariophyta* sp. *Fuel* 183:9–19.

Huntley, M. E., and D. G. Redalje. 2007. CO_2 mitigation and renewable oil from photosynthetic microbes: A new appraisal. *Mitigation and Adaptation Strategies for Global Change* 12:573–608.

Iqbal, S. Z., N. Malik, K. Rehman, and M. R. Asi. 2017. Processing techniques of algae-based materials. In *Algae Based Polymers, Blends, and Composites*, ed. K. M. Zia, M. Zuber, and M. Ali, 671–686. Amsterdam, the Netherlands: Elsevier.

Kumar, M., A. O. Oyedun, and A. Kumar. 2018. A review on the current status of various hydrothermal technologies on biomass feedstock. *Renewable and Sustainable Energy Reviews* 81:1742–1770.

Kumar, S., and R. B. Gupta. 2009. Biocrude production from switchgrass using subcritical water. *Energy & Fuels* 23:5151–5159.

Liu, A., Y. Park, Z. Huang, B. Wang, R. O. Ankumah, and P. K. Biswas. 2006. Product identification and distribution from hydrothermal conversion of walnut shells. *Energy & Fuels* 20:446–454.

Martinez, M. E., J. M. Jimenez, and F. El Yousfi. 1999. Influence of phosphorus concentration and temperature on growth and phosphorus uptake by the microalga *Scenedesmus obliquus*. *Bioresource Technology* 67:233–240.

Milano, J., H. C. Ong, H. H. Masjuki, W. T. Chong, M. K. Lam, P. K. Loh, and V. Vellayan. 2016. Microalgae biofuels as an alternative to fossil fuel for power generation. *Renewable and Sustainable Energy Reviews* 58:180–197.

Mohan, S. V., J. A. Modestra, K. Amulya, S. K. Butti, and G. Velvizhi. 2016. A circular bioeconomy with biobased products from CO_2 sequestration. *Trends in Biotechnology* 34:506–519.

Muller-Feuga, A. 2000. The role of microalgae in aquaculture: Situation and trends. *Journal of Applied Phycology* 12:527–534.

Murphy, C. F., and D. T. Allen. 2011. Energy-water nexus for mass cultivation of algae. *Environmental Science and Technology* 45:5861–5868.

Najafi, G., B. Ghobadian, and T. F. Yusaf. 2011. Algae as a sustainable energy source for biofuel production in Iran: A case study. *Renewable and Sustainable Energy Reviews* 15:3870–3876.

Nanda, S., A. K. Dalai, and J. A. Kozinski. 2016a. Supercritical water gasification of timothy grass as an energy crop in the presence of alkali carbonate and hydroxide catalysts. *Biomass and Bioenergy* 95:378–387.

Nanda, S., A. K. Dalai, F. Berruti, and J. A. Kozinski. 2016b. Biochar as an exceptional bioresource for energy, agronomy, carbon sequestration, activated carbon and specialty materials. *Waste and Biomass Valorization* 7:201–235.

Nanda, S., A. K. Dalai, I. Gökalp, and J. A. Kozinski. 2016c. Valorization of horse manure through catalytic supercritical water gasification. *Waste Management* 52:147–158.

Nanda, S., J. Isen, A. K. Dalai, and J. A. Kozinski. 2016d. Gasification of fruit wastes and agro-food residues in supercritical water. *Energy Conversion and Management* 110:296–306.

Nanda, S., J. Mohammad, S. N. Reddy, J. A. Kozinski, and A. K. Dalai. 2014. Pathways of lignocellulosic biomass conversion to renewable fuels. *Biomass Conversion and Biorefinery* 4:157–191.

Nanda, S., M. Gong, H. N. Hunter, A. K. Dalai, I. Gökalp, and J. A. Kozinski. 2017a. An assessment of pinecone gasification in subcritical, near-critical and supercritical water. *Fuel Processing Technology* 168:84–96.

Nanda, S., P. Mohanty, K. K. Pant, S. Naik, J. A. Kozinski, and A. K. Dalai. 2013. Characterization of North American lignocellulosic biomass and biochars in terms of their candidacy for alternate renewable fuels. *Bioenergy Research* 6:663–677.

Nanda, S., R. Azargohar, A. K. Dalai, and J. A. Kozinski. 2015a. An assessment on the sustainability of lignocellulosic biomass for biorefining. *Renewable and Sustainable Energy Reviews* 50:925–941.

Nanda, S., R. Rana, H. N. Hunter, Z. Fang, A. K. Dalai, and J. A. Kozinski. 2019. Hydrothermal catalytic processing of waste cooking oil for hydrogen-rich syngas production. *Chemical Engineering Science* 195:935–945.

Nanda, S., R. Rana, Y. Zheng, J. A. Kozinski, and A. K. Dalai. 2017b. Insights on pathways for hydrogen generation from ethanol. *Sustainable Energy & Fuels* 1:1232–1245.

Nanda, S., S. N. Reddy, A. K. Dalai, and J. A. Kozinski. 2016e. Subcritical and supercritical water gasification of lignocellulosic biomass impregnated with nickel nanocatalyst for hydrogen production. *International Journal of Hydrogen Energy* 41:4907–4921.

Nanda, S., S. N. Reddy, D. N. V. Vo, B. N. Sahoo, and J. A. Kozinski. 2018. Catalytic gasification of wheat straw in hot compressed (subcritical and supercritical) water for hydrogen production. *Energy Science and Engineering* 6:448–459.

Nanda, S., S. N. Reddy, H. N. Hunter, I. S. Butler, and J. A. Kozinski. 2015b. Supercritical water gasification of lactose as a model compound for valorization of dairy industry effluents. *Industrial and Engineering Chemistry Research* 54:9296–9306.

Nanda, S., S. N. Reddy, S. K. Mitra, and J. A. Kozinski. 2016f. The progressive routes for carbon capture and sequestration. *Energy Science and Engineering* 4:99–122.

Ozkan, A., K. Kinney, L. Katz, and H. Berberoglu. 2012. Reduction of water and energy requirement of algae cultivation using an algae biofilm photobioreactor. *Bioresource Technology* 114:542–548.

Rana, R., S. Nanda, A. Maclennan, Y. Hu, J. A. Kozinski, and A. K. Dalai. 2018. Comparative evaluation for catalytic gasification of petroleum coke and asphaltene in subcritical and supercritical water. *Journal of Energy Chemistry*. doi:10.1016/j.jechem.2018.05.012

Rana, R., S. Nanda, J. A. Kozinski, and A. K. Dalai. 2017. Investigating the applicability of Athabasca bitumen as a feedstock for hydrogen production through catalytic supercritical water gasification. *Journal of Environmental Chemical Engineering* 6:182–189.

Rawat, I., R. R. Kumar, T. Mutanda, and F. Bux. 2011. Dual role of microalgae: Phycoremediation of domestic wastewater and biomass production for sustainable biofuels production. *Applied Energy* 88:3411–3424.

Reddy, S. N., S. Nanda, A. K. Dalai, and J. A. Kozinski. 2014. Supercritical water gasification of biomass for hydrogen production. *International Journal of Hydrogen Energy* 39:6912–6926.

Reddy, S. N., S. Nanda, and J. A. Kozinski. 2016. Supercritical water gasification of glycerol and methanol mixtures as model waste residues from biodiesel refinery. *Chemical Engineering Research and Design* 113:17–27.

Renuka, N., A. Sood, S. K. Ratha, R. Prasanna, and A. S. Ahluwalia. 2013. Evaluation of microalgal consortia for treatment of primary treated sewage effluent and biomass production. *Journal of Applied Phycology* 25:1529–1537.

Rosenberg, J. N., G. A. Oyler, L. Wilkinson, and M. J. Betenbaugh. 2008. A green light for engineered algae: Redirecting metabolism to fuel a biotechnology revolution. *Current Opinion in Biotechnology* 19:430–436.

Saber, M., A. Golzary, M. Hosseinpour, F. Takahashi, and K. Yoshikawa. 2016. Catalytic hydrothermal liquefaction of microalgae using nanocatalyst. *Applied Energy* 183:566–576.

Sasaki, M., T. Adschiri, and K. Arai. 2004. Kinetics of cellulose conversion at 25 MPa in sub-and supercritical water. *AIChE Journal* 50:192–202.

Savage, P. E., R. B. Levine, and C. M. Huelsman. 2010. Hydrothermal processing of biomass: Thermochemical conversion of biomass to liquid. In *Fuels and Chemicals*, ed. M. Crocker, 192–215. Cambridge, UK: RSC Publishing.

Singh, K., D. Kaloni, S. Gaur, S. Kushwaha, and G. Mathur. 2017. Current research and perspectives on microalgae-derived biodiesel. *Biofuels*. doi:10.1080/17597269.2017.1278932

Song, W., S. Wang, Y. Guo, and D. Xu. 2017. Bio-oil production from hydrothermal liquefaction of waste *Cyanophyta* biomass: Influence of process variables and their interactions on the product distributions. *International Journal of Hydrogen Energy* 42:20361–20374.

Toor, S. S., L. Rosendahl, and A. Rudolf. 2011. Hydrothermal liquefaction of biomass: A review of subcritical water technologies. *Applied Energy* 36:2328–2342.

Wang, B., Y. Huang, and J. Zhang. 2014. Hydrothermal liquefaction of lignite, wheat straw and plastic waste in sub-critical water for oil: Product distribution. *Journal of Analytical and Applied Pyrolysis* 110:382–389.

Wang, W., Y. Xu, X. Wang, B. Zhang, W. Tian, and J. Zhang. 2018. Bioresource Technology Hydrothermal liquefaction of microalgae over transition metal supported TiO_2 catalyst. *Bioresource Technology* 250:474–480.

11 Life-Cycle Assessment of Biofuels Produced from Lignocellulosic Biomass and Algae

Naveenji Arun and Ajay K. Dalai

CONTENTS

11.1 Introduction 177
11.2 Life-Cycle Assessment Methodology 178
11.3 Life-Cycle Assessment of Biomass-Based Biofuels 180
11.4 Influence of Fertilizer Usage on Environmental Credibility 181
11.5 Impact of By-products and Co-products on Environmental Credibility 181
11.6 Life-Cycle Assessment of Algae-Based Biofuels 182
11.7 Conclusions 184
Acknowledgments 184
References 184

11.1 INTRODUCTION

In the twenty-first century, the primary energy concern is development of a sustainable energy sector that is commercially and technologically feasible (Requena et al. 2011). In the United States, the transportation sector accounts for one-third of the total CO_2 emissions. Bioenergy has the potential to meet the basic energy requirements without mandatory changes in energy storage facilities. Development and usage of biomass as alternate feedstock for biofuels production gained importance in 1990s because of the colossal availability of biomass (Zhang et al. 2014). Presently, the focus on food crops as feedstocks is diminishing and lignocellulosic bioenergy crops are gaining attention as promising feedstocks (Nanda et al. 2015).

The European Union (EU) directive 28/2009EC has set standards for the promotion of renewable energy resources and these standards promote the usage of renewable fuels up to 10% in the transportation sector by 2020 (Luque et al. 2010). It has been estimated that contribution of biomass energy will be 60% of the overall energy from renewable resources by 2020. The Renewable Fuel Standard (RFS2) program and the Environmental Impact Statement Assessment (EISA) mandate the use of cellulosic biomass for alternative fuels production because they can aid in the reduction of life-cycle emissions by 20%–60% compared to the emissions from petroleum baseline fuels. Figure 11.1 indicates the roadmap for biofuel production technologies (Luque et al. 2008). Moreover, for the next 20 years, it is projected that the overall biofuels share in the Canadian fuel scenario will increase in comparison to its past contribution. As the need for biofuels is rising, the portion of agricultural areas used for biofuels crop production will increase resulting in the usage of arable lands which will cause food versus fuel issues.

Presently, commercial use of generic biomass and non-edible oils to produce bio-oils that can be further hydrotreated to produce jet fuels and synthetic fuels are in progress. Biomass for biofuel

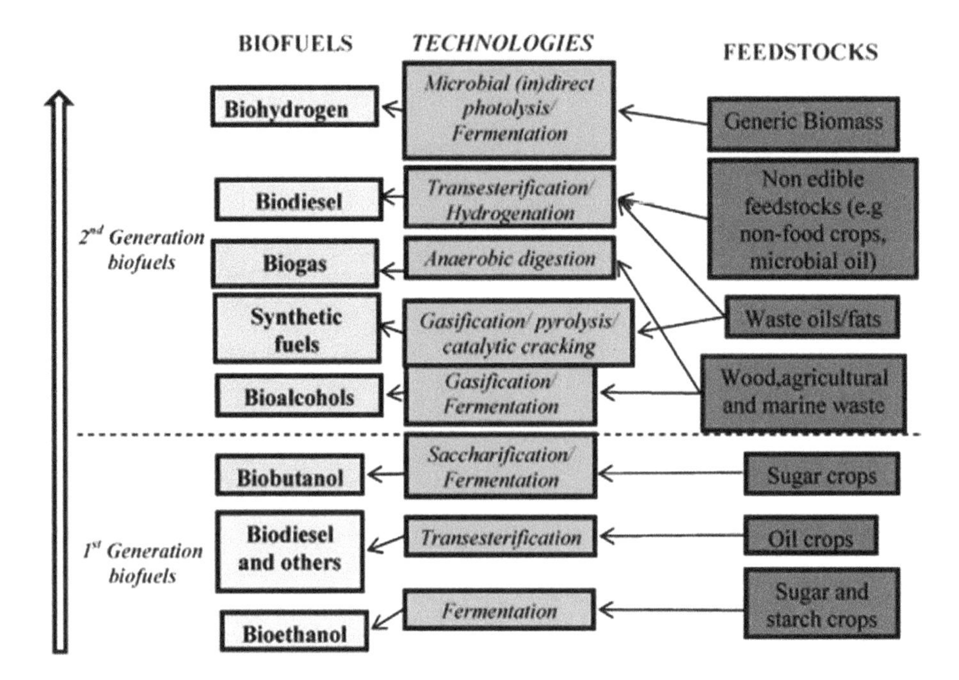

FIGURE 11.1 Roadmap of biofuel production technologies from different feedstocks. (Luque, R. et al., *Energy Environ. Sci.*, 1, 542–564, 2008. Reproduced by permission of The Royal Society of Chemistry.)

production can be broadly classified as herbaceous, woody, energy crops and green-wastes (Nanda et al. 2013, 2014). It is important to find commercial values for co-products formed during biofuels production and this is the only means to make the green fuels 'eco-friendly' on a commercial scale. Before justifying the commercialization of alternative fuels to meet present energy demands, justifications based on their environmental impacts are essential; hence, life-cycle assessment (LCA) is crucial. The environmental load of a process and its products can be evaluated and assessed using LCA from its cradle to grave. Various environmental aspects such as land usage, ozone layer depletion, acidification, eutrophication, and endemic air pollution are included in the LCA and it is a method of comparing biofuels and fossil fuels based on their energy efficiency, impact on the environment, and economic parameters. LCA of biofuel industries are generally data intensive and various LCA models have been applied to study the biofuel sector. Most often, LCA analyses are performed based on the standard norms provided by ISO 14040-43.

Lignocellulosic biomass such as pinewood or switchgrass is being cultivated specifically to produce biofuels and, in such cases, the biomass treatment plants are usually located close to the biomass cultivation lands. Usually, pyrolysis of biomass is carried out to produce bio-oil. Fast and slow pyrolysis of biomass to yield bio-oil requires a high reaction temperature (>300°C) and can contribute slightly to the emission of greenhouse gases (GHG). Moreover, hydrotreatment of bio-oil to produce biofuels requires the use of hydrogen. The by-products (water, CO_2, and CO) may contribute to GHG emissions if not used efficiently. Biofuels produced from plants are usually transported to the end stations for final disposal to customers. Transportation of biofuels through locomotives such as trucks and railways, directly and indirectly, contributes to considerable GHG emissions.

11.2 LIFE-CYCLE ASSESSMENT METHODOLOGY

Over the years, LCA has gained importance as a powerful assessment technique to analyze the environmental burdens caused by different operations in biofuel production industries. LCA calculations are based on the "well-to-wheels" approach incorporating all the emissions from the procurement of feedstocks to the end usage of biofuels (Figure 11.2).

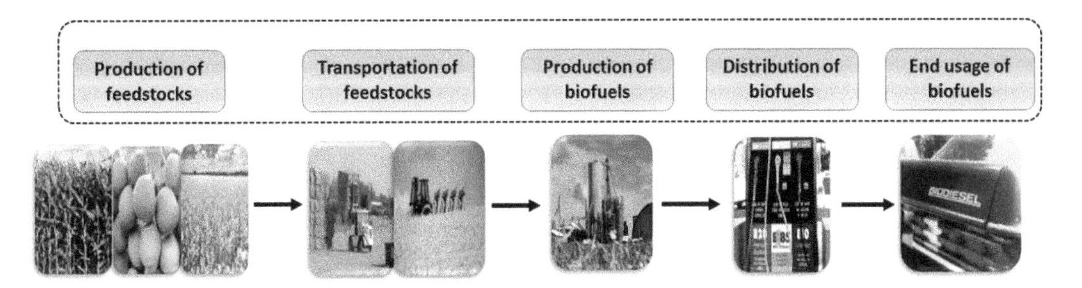

FIGURE 11.2 Cradle to grave analysis of biofuels production process.

The LCA quantitatively predicts the influence of a process on the environment under different factors such as global warming, the rate of acidification, and eutrophication. Some major activities involved in LCA are:

1. Definition of the goal and identification of the scope of the assessment
2. Collection of inventory data on materials and energy flows, emissions, and release of waste in the process
3. Testing the performance of LCA
4. Interpretation of the analyzed data and development of decisions based on the life-cycle, sensitivity, and uncertainty analysis

The common factors that are analyzed during LCA are the impact on climate, emission of pollutants and their impact on the atmosphere, water resources, land usage, human health, and ecology. The Energy Policy Act of 2005 put forth renewable fuel standard (RFS) program to set biofuels volume requirements. In 2007, the program was later revised and expanded under the Energy Independence and Security Act (EISA). Understanding the sustainability of newly developed process is of great concern for commercialization of novel laboratory-scale processes (Figure 11.3).

Sustainability studies deal with the assessment of economic feasibility, environmental impact, potential risks, and the advantages of the green chemistry process. For qualitative and quantitative LCA, many packages are available such as e-factor2, Ecoscale4, GME3, ProSuite5, BASF eco-efficiency6, and sustainability consortium Open IO7. Most of these packages require a considerable amount of data input for accurate analysis and directly affect time, resource management, and cost investment.

Major tasks and challenges for LCA of biofuels are (Kauffman et al. 2011):

1. Updated information on farming of feedstock by farmers, availability of different feedstock, and land usage
2. Prediction of practices and technologies involved in biofuel production
3. Characterization of tailpipe emissions and understanding their health consequences
4. LCA involving time as an independent and significant factor
5. Assessment of transitions and end states in the process
6. Accounting uncertainty and variability during analysis

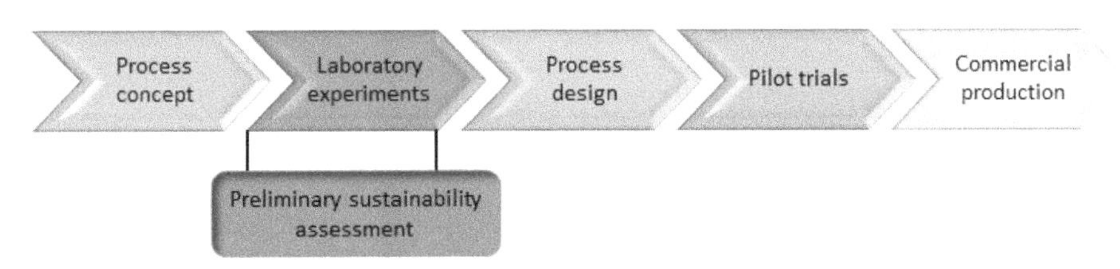

FIGURE 11.3 Development of process and methodology application. (Patel, A. D. et al., *Energy Environ. Sci.*, 5, 8430–8444, 2012. Reproduced by permission of The Royal Society of Chemistry.)

Defining the system boundaries is crucial for defining the objective of LCA and assessments performed by attributional or consequential methodologies. In attributional methodology, supply-chain interactions are used to correlate the production rate of a product and the GHG emissions associated with the process. The consequential methodology illustrates the change in GHG emissions with respect to the change in demand for a product and it indicates the direct and indirect effects within a system boundary. During LCA, optimization of production factors and maximization of profit and availability of resources are important. Production factors include goods related to land, labor, and capital that are used for production. Land includes flora, fauna, water, air, soil, and minerals. Labour includes manpower—the number of people available as labours in a province or country. Capital includes constructions, trucks, machinery, and equipment for the process.

11.3 LIFE-CYCLE ASSESSMENT OF BIOMASS-BASED BIOFUELS

Table 11.1 illustrates the major categories analyzed during the LCA and the source that contributes to the factors (Luque et al. 2008). During "cradle to grave" analysis of biofuels production processes, it is evident that the process has a considerable environmental impact in all these categories. During cultivation of biomass, usage of fertilizers, chemicals, and solvents contributes to ozone layer depletion, global warming, acidification, ecotoxicity, and eutrophication.

Rigorous steps were taken by the United States such as the U.S. Energy Policy Act 2005 to curb the GHG emission and promote commercialization of processes for alternative fuels production. The study by Searchinger et al. (2008) indicates that the promotion of first-generation fuels such as corn to ethanol can cause increase in the GHG emissions by two-folds in a span of 30 years rather that reducing them if the land is used for the cultivation of corn plants after the clearance of forests and grasslands. Clearance of grasslands for the growth of these agricultural crops causes a shift in carbon balance and, in fact, increases carbon emissions. Liang et al. (2013) studied the economic and environmental impact of biofuel production processes using different feedstocks such as soybean, jatropha, castor oil from vegetables, algae, waste extraction, and waste cooking oil. In North America, the most promising feedstocks are corn, soybean, and canola oil (Azargohar et al. 2013; 2018). However, the debate over food versus fuel on the choice of feedstocks is well

TABLE 11.1
Major Sources of Environmental Burdens for Each Impact Category

Impact Category	Major Sources of Environmental Burdens
Ozone layer depletion	Production of soil disinfectant (sweet potato) Mining of fossil fuels for fertilizer production (other crops)
Global warming	N_2O from soil
Acidification	Ammonia from compost
Human toxicity (cancer)	Arsenic emissions from fertilizer production and the operation of machinery
Human toxicity (chronic diseases)	Mercury emissions from fertilizer production (all crops examined) and benzene emissions from the operation of machinery (sweet potato)
Terrestrial ecotoxicity	Fertilizer production-related pyrethroid discharge
Eutrophication	Ammonia from compost
Energy resources	Fossil fuels for the production of fertilizer and fuels (all crops examined), as well as plastic (sweet potato)

Source: Luque, R. et al., *Energy Environ. Sci.*, 1, 542–564, 2008.

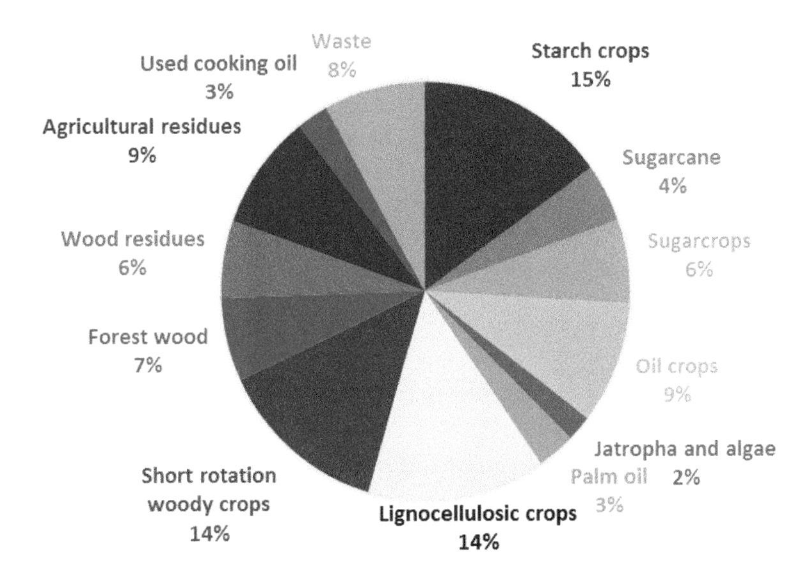

FIGURE 11.4 Choice of biomass raw materials for life-cycle assessment. (From Cherubini, F. and Strømman, A. H., *Bioresour. Technol.*, 102, 437–445, 2011.)

known. In Canada, green seed canola oil, the non-edible form of canola oil is gaining attention as a promising feedstock for biofuel production. Figure 11.4 illustrates the choice of biomass raw materials for bioenergy production (Cherubini and Strømman 2011).

11.4 INFLUENCE OF FERTILIZER USAGE ON ENVIRONMENTAL CREDIBILITY

Though biofuels are environmentally friendly, production and transportation of biomass (feedstock) requires fertilizers, mechanical equipment, and motors that presently use fossil fuels such as kerosene and gasoline. Fundamentally, they have deleterious effects on the environmental credibility of biofuels but there are certain ways to increase the credibility. Genetic modification of crops to produce seeds with a higher yield of fatty acids or increasing the photosynthetic ability and nitrogen-absorbing potential of plants to increase the overall yield of seeds are some techniques that can be of interest to the present generation researchers. It is highly recommended to promote the production of biofuels from waste biomass and crops grown on croplands that are considered waste or non-productive in terms of soil nutrients and mineral contents. In this way, it is easier to minimize the carbon debt (Fargione et al. 2008).

11.5 IMPACT OF BY-PRODUCTS AND CO-PRODUCTS ON ENVIRONMENTAL CREDIBILITY

To date, biomasses are being transported using railways and trucks which predominantly run on fossil fuel resources. Research on the development of a commercial-scale mobile biomass processing unit can aid in increasing the overall environmental credibility of the process. Co-processing or altering the present hydrotreating unit to accommodate biomass feedstocks will contribute positively to the overall environmental impact of these futuristic biofuel production processes. Production of bioethanol from corn leads to two major environmental concerns. Being an edible feedstock, the usage of corn is not recommended because it is more important to meet the dietary needs of the people than to fuel the energy needs of the future. Moreover, blending bioethanol in excess with gasoline (>15% by volume) can cause severe engine damage (Lin et al. 2013).

For assessing the sustainability of biofuels, LCA is the most standard framework and it was concluded that the production of sunflower seeds has more environmental impacts than the production of rapeseed or soybean seed. This fact is attributed to the large use of land by sunflower compared to rapeseed and soybean plants. Larger land use results in increased usage of insecticides and pesticides. Canola oil is predominantly produced in the prairie provinces of Canada, especially Saskatchewan and Alberta. However, in the United States, corn and sunflower oils may be potential feedstocks. However, these oils lead to food versus fuel issues. First-generation biofuels tend to have lesser environmental impacts than the fossil fuels. However, sustainability of first and second-generation fuels can be a major challenge in the commercialization of these fuels.

11.6 LIFE-CYCLE ASSESSMENT OF ALGAE-BASED BIOFUELS

Recently, algae-derived biofuels have been gaining attention as they can help to overcome the problems associated with the usage of first and second-generation biofuels. Third-generation fuels involve feedstocks such as algae, lignocellulosic biomass, and waste vegetable oil as feedstocks to produce alternative fuels. The usage of microalgae as a feedstock to produce biodiesel, bioethanol, and biohydrogen on a commercial scale is already in progress. From the study by Singh and Olsen (2011), it was concluded that the usage of algal biomass to produce biodiesel, bioethanol, and biohydrogen is a commercially viable process if process efficiency, operational cost, and the reuse of lipid mass are optimized. Alvarado-Morales et al. (2013) performed LCA to compare the production of biofuels from brown seaweed based on two cases. In Scenario 1, brown seaweed (*Laminaria digitata*) was used to produce bio-gas. In contrast, in Scenario 2, the same algal species was used to produce bio-gas and bioethanol. The environmental impacts from both processes were compared. The two scenarios were compared on environmental impact factors such as acidification, global warming, and terrestrial eutrophication. It was concluded that the Scenario 1 was more energy efficient than Scenario 2 because of the high cost and energy intensive bioethanol separation process associated with Scenario 2. However, cost and profits involved with the commercial use of bioethanol should be incorporated to completely prove that Scenario 1 is more energy efficient than Scenario 2.

Some studies have shown that algae-derived biofuels have negative net economic benefits and energy yields. However, technological improvements are essential to make the net energy yield positive (Clarens et al. 2010; Stephenson et al. 2010). It should be noted that algae-based biofuels require fewer freshwater resources and hence can have a major positive impact on environment. Hence, in the long-term, promotion of algae-based biofuels is highly recommended. Microalgae or seaweed has shown higher biomass yield, photosynthetic efficient, solar energy storage, and nutrient assimilation. In an algae-based biofuel production process, extraction of oil is usually carried out using three processes: dry extraction, wet extraction, and secretion (Figure 11.5). It is well established that the secretion methodology is economical in laboratory-scale analysis. Based on the advantages and disadvantages of the secretion method in small-scale studies, the pilot-plant experimentation and demonstration-scale studies are still in progress. The study by Woertz et al. (2014) focused on the LCA of GHG emissions from microalgae biodiesel using the CA-GREET model. In comparison to fossil fuels, the GHG emissions using algal biofuel system was 70% lower and met the requirements of the EPA RFS2. It also conformed to the regulation of European Union Renewable Energy Directive. Large-scale production of algae for biofuel production is economically challenging and it is feasible only based on certain key assumptions related to its sustainable growth.

Globally, research on genetic modification of algae to increase their biomass production rate and oil content is in progress and this is a promising area for bioenergy researchers. Lardon et al. (2009 investigated he production of algae-based biodiesel using *Chlorella vulgaris* as the microalgae species. It is very evident that *Chlorella* is sensitive to the nitrogen content in its growing condition. In a nitrogen-deprived environment, *Chlorella* tends to accumulate lipids and carbohydrates, which in turn affects the yield of oil (Table 11.2). Wet and dry methods of extraction were compared for their environmental impact and, finally, the extracted oil was processed in a transesterification unit to produce biodiesel. It is documented that Eustigmatophyte *Nannochloropsis* could be a promising feedstock for production of sufficient lipids.

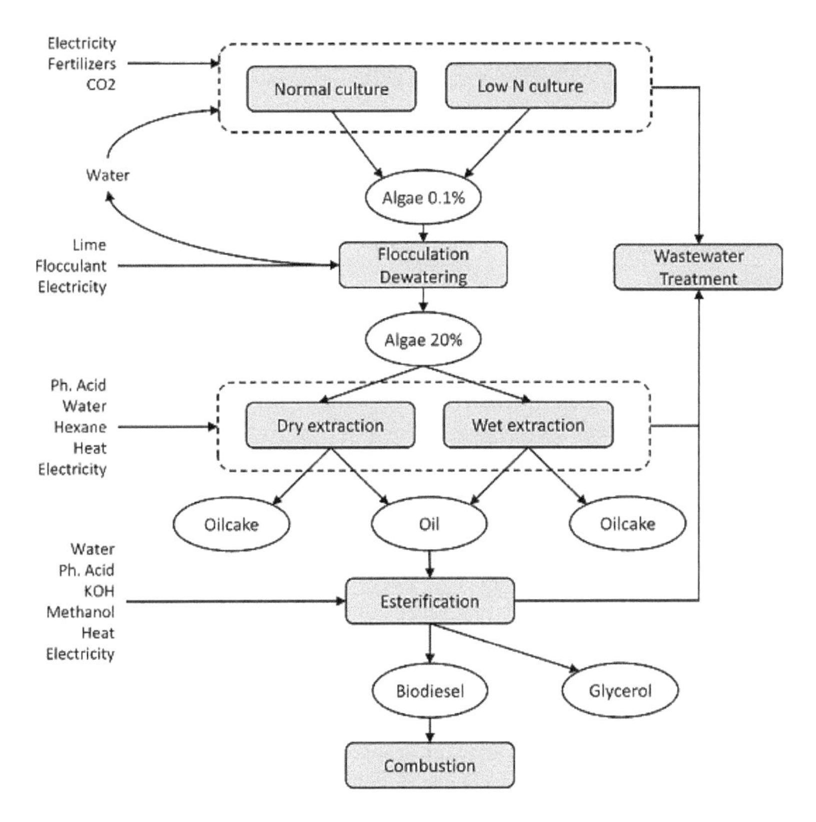

FIGURE 11.5 Process chain overview for biofuel production from algae. (Reprinted with permission from Lardon, L. et al., *Environ. Sci. Technol.*, 43, 6475–6481, 2009. Copyright 2009 American Chemical Society.)

TABLE 11.2
Composition and Culture Parameters of *Chlorella vulgaris*

Parameter	Normal	Low Nitrogen
Protein (g/kg)	282	67
Lipid (g/kg)	175	385
Carbohydrates (g/kg)	495	529
Lower heating value (MJ/kg)	17.5	22.6
C (g/kg)	480	538
N (g/kg)	46	10.9
P (g/kg)	9.9	2.4
K (g/kg)	8.2	2
Mg (g/kg)	3.8	0.9
S (g/kg)	2.2	0.5
CO_2 (kg/kg)	1.8	2
Growth rate (per day)	0.99	0.77
Productivity (g/m^2/day)	24.75	19.25

Source: Reproduced with permission from Lardon, L. et al., *Environ. Sci. Technol.*, 43, 6475–6481, 2009. Copyright 2009 American Chemical Society.

In most of the biofuel industries, considerable amounts of by-products are produced. For example, soy oil is obtained from soybeans and soy meal is obtained as by-product. LCA involving these by-products is essential, as the prime objective is to optimize energy consumption in an industry and minimize the GHG emissions. Displacement, substitution, or system boundary expansions method recommended by ISO 14040 can be used for the identification and quantification of energy from by-products obtained in a biofuel industry.

11.7 CONCLUSIONS

The production of biofuels from algae needs considerable attention from the research community before it can be commercialized. The use of algae for biofuels production is reported to be energy intensive in comparison to the other feedstocks. The literature review proved that techno-economic and the well-to-wheel analysis of biofuel production process, especially hydrotreament and micro-algae biofuel processes are very crucial for the commercialization of futuristic fuel processing. The shift from the first-generation to third-generation feedstocks looks positive but exergy analysis is very crucial to justify the fact that biofuels are actually "beneficial bio-based fuels."

It is evident that commercialization of a new era of bioenergy processes is challenging and there are considerable levels of risk. For example, use of biofuels may require engine alteration, increased usage of additives during processes, and pose challenges in finding alternate uses for the by-products. Conclusively, GHG savings in a biofuel industry can be enhanced if basic inputs such as hydrogen and electricity are produced from bio-based resources instead of fossil fuel resources. Artificial or anthropogenic systems such as microalgae production require the addition of insecticides, pesticides, water, and considerable energy. However, the residual natural resources such as forest and agricultural biomass require less usage of insecticides and pesticides. The production of biofuels from algal biomass is gaining importance in this decade and hence it is too early to comment on the optimized production route for the synthesis of biofuels from algal biomass on a commercial scale.

ACKNOWLEDGMENTS

The financial support from BioFuelNet Canada is greatly acknowledged.

REFERENCES

Alvarado-Morales, M., A. Boldrin, D. B. Karakashev, S. L. Holdt, I. Angelidaki, and T. Astrup. 2013. Life cycle assessment of biofuel production from brown seaweed in Nordic conditions. *Bioresource Technology* 129:92–99.

Azargohar, R., S. Nanda, B. V. S. K. Rao, and A. K. Dalai. 2013. Slow pyrolysis of deoiled Canola meal: Product yields and characterization. *Energy and Fuels* 27:5268–5279.

Azargohar, R., S. Nanda, K. Kang, T. Bond, C. Karunakaran, A. K. Dalai, and J. A. Kozinski. 2019. Effects of bio-additives on the physicochemical properties and mechanical behavior of canola hull fuel pellets. *Renewable Energy* 132:296–307.

Cherubini, F., and A. H. Strømman. 2011. Life cycle assessment of bioenergy systems: State of the art and future challenges. *Bioresource Technology* 102:437–445.

Clarens, A. F., E. P. Resurreccion, M. A. White, and L. M. Colosi. 2010. Environmental life cycle comparison of algae to other bioenergy feedstocks. *Environmental Science & Technology* 44:1813–1819.

Fargione, J., J. Hill, D. Tilman, F. Polasky, and P. Hawthorne. 2008. Land clearing and the biofuel carbon debt. *Science* 319:1235–1238.

Kauffman, N., D. Hayes, and R. Brown. 2011. A life cycle assessment of advanced biofuel production from a hectare of corn. *Fuel* 90:3306–3314.

Lardon, L., A. Helias, B. Sialve, J. P. Steyer, and O. Bernard. 2009. Life-cycle assessment of biodiesel production from microalgae. *Environmental Science & Technology* 43:6475–6481.

Liang, S., M. Xu, and T. Zhang. 2013. Life cycle assessment of biodiesel production in China. *Bioresource Technology* 129:72–77.

Lin, C. S. K., L. Pfaltzgraff, L. Herrero-Davila, E. B. Mubofu, S. Abderrahim, J. H. Clark, A. Koutinas et al. 2013. Food waste as a valuable resource for the production of chemicals, materials and fuels. Current situation and global perspective. *Energy and Environmental Science* 6:426–464.

Luque, R., J. C. Lovett, B. Datta, J. Clancy, J. M. Campelo, and A. A. Romero. 2010. Biodiesel as feasible petrol fuel replacement: A multidisciplinary overview. *Energy and Environmental Science* 3:1706–1721.

Luque, R., L. Herrero-Davila, J. M. Campelo, J. H. Clark, J. M. Hidalgo, D. Luna, J. M. Marinas, and A. A. Romero. 2008. Biofuels: A technological perspective. *Energy and Environmental Science* 1:542–564.

Nanda, S., J. Mohammad J., Reddy S.N., Kozinski J.A., and Dalai A.K. 2014. Pathways of lignocellulosic biomass conversion to renewable fuels. *Biomass Conversion and Biorefinery* 4; 157–191.

Nanda, S., P. Mohanty, K. K. Pant, S. Naik, J. A. Kozinski, and A. K. Dalai. 2013. Characterization of North American lignocellulosic biomass and biochars in terms of their candidacy for alternate renewable fuels. *Bioenergy Research* 6:663–677.

Nanda, S., R. Azargohar, A. K. Dalai, and J. A. Kozinski. 2015. An assessment on the sustainability of lignocellulosic biomass for biorefining. *Renewable and Sustainable Energy Reviews* 50:925–941.

Patel, A. D., K. Meesters, H. Uil, E. Jong, K. Blok, and M. K. Patel. 2012. Sustainability assessment of novel chemical processes at early stage: Application to biobased processes. *Energy and Environmental Science* 5:8430–8444.

Requena, J. F. S., C. Guimaraes, S. Q. Alpera, E. R. Gangas, S. Hernandez-Navarro, L. M. N. Gracia, J. Martin-Gil, and H. F. Cuesta. 2011. Life Cycle Assessment (LCA) of the biofuel production process from sunflower oil, rapeseed oil and soybean oil. *Fuel Processing Technology* 92:190–199.

Searchinger, T., R. Heimlich, R. A. Houghton, F. Dong, A. Elobeid, J. Fabiosa, S. Tokgoz, D. Hayes, and T. Tu. 2008. Use of U.S. croplands for biofuels increases greenhouse gases through emissions from land-use change. *Science* 319:1238–1240.

Singh, A., and S. I. Olsen. 2011. A critical review of biochemical conversion, sustainability and life cycle assessment of algal biofuels. *Applied Energy* 88:3548–3555.

Stephenson, A. L., E. Kazamia, J. S. Dennis, C. J. Howe, S. A. Scott, and A. G. Smith. 2010. Life-cycle assessment of potential algal biodiesel production in the United Kingdom: A comparison of raceways and air-lift tubular bioreactors. *Energy and Fuels* 24:4062–4077.

Woertz, I. C., J. R. Benemann, N. Du, S. Unnasch, D. Mendola, B. G. Mitchell, and T. J. Lundquist. 2014. Life cycle GHG emissions from microalgal biodiesel–A CA-GREET Model. *Environmental Science & Technology* 48:6060–6068.

Zhang, T., X. Xie, and Z. Huang. 2014. Life cycle water footprints of nonfood biomass fuels in China. *Environmental Science & Technology* 48:4137–4144.

12 Synthetic Crude Processing

Impacts of Fine Particles on Hydrotreating of Bitumen-Derived Gas Oil

Rachita Rana, Sonil Nanda, Ajay K. Dalai, Janusz A. Kozinski, and John Adjaye

CONTENTS

12.1 Introduction 187
12.2 Upgrading of Canadian Athabasca Bitumen 190
12.3 Hydrotreating Process 192
12.4 Hydrotreating Catalyst and Catalyst Fouling 195
12.5 Organic Coated Solids in the Bitumen Feed 196
12.5.1 Fines 197
12.5.2 Asphaltenes 198
12.6 Mechanism of Fine Particle Deposition 199
12.7 Impacts of Hydrotreating Reaction Conditions on Particle Deposition 200
12.8 Industrial Challenges 201
12.9 Conclusions 203
Acknowledgments 203
References 203

12.1 INTRODUCTION

The U.S. Energy Information Administration reports that the demand for oil will globally escalate from 95 million barrels per day in 2016 to 105 million barrels per day by 2030 (USEIA 2016). The evident depletion rate of fossil fuel resources is not expected to meet this growing energy demand, which reinforces the relevance of exploring the unconventional crude oil resources (Vosoughi et al. 2016). Canada has the second largest oil reserve in the world, which primarily includes the oil sands in Athabasca basins that cover the area of northeastern Alberta and Saskatchewan (Dunbar 2009). Canada has stockpiled more than 79.8 million tons of petroleum coke in open-pit oil sands mines and tailing ponds in Alberta because of the prospect of coking of bitumen to release oil (Rana et al. 2018a). Figure 12.1 shows the top ten countries with highest crude oil reserves.

The first crudely extracted product from oil sands is called bitumen. Bitumen is a highly viscous mixture of hydrocarbons, silica sand, and clay minerals, which are extracted directly from the oil sands (Rana et al. 2018b). The ultra-fine clays and organic components contribute to the fine particles that remain entrained in the bituminous product streams. These product streams are further upgraded in an atmospheric distillation unit where virgin light gas oil and topped bitumen are obtained. The topped bitumen is further upgraded in the secondary upgrading units that include vacuum distillation, cokers, and hydrocrackers. All these units yield light gas oil (LGO) and heavy gas oil (HGO), which inherently entrain fine particles (<44 μm) (Iliuta et al. 2003a). The gas oils

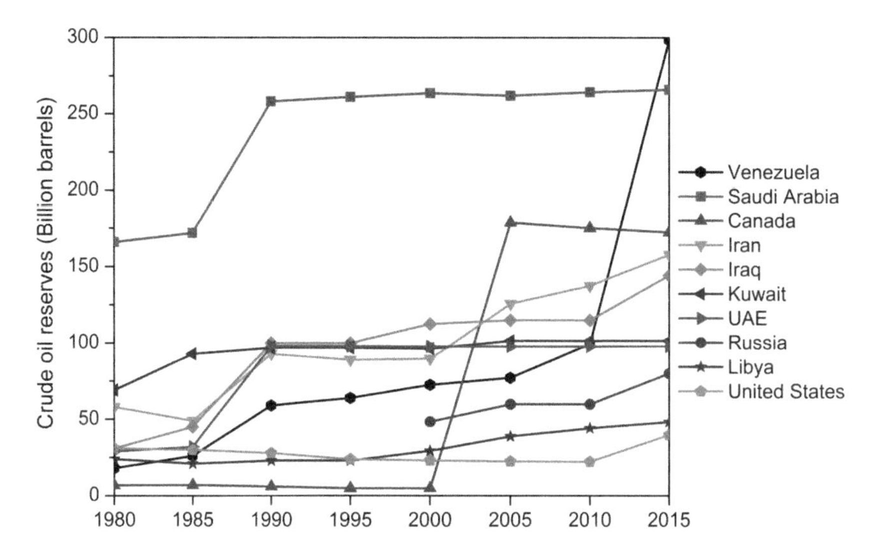

FIGURE 12.1 Top ten countries with highest reserves of crude oil. (Data source: USEIA 2017.)

are further refined in the hydrotreating unit at high temperature (355°C–395°C) and pressure (1000–1400 psig) to remove sulfur and nitrogen as well as to saturate the aromatics to produce synthetic crude oil. This process helps in making the crude oil compatible with the stringent environmental regulations (Jones and Pujado 2006).

When bitumen-derived gas oils are hydrotreated, the entrained fines (clay particles that surface-adsorb organic coating) that are less than 20 μm pass through the guard-bed filters and enter the hydrotreating unit with the feed stream. Out of the several challenges faced during the synthesis of synthetic crude, the entrainment of these fines that deposit over the catalyst bed and cause a sudden pressure-drop in the reactor is a significant one (Iliuta and Larachi 2005). When the reactor pressure drops due to catalyst fouling, the ripened bed is replaced, which leads to numerous operational problems and economical losses to the refining industry. Hence, it is significant to strategically study and analyze the possible measures to address this problem.

The nature of fine particles strongly impacts their interaction with the catalyst (Wang et al. 1999). Fine particles in the bitumen-derived gas oils are found to be like the asphaltene-coated kaolin. Kaolin (alumino silicate) is like naturally occurring clay and asphaltenes are the organic particles that are adsorbed on clay surface which form fine particles. Figure 12.2a and b illustrate the chemical structures of kaolin and asphaltene, respectively. At the reaction conditions, the organic coating (asphaltene) of the clay particle (kaolin) desorb, which causes the clay particles to drop on the catalyst bed leading to their substantial accumulation (Wang et al. 1999). Furthermore, the hydrotreating products such as ammonia, hydrogen sulfide, and water can also contribute to the deposition of fine particles on the catalyst bed. However, the water released during hydrodeoxygenation (HDO) displaces asphaltene molecules from the surface of asphaltene-coated kaolin. This process exposes the solid surface to more hydrophilic areas, thus increasing the tendency of fine particles to deposit on the catalyst bed (Wang et al. 2001).

The fine particle deposition also is believed to be significantly affected by the process conditions within the reactor such as temperature, pressure, and reaction time. In addition, the variation in particle loading of the feed and its particle size can be of interest in determining the impact of fine particle deposition. The fines deposition can lead to significant pressure drops in the reactor, thus influencing the operations. The incidence of particles agglomeration was evaluated based on the increase in pressure drop as a function of time by Iliuta et al. (2003a, 2003b). Several models have been derived for the pressure build-up and dynamics of the fine particles in the packed column (Gray et al. 2002; Iliuta et al. 2003b). Fine particle deposition requires its migration to the catalyst surface followed by catalyst-fines interaction that leads to their attachment to the catalyst

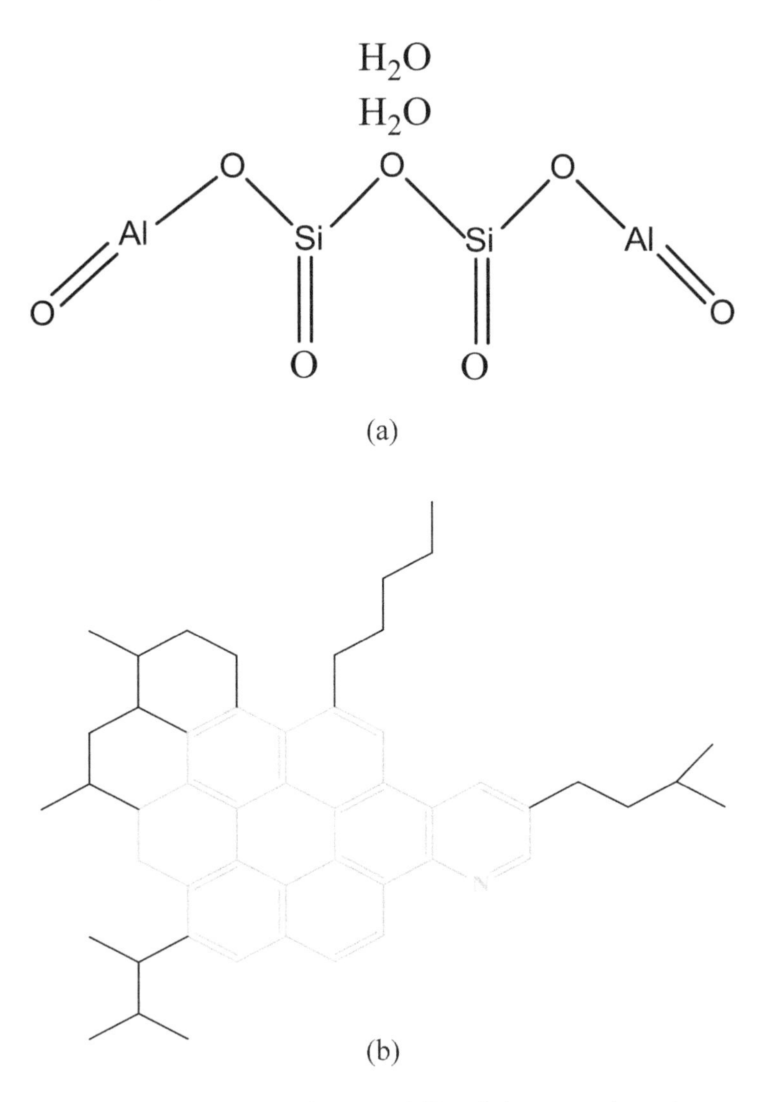

FIGURE 12.2 Chemical structure of (a) kaolin clay and (b) asphaltene organic coating.

(Wang et al. 1999). Although the mechanism is still unclear, it is believed that the process conditions can be optimized to hinder the interaction between the catalyst and the fine particles.

Today, the escalating energy demand and competitive oil market place emphasis on supplying fuel at the most competent prices. North America lacks sweet crude, unlike the Middle East; hence, to stand as a potential oil supplier the North America must develop resources and techniques to curb the cost of oil sand extraction, production, and processing. The classical approach to address these issues is to develop catalysts with high activity and selectivity as proposed by several authors (Badoga et al. 2014; Ferdous et al. 2004, 2007). The best hydrotreating conditions for various oil streams also have been established as a tactical solution (Botchwey et al. 2004). However, issues related to catalyst fouling still have significant room for improvement. This chapter provides a detailed insight into the process of fine particle deposition on the catalyst surface from the initial stage of bitumen upgrading to the final stage of hydrotreating the process streams. The purpose of this chapter is to understand the origin, behavior, and nature of these fine particles at various stages of hydrotreating and finally understanding the impact of hydrotreating on fines deposition.

12.2 UPGRADING OF CANADIAN ATHABASCA BITUMEN

Further to the mining of oil sands, bitumen extraction and upgrading are the two major steps in oil refining. The process of bitumen extraction and crude oil hydroprocessing is outlined in Figure 12.3. Bitumen is a black, viscous, sticky mixture of organic liquids (hydrocarbons) that are composed of highly condensed chemical compounds. It is extracted as a by-product or residue of the fractional distillation of crude oil (Banerkke 2012; Jones and Pujado 2006). It can be considered a form of petroleum with high boiling point and viscosity (Zhao et al. 2002). A few important chemical properties of bitumen and its subsequent HGO and LGO fractions are given in Table 12.1.

Liu et al. (2005) studied the processing of oil sand ores in Alberta and stated that the physical and chemical properties of the ore dictate the extent of bitumen liberation at the same extent as fine particle deposition. In their study, poor processing ores showed surface properties such as zeta potential along with higher induction time for bitumen and air bubble, as well as colloidal properties between bitumen and silica. The slime coating (adsorption of fines to bitumen) was observed to be the main reason for poor processing of the high fine ores. Table 12.2 shows the elemental composition of bitumen and bitumen-derived gas oil.

Zhao et al. (2002) compared the Canadian Athabasca bitumen with conventional and heavy crudes. In their experiment, they prepared the narrow-cut fractions of bitumen pitch by the technique of supercritical fluid extraction with pentane. The temperature was less than the thermal process. Along with Canadian Athabasca bitumen, Saudi Arabian light crude oil, Venezuelan heavy oil, and Chinese Daquing conventional crude were also investigated. From several characterization studies, it was reported that the end-cuts from Athabasca bitumen contained more solids (7 wt/wt%) than the other crudes. Nanosized alumino silicate clay particles reportedly found in Athabasca bitumen were non-uniformly covered with polar and aromatic organic matter insoluble in toluene (Zhao et al. 2002).

Selucky et al. (1977) reported the results of a detailed study of maltenes, the de-asphalted bitumen from Fort McMurray, Alberta. Bitumen was extracted from the sand using Soxhlet extraction with benzene as the extracting solvent. Asphaltene was precipitated using pentane in a centrifuge at 2800 rpm after 12 hours under nitrogen atmosphere. The de-asphalted bitumen was set for column

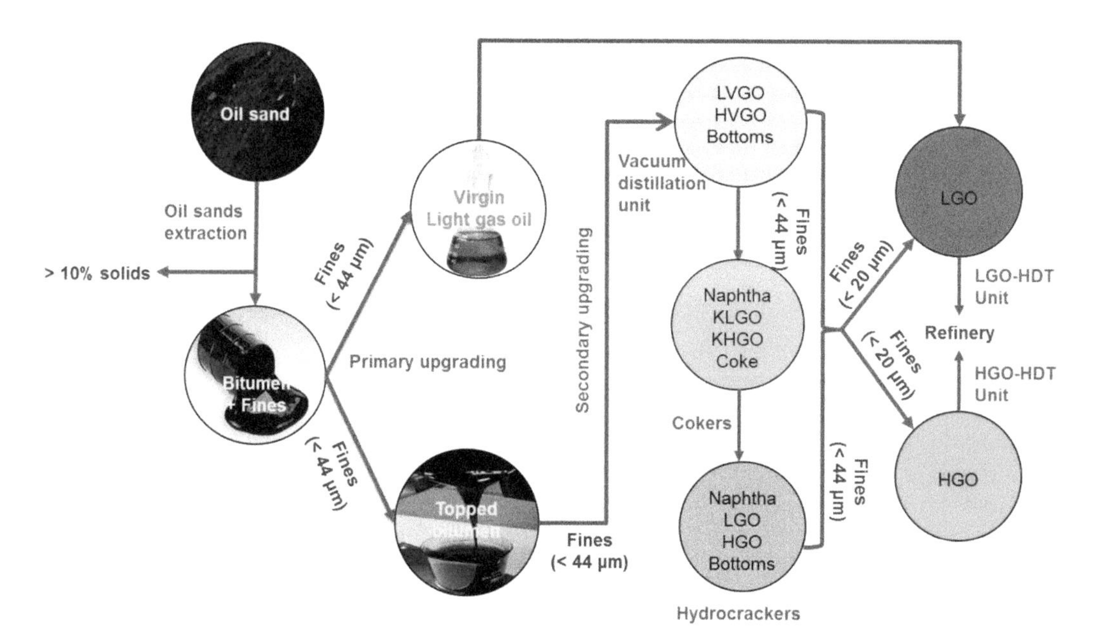

FIGURE 12.3 Flow diagram for bitumen-derived gas oil upgrading. *Abbreviations*: light vacuum gas oil, LVGO; heavy vacuum gas oil, HVGO; coker light gas oil, KLGO; coker heavy gas oil, KHGO; light gas oil, LGO; heavy gas oil, HGO; hydrotreating, HDT.

TABLE 12.1
Physical Properties of Heavy Oils and Bitumen

Fraction	Viscosity (mPa.s)	Density (g/cm^3)	API Gravity (°)	Boiling Point (°C)	Nitrogen (ppm)	Sulfur (ppm)
Bitumen	>10^5	>1.0	<10	525–675	4222	10^6
Heavy gas oil	10^2–10^5	0.9–1.0	10–20	350–650	3000	40370
Light gas oil	1–2	0.9	27	200–450	1500	28500

Source: Gray, M.R., *Upgrading Petroleum Residues and Heavy Oils*, Marcel Dekker, New York, 1994; Jechura, J., Refinery feedstocks and products-properties and specifications, Colorado School of Mines, https://inside.mines.edu/~jjechura/Refining/02_Feedstocks_&_Products.pdf, (accessed May 20, 2017), 2016; The Engineering ToolBox, Classification of Gas Oil, http://www.engineeringtoolbox.com/classification-gas-oil-d_165.html, (accessed March 26, 2017), 2017.

TABLE 12.2
Elemental Composition of Athabasca Bitumen and Bitumen-Derived Gas Oil

Element	Athabasca Bitumen	Gas Oil
Carbon	82–83 wt%	83–87 wt%
Hydrogen	10.1–10.2 wt%	10–14 wt%
Nitrogen	3000–5000 ppm	3000–1500 ppm
Sulfur	4–6 wt%	0.05–6 wt%
Oxygen	<1.0 wt%	0.05–1.1 wt%
Vanadium	180–250 ppm	10–170 ppm
Nickel	60–90 ppm	10–30 ppm

Source: Banerkke, L., *Oil Sands, Heavy oil and Bitumen: From Recovery to Refinery*, 1st ed, Pennwell Corporation, Tulsa, Oklahoma, 2012; Ferdous, D., Surface morphology of nickel-molybdenum catalyst supported on gamma-aluminum oxide: Impact on hydroprocessing of heavy gas oil derived from Athabasca bitumen – Thesis, University of Saskatchewan, Canada, 2003.

chromatography on silica, after which alumina and silver nitrate were used for the column chromatography of pentane eluate from silica. The straight chains of paraffins were separated on molecular sieves non-adduct. The extracted oil sands yielded 16.6% asphaltene and the de-asphalted oil content was 83.4%.

The chemical composition of Athabasca bitumen has been reported by numerous authors (Luo and Gu 2007; Rajagopalan 2010; Sparks et al. 2003; Strausz et al. 2011). Several studies have been conducted to understand the chemical composition of distillable aromatic fractions. An extensive fractionation of the bitumen was carried out by Strausz et al. (2011), which involved the extractions, precipitation, molecular distillation, complexation, adsorption, and adduction chromatography to separate asphaltene from maltene. Chromatographic method makes the isolation of an aromatic component possible and then it can further be separated to mono-aromatic, di-aromatic, tri-aromatic, and poly-aromatic sub-fractions. The sample was found to contain 38.6% aromatics, out of which 76.4% of the aromatic fraction was distillable at 10^{-3} Torr and 240°C.

The bitumen sample from Fort McMurray, Canada (from 18 meters below the ground surface) was collected by Strausz et al. (2010), which was from Syncrude High Grade (SHG) oil sand with

12% bitumen content. Soxhlet extraction with dichloromethane as the extracting solvent was used to extract the bitumen. Strausz et al. (2010) studied the chemical composition of the saturated fraction that was sourced from the Athabasca bitumen. This bitumen yielded 17% of asphaltene of the total bitumen content. Fluorescence micro-spectrophotometry, gravimetric composition analysis, and carbon number maxima were used to determine the concentration distribution of the alkanes (mono-hexacyclic) based on carbon number. The biodegradation severity was in the following order: Grosmont > Athabasca > Peace River > Lloyddminster. The distillable fractions and non-distillable fractions were different in molecular size and chemical composition. Nuclear magnetic spectroscopy of the distillable sub-fraction indicated a low ratio of mid-chain methylene over the end chain methyl resonance. This result manifested the deficiency of long alkyl chains in the distillable saturated sub-fractions. It was suggested that, more than the distillable fraction, the non-distillable fraction contained the aromatic carbon, although in trace amounts.

A study on the extraction measurements and physical properties of Athabasca bitumen and the light hydrocarbon stream was conducted by Nourozieh et al. (2011). Liquid upgrading process was used for upgrading bitumen by using ethane as the solvent. The process proceeded in two phases: solvent-enriched phase and bitumen-enriched phase. The parameters affecting the extraction yield were pressure, temperature, and solvent-to-bitumen ratio. The reaction was carried out at 21.6°C and 725–1305 psig at four different ethane concentrations. The study supported that the extraction yield increased at higher pressures. The solvent-to-bitumen ratio variations showed that with an increased concentration of ethane more light components were extracted. The same experiment carried out with propane as the solvent gave better yields than that ethane at similar conditions. This result was owing to the nature of propane that enabled it to extract more components to solvent-enriched phase as compared to ethane.

The advances in process technologies that are applicable for upgrading of heavy oils and residues were reported by Rana et al. (2007). Two types of residues generated are atmospheric residue (AR > 343°C) and the vacuum residue (VR > 565°C). It is reported that at a lower temperature with constant solvent composition and pressure, the yield increases but the quality degrades. Gasification is an upgrading technique that causes complete cracking of residues into combustible syngas (H_2, CO, and traces of CH_4, C_2H_6, and C_{2+}). Delayed coking is another upgrading technique that leads to complete rejection of metals and carbon while partially converting the feed to liquid product (naphtha and diesel). On the other hand, fluid coking is a better technique than delayed coking with the advantage of a slightly improved liquid yield along with economic benefits. Flexicoking is an extended form of fluid coking where excess coke is converted to syngas at about 1000°C – a temperature able to burn all the coke.

Visbreaking is an established process that can be applied to atmospheric residues, vacuum residues, and solvent de-asphalter pitch to mildly improve their viscosity. It is popularly used to improve the refinery net distillate yield. The thermal process, used to produce large amounts of low-value by-products, requires extensive processing of its liquid yields. For this reason, the catalytic residue processes are more popular. The fluid catalytic cracking (FCC) converts a significant amount of the heavier fractions of oil to high-octane gasoline components. In hydroprocessing of the residues, a substantial amount of hydrogen is consumed. It has a high product selectivity and better product yield (near 85%) (Rana et al. 2007). For the final upgrading of bitumen to make it commercially ready product, it should be processed to lighter hydrocarbons (Aoyagi et al. 2003). The yield is generally termed synthetic crude oil, and it necessitates primary and secondary upgrading processes. After primary and secondary upgrading, the LGO and the HGO streams are sent to the hydrotreating units for further processing.

12.3 HYDROTREATING PROCESS

Hydrotreating is a catalytic hydrogenation process used in refining or purification of fuel and other by-products. The emphasis of hydrotreating is on improving the quality of the final product. Hydrotreating uses hydrogen at high pressure and temperature in the presence of a metal

TABLE 12.3
Key Factors Affecting Hydrotreating Activity

Feed	Catalyst	Process
Sulfur	Support	Reactor design
Nitrogen	Active metal	Catalyst bed configuration
Aromatics	Promotor	Process conditions
Entrained particles	Preparation method	Pressure drop
Boiling range	Active site mechanism	Catalyst deactivation

catalyst for the removal of unwanted constituents of the crude oil. A representation of the hydrotreating reaction is given in Equation 12.1. Table 12.3 shows the key factors affecting hydrotreating.

$$\text{Feed} \overset{H_2,\ \text{Catalyst}}{\rightarrow} H/C + H_2S + NH_3 + H_2O \tag{12.1}$$

The major objectionable components that are desired to be removed during hydrotreating are sulfur, nitrogen, olefins, and aromatics. The addition of hydrogen helps burn the olefins as clean-burning paraffins. Naphtha and other light materials are treated in the catalytic reforming units, whereas the heavy material is further treated to meet the fuel quality standards for use as commercial fuel oil. This heavy stream has a high content of unwanted sulfur, nitrogen, metals, unsaturates, and other hetero-compounds. Hydrotreating plays a significant role in upgrading the heavy stream to environmental and commercial industry standards (Gary et al. 2007; Satterfield 1996).

The removal of sulfur in the form of H_2S is termed as hydrodesulfurization (HDS). Figure 12.4 shows some of the common sulfur compounds found in the feed such as thiophenes, benzothiophenes, and dibenzothiophenes. The reaction mechanism for HDS is shown in Figure 12.5. As presented in Figure 12.5, the HDS process can follow two pathways—hydrogenation or hydrogenolysis—which leads to the removal of sulfur from the aromatic structure in the form of hydrogen sulfide. Thus, organo-sulfur compounds react with hydrogen and the polluting sulfur gas is liberated in the form of hydrogen sulphide, thus making the product comply with the clean environmental norms. It is seen that organo-nitrogen compounds tend to deactivate the hydrotreating catalyst. Hence, hydrodenitrogenation (HDN) is used to remove such unwanted hetero-compounds of nitrogen in the form of ammonia.

Figure 12.6 shows some common nitro-compounds present in oil such as quinoline, acridine, indole, and cabazole. The mechanism for HDN is also shown in Figure 12.7 wherein, hydrogen reacts with the nitrogen in the aromatic structure to release nitrogen in the form of ammonia. The other reactions that take place in a hydrotreating reactor are hydrodemetallization (HDM) for the removal of metals such as arsenic, nickel, and vanadium, as well as HDO for the removal of oxygen in the form of water (Coulier et al. 2001; Delmon 1993). The pretreatment of petroleum fractions for

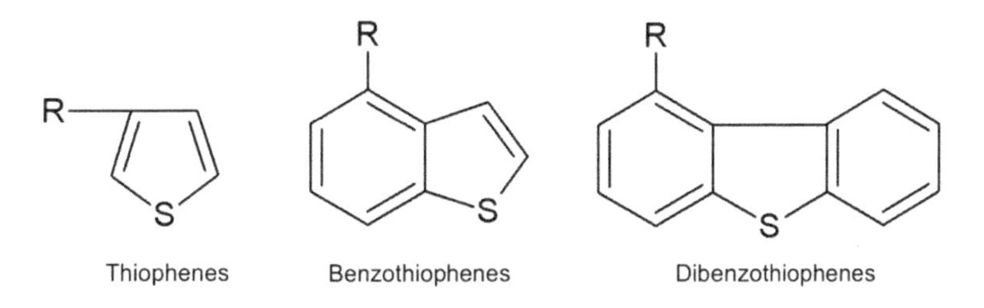

FIGURE 12.4 Structures of sulfur compounds in oil.

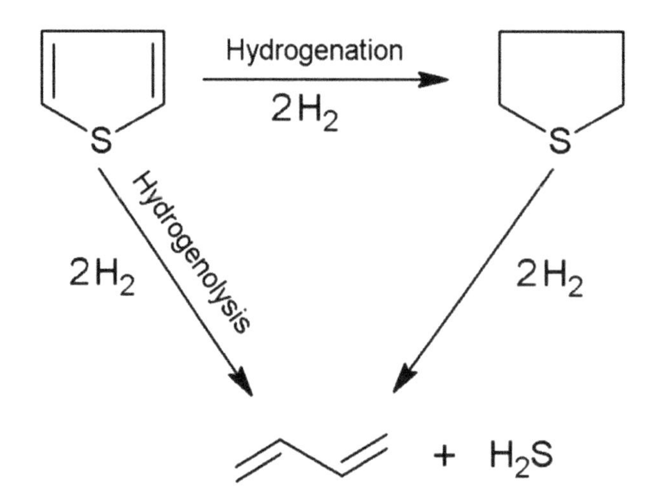

FIGURE 12.5 Hydrodesulphurization mechanisms of thiophenes.

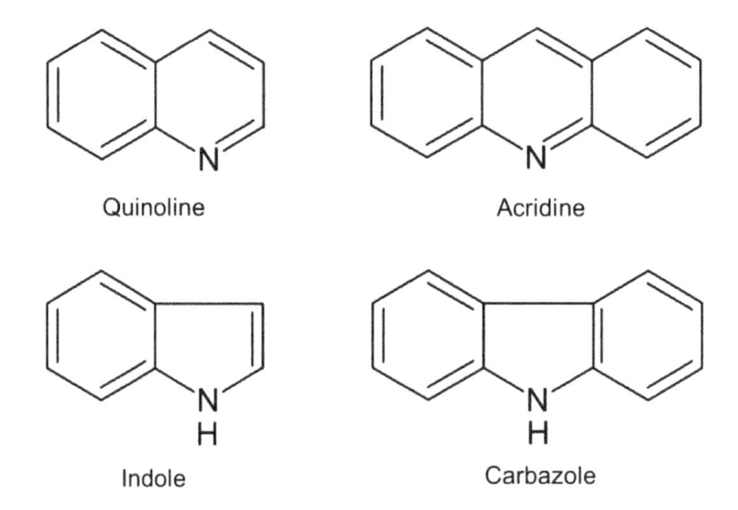

FIGURE 12.6 Structures of basic and non-basic nitrogen compounds in oil.

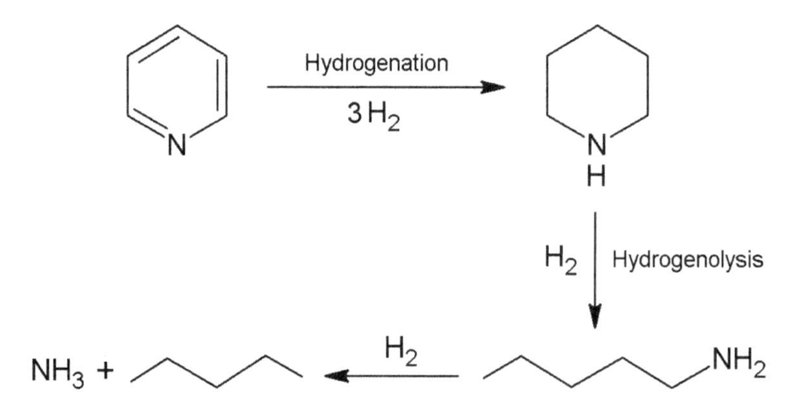

FIGURE 12.7 Hydrodenitrogenation mechanism of pyridine.

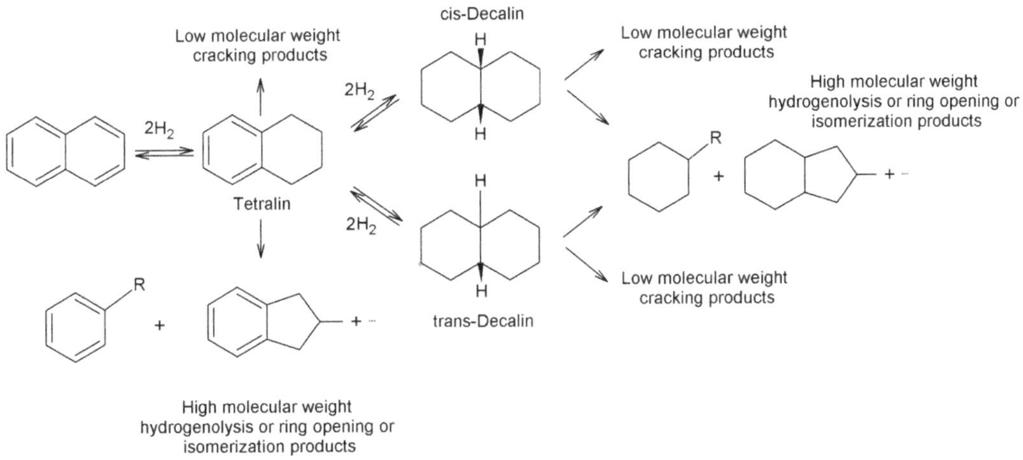

FIGURE 12.8 Hydrodearomatization mechanism of naphthalene at high pressures.

the downstream processes is essential for avoiding catalyst poisoning by hetero-atom contamination (Leffler 2000). Figure 12.8 represents the hydrodearomatization pathway of naphthalene. The saturation of aromatics improves the properties of oil products such as cetane index and smoke point (Jones and Pujado 2006).

There are various parameters that determine the hydrotreating process and it is the feedstock properties that determine the hydrotreating process parameters (Furimsky 1998). Liquid hourly space velocity (LHSV), reaction pressure, hydrogen partial pressure, reaction temperature, hydrogen-to-oil ratio, and catalysts used for hydrotreating are a few significant process parameters that determine the feedstock conversion (Ramachandran and Menon 1998). The hydrotreating catalysts play vital roles in determining the conversion, product quality, and the economics of the process (Ancheyta et al. 2005).

12.4 HYDROTREATING CATALYST AND CATALYST FOULING

Catalysts, usually used to accelerate the reaction rate, have active sites that adsorb the reacting species and desorb the product moieties. van Santen et at. (2006) explained the mechanism of catalysis in hydrotreating as a cyclic process in which the participating catalyst remains unchanged at the end of the reaction. A catalyst functions by decreasing the activation energy for a reaction without altering the equilibrium (Coulier et al. 2001; Ferdous 2003). The significance of using a catalyst in hydrotreating process lies in its applicability in removing the unwanted constituents in the feed such as sulfur, nitrogen, oxygen, and so on to an appreciable level and to increase the rate of the hydrotreating reaction for better conversion to lighter products (Ferdous et al. 2005, 2007; Furimsky and Massoth 1999).

The typical desirable properties of these catalysts are high surface area, larger number of active sites, thermal and chemical stability, selectivity, suitable shape, and pore-size (Botchwey et al. 2004). These ideal catalysts are active components of molybdenum (Mo) or tungsten (W), generally in combination with a suitable metal promoter (e.g., nickel, cobalt, and iron) and an alumina or silica support. Catalyst support is used to give mechanical strength to the catalyst and allow it to withstand extreme reaction conditions (Satterfield 1996). However, in the case of feeds with high metal content, a catalyst with high pore volume to avoid mouth plugging is required (Ancheyta et al. 2005). Apart from the feed constituents, coke formation at high reaction conditions also can contribute to poor catalytic activity that might eventually lead to catalyst deactivation (Ancheyta and Speight 2007).

The type of catalyst used during a hydrotreating reaction depends on the desired product. NiMo/γ-Al_2O_3 and CoMo/γ-Al_2O_3 are two of the most widely used commercial catalysts for hydrotreating (Papayannakos and Georgiou 1988). The active metal Mo is promoted by Ni or Co to enhance the activity and selectivity of the catalyst by reducing the metal-sulfur interaction force in MoS_2. NiMo sulfide catalysts are more suitable for hydrodenitrogenation as Ni prefers to be attached to Mo edge, thus proving higher activities for hydrogenation, whereas CoMo prefers the S-edge leading to higher HDS activities (Sun et al. 2005). Much has been experimented on the potential catalysts for hydrotreating reaction. Zeolites also have found a great deal of application in this field. There is a wide variation in the size, shape, pores, and material of the catalyst synthesized for the enhancement of the product yield and quality (Knudsen et al. 1999). For the heavy oil feeds, a few highly active catalysts reported are New Bulk Activity (NEBULA), Exxon Mobil, Super Type II Active Reaction Site (STARS), and Akzo Noble Catalyst (Rana et al. 2007).

For most of the hydrotreating reactions, the commonly used active phase is MoS_2, which is produced during the sulfidation of the catalyst. The weak interaction between the active phase and the catalyst support lead to stack formation of 2–4 layers that are defined as Type II Co-Mo-S or Ni-Mo-S (Topsøe 2007). The sulfur ion from this molecule is expelled in the form of H_2S creating sulfur vacancies that are Lewis acidic in nature, known as coordinatively unsaturated sites (CUS). This result allows the adsorption of molecules that comprise of unpaired electrons such as pyridine and dibenzothiophine. At the neighbouring sites, the homocyclic and heterocyclic splitting of hydrogen from –SH and Mo-H species become the source of hydrogen for the hydrotreating reaction. These groups have Brønsted acid characteristics (Ancheyta et al. 2005). The reaction completes by the reformation of the initial active phases by sulfur breathing (Mochida and Choi 2006).

The catalyst tends to foul after many hydrotreating reactions owing to the feed, product, or by-product concentration and conversion. In fact, catalyst deactivation due to fouling is the focus while designing and preparing catalysts for hydroprocessing of different feeds (Wardag et al. 2012). Hydrotreating catalysts are more prone to the pore mouth plugging and, hence, are often used with guard-bed catalysts with large pore diameter, high metal retention, and low surface area (Rana et al. 2007). Despite several measures, catalyst plugging is difficult to prevent during hydrotreating, but it is desirable to avoid a quick catalyst plugging during reactions. However, the catalyst is preferred to be regenerated to enhance the productivity of the process and reduce the expense of catalyst replacements (Leliveld et al. 1997). The major constituent for catalyst fouling is the coke formed during the reaction, nitrogen organics, metals, and ultra-fines in the feed (Newson 1975).

Like other crude oil processing, hydrotreating is extremely essential for the Athabasca bitumen because it contains high sulfur and nitrogen organics content (Zhao et al. 2002). However, for the hydrotreating of Athabasca bitumen one of the major challenges is catalyst fouling because of the deposition of fine particles. The amino-silicates, present in the form of asphaltenes and the humic clay that is like kaolin, are the major contents of these fines (Wang et al. 2001).

12.5 ORGANIC COATED SOLIDS IN THE BITUMEN FEED

It is a matter of foremost concern in the upgrading of Athabasca bitumen to avoid catalyst plugging due to the deposition of these fine particles on the catalyst surface and in the pores (Bensebaa 2000). Sparks et al. (2003) categorized the bitumen solids into clays, fossils, heavy minerals, and aggregates. The clay fraction comprises of alumina-silicates, which are primarily like kolinite and illite in nature. This fraction of conventional clays has an average particle size of about 3.0 μm. Clay crystallites that have a lateral extension of less than 300 nm and 10 nm or less thickness is identified particularly to create handling and technical impediments (Kotlyar et al. 1993a, 1993b). Defined as ultra-fine clays, these materials cause a major constituent in the unfiltrable solids in bitumen-derived streams and in tailings.

Other familiar components of oil sands are the fossils, which mostly originate from algae and diatoms. The fossils can be further classified as sulfurized-fossils and heavily mineralized-fossils. Sulfurized-fossils entrain in salty water and are usually hollow, while mineralized-fossils can be crystallized-pyrites. Mineralized-fossils usually fall in the range of 10–100 μm and they can be easily separated from the bitumen stream by centrifugation or conventional settling. However, the hollow sulfurized-fossils are challenging to extract because of their low density and fragmented particles that might also be the source of water-soluble salts in bitumen.

During bitumen separation, the naturally occurring hydrophobic minerals combine with other organic-rich solids and concentrate at the oil-water interface. In oil sands, the commonly found minerals are titanium dioxides and pyrites. Although these mineral particles have a high density, their relatively smaller size (500 nm) makes them difficult to separate from the bitumen stream. Quartz, clay, and heavy minerals can combine to result in aggregate formation. Usually, calcium carbonate, iron oxide, siderite, or humic materials bind these minerals together. These materials can range from being biwettable to hydrophobic depending upon their organic content. Some amounts of titanium, sulfur, and iron containing minerals also can be a part of aggregates.

Kotlyar et al. (1988) suggested that aggregates breakdown by the removal of organic contents through a plasma oxidation technique. Such particles are usually less than 44 μm but can be as large as 100 μm (Kotlyar et al. 1987a, 1987b). Liu et al. (2003) studied the hot water extraction process to recover bitumen from oil sands. It was reported that during this process bitumen is attached to air bubbles, thus the formation of bitumen-rich froth takes place. The bitumen extraction can be as high as 93% for good quality ores. However, for poor processing ores, the high content of fines (40 wt% fines in total solid content) and divalent ions cause a technical challenge for the effective bitumen extraction (Liu et al. 2003).

12.5.1 Fines

Particles in bitumen streams that are less than 44 μm in size are defined as fines. The fine particles could originate from iron sulfides in the upstream equipment, as organic precipitates, or as naturally occurring fine clays in oil sands bitumen and in-situ coke fines. There is a large scope for the study of these fines agglomeration and for planning and suggesting strategies against reactor plugging. The clay crystallites with a size of less than 300 nm and thickness of less than 10 nm are reported to be particularly problematic materials (Kotlyar et al. 1999). These are the major contributor to non-filterable ultra-fines found in the bottom tailings and in the bitumen feed stream (Sparks et al. 2003).

In an undisturbed state, these fines generally cover the industrial solid surface (catalyst), whereas when the fluid saturating the catalyst pore is set in motion these fines are entrained and later deposited at a convenient accumulation site leading to reactor plugging (Gruesbeck 1982). The major content of these fines is the clay particles that are called humic clays, which are found to be like asphaltene-coated kaolin particles (Wang et al. 2001). The bitumen feed entering the hydrotreating reactor is pretreated using techniques such as desalting, distillation, and filtration to eliminate most of the unwanted particles. However, the ultra-fines (particle diameter <20 μm) can persist in the feed stream owing to their non-filterable size (Wang et al. 1999).

Kaolinite is the most abundant clay in oil sands. It is a layered alumino-silicate with a tetrahedral sheet that has oxygen-linking atoms binding it to the octahedral sheet of alumina in the ore. The adsorption of asphaltenes on the kaolin surface has a relevant impact on the interactions of the fines. The major interacting forces are van der Waals forces, double layer repulsive forces, and electrostatic forces (Mendoza et al. 2009; Murgich 2002; Wang et al. 1999). The deposition of particles to a substrate is governed by the size of the particles. Fine particles can clog the catalyst pores, whereas the larger particles can settle on the surface by the physical forces arising from fluid drag and gravity (Elimelech and O'Melia 1990).

Some of the important factors that govern the fines deposition are particle size, fluid velocity, physical properties of the fluid, and porous solid (Gruesbeck 1982). It is essential to study where exactly on the catalyst these fines entrain and deposit during the hydrotreating reaction. The principle for entrainment of fine particles was studied and reported along with the particle size determining the fraction of fluid pathway which was found to be plug type (Gruesbeck 1982). It is suggested that the clay particles in the bitumen feed tend to adsorb asphaltenes and form agglomerates. The adsorption of the asphaltene layer over the clay particles is also non-uniform and patchy (Sparks et al. 2003). Fine particles are generally coated with asphaltenes, which make it noteworthy to study the role of asphaltenes in hydrotreating of bitumen streams and to understand the surface chemistry that comes into play when asphaltene coats the clay particles that are deposited on the catalyst surface.

12.5.2 Asphaltenes

Asphaltenes are the target molecules responsible for deactivating the hydrotreating catalyst in either fixed or moving bed operations. The actual chemical structure of asphaltene is not well understood but several structures have been proposed. It exists as a common uncertain molecule in heavy oil and bitumen (Liu et al. 2003). Asphaltenes can be considered as large aromatic sheets with large molecular weight, layered on one another to form unit cell and larger asphaltene molecules. These asphaltenes cover the fine particles associated with the bitumen feed (Rana et al. 2007). Asphaltenes from the same source are expected to have the same molecular dimension, although the chemical structure may differ if the sources and unit sheets differ (Gawel et al. 2005).

Asphaltenes are formed by aromatic compounds, having a π-π interaction, which undergo an acid-base reaction and are self-associated by hydrogen bonds (Chang and Fogler 1994a, 1994b). Asphaltenes are tightly bound to bitumen solids as evidenced by high nickel and vanadium concentration detected as compared to ultra-fines from fine tailings. In asphaltenes, they occur as chelates or porphyrins of nickel and vanadium, along with the occasional presence of iron and sulfur (Kotlyar et al. 1999). The concept of aggregates and micelle formation in case of asphaltene has been studied. A micelle is a reversible aggregate formed in a polar environment that remains constant in size for the given environmental constraints. The term critical micelle concentration (CMC) is used for the condition when the concentration of the asphaltene molecules is such that it begins to self-aggregate in a certain polar media. In a paraffinic environment (non-polar media) larger, non-dissolvable asphaltene agglomerates are formed (Haraguchi et al. 2001).

Asphaltene adsorption on solid substrate is essential to study due to its effects on the wettability of the minerals that might affect fluid permeability in rock reservoirs. Wang et al. (1999) studied the interaction of asphaltene and kaolin along with their surface chemistry. Asphaltene-coated kaolin gives deep-bed filtration due to steric repulsion while without asphaltene coating, the kaolin clay particles gave cake filtration. It is difficult to comment if an interaction would take place between asphaltene molecules or the aggregates formed (Haraguchi et al. 2001).

An artificial model of fine particles to study the effect and filtration trend of fines on catalyst surface and within the pores was developed by Wang et al. (1999). The fine particles introduced in the bitumen feed were asphaltene-coated kaolin. The surface chemistry and an illustration of the deposition mechanism is shown in Figure 12.9. It was suggested that the adsorbed organics on the clay were desorbed in water and led to the flocculation of the clay particles on the collector (catalyst) surface (Wang et al. 1999, 2001). Asphaltene coated kaolin is commonly used in fine particles studies. The trickle bed reactor is used commercially owing to the wide range of the operating condition to which it can cater. However, it shows complex behavior due to hydrodynamic characteristics that arise from the gas-liquid flow in the packed bed (Wang 2000).

The effect of hydrotreating product in deposition of fine particles on the packed column of the reactor was reported (Wang 2000). Fine particles that cause reactor plugging in the reactor bed at hydrotreating conditions of temperature and pressure are also studied. Ammonia and quinoline were found to have no effect on the particle deposition in the reactor. From this result, it could be concluded that HDN did not affect the particle behaviour, whereas HDS, HDO, and HDM had some noticeable effects on the particle deposition owing to the hydrogen sulfide, water, and metal oxides formation (Wang 2000).

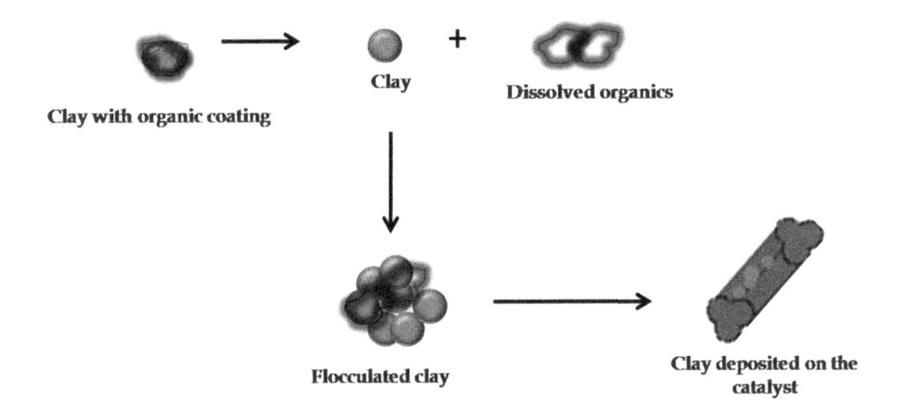

FIGURE 12.9 An illustrative asphaltene-kaolin surface interaction model.

Fine particles and asphaltenes play significant roles in hydrotreating reaction due to their nature of plugging the catalyst and affecting the activity. However, the reactor used in hydrotreating and the operational problems associated with catalyst fouling due to fine particles deposition are essential to understand to provide a better incentive to the Athabasca bitumen upgrading.

12.6 MECHANISM OF FINE PARTICLE DEPOSITION

The deposition of fine particles onto a substrate is influenced by their size. Smaller fine particles can plug the pores of the catalyst, whereas the larger ones settle on the surface due to the physical forces arising from gravity and fluid drag (Elimelech and O'Melia 1990). Figure 12.10 shows a conceptualized diagram of particle deposition in trickle bed hydrotreater. The forces of attraction that majorly govern the fine particles and catalyst interaction are van der Waals forces and double layer repulsion (Chowdiah et al. 1981; Elimelech and O'Melia 1990).

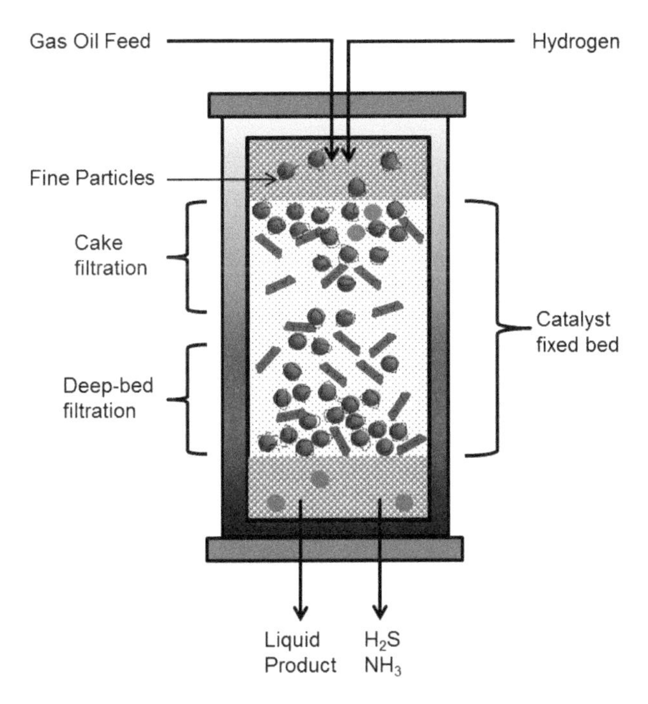

FIGURE 12.10 Particle deposition illustration in a trickle bed hydrotreating reactor.

The effect of flocculation time on the structure and size of model clay fines (kaolinite mix) was studied by Vaezi et al. (2011). The study led to the understanding that the flocculation kinetics depends on aggregate size density. These aggregates grow rapidly to form enormous open structure in the early stages and the aggregate density tends to increase slightly at prolonged flocculation times.

The particle capture in a packed-bed reactor depends on the size of the particles, catalyst pore diameter, deep-bed, and granular filtration (Narayan et al. 1997). In a study by Narayan et al. (1997), the pressure-drop increased with fines deposition in all cases, but the intensities varied. The efficiency of trapping the fines was reported to change with deposition (i.e., an increase in deposition is noticed as the bed "ripens") (Zamani and Maini 2009).

The three main phases of the particle deposition on the catalyst were conceptualized as follows (Choo and Tien 1995):

1. The smooth cover that includes the trapped particles on the surface of the collector (catalyst)
2. Multi-layer deposition of the particles
3. Clogging of the particles at the catalyst pores in the bed wherein straining (cake filtration) acts as a dominant mechanism

A complex situation may arise when the gas and liquid phases flow concurrently downwards through the packed bed of solid particles. The prevailing flow regime is the function of the followings: (1) gas and liquid flow rates, (2) reactor dimensions, (3) particle size and shape, (4) packing, and (5) thermo-physical properties of gas and liquid (Choo and Tien 1995).

Asphaltene-coated kaolin was mixed in a bitumen-extracted gas oil feed to substitute the fine particles, and the deposition trend was reported by Wang et al. (1999, 2001). Asphaltene-coated kaolin gave deep-bed filtration, whereas non-coated kaolin gave cake filtration. The steric resistance against multilayer deposition acting in case of the asphaltene-coated kaolin led to a greater resistance to deposition than in the case of non-coated kaolin. In the specific case of asphaltene-coated kaolin, the study showed that the deposition increased with the progress of the catalytic reaction. In the high-temperature zones of the reactor, the efficiency of filtration is higher as compared to the low-temperature zones. This result explains the greatest deposition of fine particles at the exit of the reactor. As the reaction proceeds, the ratio of the liquid to gas reduces, which further enhances the particle deposition (Wang et al. 1999). Variables such as reaction kinetics, heat transfer, mass transfer, bed porosity, catalyst shape and size, wettability, interfacial tension, gas and liquid flow rates, and viscosities have considerable effects on the operation of the reactor (Zhou et al. 2009).

12.7 IMPACTS OF HYDROTREATING REACTION CONDITIONS ON PARTICLE DEPOSITION

The use of trickle bed reactors is quite common for hydrotreating of bitumen-derived products. The hydrotreating reaction conditions require high pressure (870–1740 psig H_2) and temperature (355°C–395°C) and the commonly used catalyst NiMo/γ-Al_2O_3. There have been relevant studies explaining various steps of the hydrotreating process. However, not much has been reported on the trend of the fine particle deposition despite the operational problems related to it. Little work has been done to understand the filtration of these fine particles and how they affect the pressure drop (Gray et al. 2002).

Even small quantities of fine particles could increase the pressure drop in the packed column (Choo and Tien 1995). It was reported by Iliuta et al. (2003a, 2003b) that the occurrence of fines deposition could be suitably evaluated by the rise in pressure drop as a function of time and in terms of plugging for local porosity and specific velocity versus bed depth. Larachi et al. (1995) studied the three-dimensional mapping of solids flow fields in multiphase reactors with the help of radioactive particle trapping. Such studies help in understanding the dynamics of solid particles in the packed column that runs in multiphase flows. The impact of fine particles in terms of pressure drop as a function of time was analyzed along with the plugging pattern by Ranade et al. (2011). The following are some of the concluding remarks from their study:

1. Two-phase (liquid feed and hydrogen) pressure drop and the volume-average specific deposits increases with an increase in the flow rate of liquid
2. The particle deposition is independent of the density of the gas used
3. Two-phase pressure drop ratio is inversely proportional to the liquid velocity
4. Two-phase pressure drop ratio is lowered if the concentration of the inlet fines is decreased
5. Two-phase pressure drop ratio rises with an increase in the diameter of the inlet fines

Ranade et al. (2011) also suggested that in a trickle-bed reactor, two-phase (liquid feed and hydrogen) pressure drop along the length of the bed is a function of the following:

1. Reactor hardware such as, diameter of the reactor column, size and shape of the fine particles, and internal assembly
2. Operating parameters (e.g., gas-liquid flow rates)
3. Properties of the fluid such as density and viscosity of the flowing fluid, surface tension, and surface characteristics
4. Operating parameters like temperature and pressure

The pressure drop for a single-phase flow in a packed bed reactor is calculated using the Ergun equation (Equation 12.2) for the liquid phase as follows. In the equation, ε is the porosity and Re_L is the Reynold's number of the liquid. ε_1, ε_2 will depend on the properties of the bed packing material (Gray et al. 2002).

$$\frac{\Delta P_L}{L} = \left[\frac{\mu^2}{\rho_L dp^3}\right]\left[\frac{\varepsilon}{1-\varepsilon}\right]^3 Re_L\left(\varepsilon_1 + \varepsilon_2 Re_L\right) \tag{12.2}$$

Hydrotreating conditions (i.e., catalyst, temperature, and hydrogen concentration) have significant effects on fine particles deposition (Wang 2000). Wang et al. (2001) reported the particle-particle interaction between the fines and its impacts on the surface chemistry. The experiments conducted in a batch reactor contributed to understanding the filtration trends and the mechanism of kaolin and asphaltene behavior in a suspension. The impact of process conditions can be an essential parameter for fines deposition. Rana et al. (2017) studied the impact of process parameters such as temperature, pressure, and particle loading on the deposition of fines in the catalyst bed. It was reported that particle deposition increased with an increase in particle loading. However, the impact of temperature on particle deposition also was also found significant with higher temperatures (381°C) giving high bed deposition.

12.8 INDUSTRIAL CHALLENGES

There are several challenges in the hydroprocessing of bitumen-derived oil that remain untouched. Significant amounts of bitumen are transported to the industrial sites from the exploration site. The transportation cost of the extremely viscous product is high because a diluent is added to it to make it transportable from the reservoir. Thus, the prices of bitumen are usually varying due to the total exploration and transporting cost because of its viscous nature. There is plenty of scope to establish a cost-effective method for bitumen exploration and transportation that can affect the overall bitumen-derived oil production prices. The production and total cost

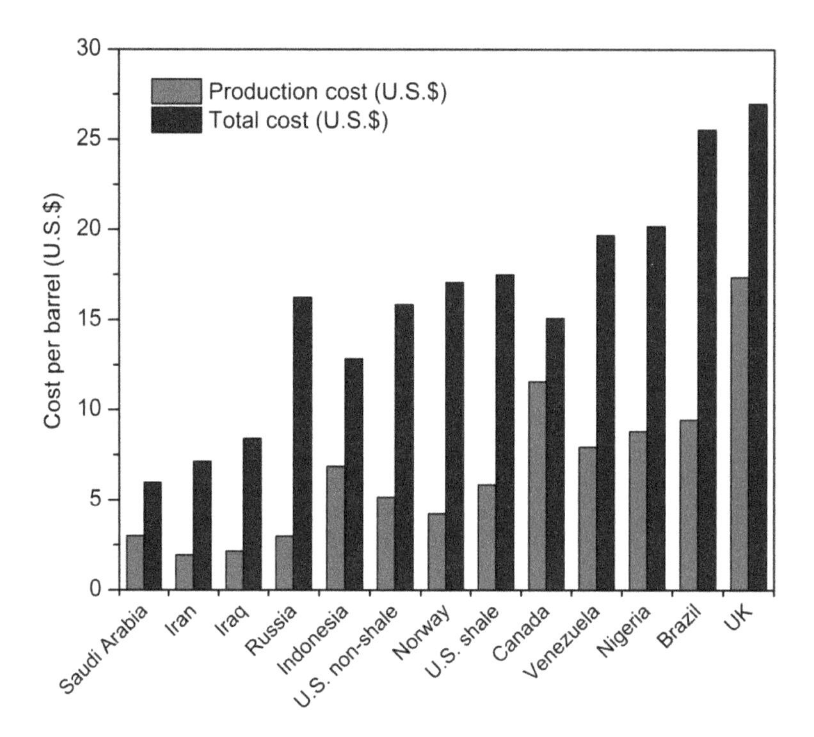

FIGURE 12.11 Current cost of producing a barrel of oil and gas. (From The Wall Street Journal, Barrel Breakdown. http://graphics.wsj.com/oil-barrel-breakdown/, 2016.)

of fuel from different sources is summarized in Figure 12.11. As seen from the figure, although the total cost per barrel of oil in Canada (U.S. $26.64) is competitive, its production cost is the second highest (U.S. $11.56) after that of the UK (U.S. $17.36) (The Wall Street Journal 2016). Hence, it is a challenge to bring down the production cost of Canadian oil by making the process more efficient.

The upgrading of bitumen possesses many limitations due to the presence of impurities in high concentration. The inherently entrained impurities that are present in bitumen make it extremely challenging for the oil industries to process the bitumen-derived oil in compliance with stringent environmental regulations. The impurities in the unconventional crude (3,000 ppm nitrogen and 40,370 ppm sulfur) are at higher levels than that of the conventional crude (900 ppm nitrogen and 5,000 ppm sulfur). Therefore, there are some challenges to improving the catalytic efficiencies to remove impurities from crude with minimum capital investment. Studies of synthesizing effective catalysts for better activity, higher surface area, greater mechanical strength, and stability with low cost are being conducted. There are developments in this field with established conditions for hydrotreating of the oil but when the cost is compared globally, the unconventional crude still lags.

In addition to the previously stated disadvantages, another problem with hydrotreating of bitumen-derived oil is the catalyst bed plugging due to particle deposition. The theoretical models for particle deposition and pressure drop in the reactor have been derived. However, the available sources lack information on the practical verification of the derived models. In addition, the experimental data on the impact of catalyst properties, hydrotreating process parameters, and chemistry of catalyst-fines interaction is insufficient in available literature. The fate of fine particles in treated gas oil is essential to govern the economics of the bitumen-upgrading process. Hence, it becomes imperative to explore this neglected area of research and establish optimal operating conditions, catalysts, and reactor designs to minimize the fine particle deposition without affecting bitumen conversions.

12.9 CONCLUSIONS

Oil has been one of the major global sources of energy and continues to be so. With advances in technology, the depletion of conventional oil resources have posed a challenge to the environment for its energy needs, thus, bringing the unconventional crude to light and placing emphasis on its exploration. Several techniques have been established such as vapour extraction (VAPEX), Steam Assisted Gravity Drainage (SAGD), and so on for the exploration and extraction of unconventional oil. Considering the high impurities in the unconventional oil, it becomes difficult to produce synthetic crude oil from bitumen with similar standards as the conventional crude (sweet crude).

Canadian oil sands are a vast resource of unconventional heavy oil with surplus to meet the growing energy demands. However, the presence of impurities, high viscosity of bitumen, and high sulfur and nitrogen content makes the processing of Canadian oil sands expensive. This review presents the underlying principles involved in bitumen extraction and hydrotreating of the oil streams along with a detailed discussion about the nature, interactions, and problems faced by the industry due to catalyst-fines interaction.

There have been studies that report pressure-drop in the hydrotreating reactor and the deposits on the catalyst bed are the major contributor to this case. The incidence of fine particle deposition is essentially correlated to the pressure-drop and hence it becomes extremely important to understand the catalyst-fines interaction and the mechanism of particle deposition at the physical level and the chemical level. This chapter focused on the compilation of literature that is available related to the challenges during hydrotreating due to fine particle deposition. It is believed that addressing this problem of catalyst plugging and pressure-drop due to particle deposition can affect the production cost of oil production and help in making industrial hydrotreating more efficient and cost effective.

ACKNOWLEDGMENTS

The authors would like to acknowledge the Natural Science and Engineering Research Council of Canada (NSERC), MITACS Canada and Syncrude Ltd. for funding this research.

REFERENCES

Ancheyta, J., and J. G. Speight. 2007. *Hydroprocessing of Heavy Oils and Residua*. Boca Raton, FL: Taylor & Francis Group.

Ancheyta, J., M. S. Rana, and E. Furimsky. 2005. Hydroprocessing of heavy petroleum feeds: Tutorial. *Catalysis Today* 109:3–15.

Aoyagi, K., W. McCaffrey, and M. R. Gray. 2003. Kinetics of hydrocracking and hydrotreating of coker and oilsands gas oils. *Petroleum Science and Technology* 21:997–1015.

Badoga, S., A. K. Dalai, J. Adjaye, and Y. Hu. 2014. Combined effects of EDTA and heteroatoms (Ti, Zr, and Al) on catalytic activity of SBA-15 supported NiMo catalyst for hydrotreating of heavy gas oil. *Industrial & Engineering Chemistry Research* 53:2137–2156.

Banerkke, L. 2012. *Oil Sands, Heavy oil and Bitumen: From Recovery to Refinery*, 1st ed. Tulsa, Oklahoma: Pennwell Corporation.

Bensebaa, F. 2000. Organic coated solids in Athabasca bitumen: Characterization and process implications. *Canadian Journal of Chemical Engineering* 78:610–616.

Botchwey, C., A. K. Dalai, and J. Adjaye. 2004. Two-stage hydrotreating of Athabasca heavy gas oil with interstage hydrogen sulfide removal: Effect of process conditions and kinetic analyses. *Industrial & Engineering Chemistry Research* 43:5854–5861.

Chang, C. L., and H. S. Fogler. 1994a. Stabilization of asphaltenes in aliphatic solvents using alkylbenzene-derived amphiphiles: 1. Effect of the chemical structure of amphiphiles on asphaltene stabilization. *Langmuir* 10:1749–1757.

Chang, C. L., and H. S. Fogler. 1994b. Stabilization of asphaltenes in aliphatic solvents using alkylbenzene-derived amphiphiles: 2. Study of the asphaltene–amphiphile interactions and structures using Fourier transform infrared spectroscopy and small-angle X-ray scattering techniques. *Langmuir* 10:1758–1766.

Choo, C. U., and C. Tien. 1995. Analysis of the transient behavior of deep-bed filtration. *Journal of Colloid and Interface Science* 169:13–33.

Chowdiah, P., D. T. Wasan, and D. Gidaspow. 1981. Electrokinetic phenomena in the filtration of colloidal particles suspended in nonaqueous media. *AIChE Journal* 27:975–984.

Coulier, L., V. H. J. de Beer, J. A. R. van Veen, and J. W. Niemantsverdriet. 2001. Correlation between hydrodesulfurization activity and order of Ni and Mo sulfidation in Planar silica-supported NiMo catalysts: The influence of chelating agents. *Journal of Catalysis* 197:26–33.

Delmon, B. 1993. New technical challenges and recent advances in hydrotreatment catalysis. A critical updating review. *Catalysis Letters* 22:1–32.

Dunbar, R. B. 2009. *Canada's Oil Sands – A World-Scale Hydrocarbon Resource*. Strategy West Inc, Calgary, Canada.

Elimelech, M., and C. R. O'Melia. 1990. Effect of particle size on collision efficiency in the deposition of Brownian particles with electrostatic energy barriers. *Energy and Fuels* 6:1153–1163.

Ferdous, D. 2003. Surface morphology of nickel-molybdenum catalyst supported on gamma-aluminum oxide: Impact on hydroprocessing of heavy gas oil derived from Athabasca bitumen. Thesis, University of Saskatchewan.

Ferdous, D., A. K. Dalai, and J. Adjaye. 2004. A series of NiMo/Al_2O_3 catalysts containing boron and phosphorus. *Applied Catalysis A: General* 260:153–162.

Ferdous, D., A. K. Dalai, J. Adjaye, and L. Kotlyar. 2005. Surface morphology of NiMo/Al_2O_3 catalysts incorporated with boron and phosphorus: Experimental and simulation. *Applied Catalysis A: General* 294:80–91.

Ferdous, D., N. N. Bakhshi, A. K. Dalai, and J. Adjaye. 2007. Synthesis, characterization and performance of NiMo catalysts supported on titania modified alumina for the hydroprocessing of different gas oils derived from Athabasca bitumen. *Applied Catalysis B: Environmental* 72:118–128.

Furimsky, E. 1998. Selection of catalysts and reactors for hydroprocessing. *Applied Catalysis A: General* 171:177–206.

Furimsky, E., and F. E. Massoth. 1999. Deactivation of hydroprocessing catalysts. *Catalysis Today* 52:381–495.

Gary, H., E. Handewerk, and J. Kaiser. 2007. *Petroleum Refining: Technology and Economics*. Boca Raton, FL: Taylor & Francis Group.

Gawel, I., D. Bociarska, and P. Biskupski. 2005. Effect of asphaltenes on hydroprocessing of heavy oils and residua. *Applied Catalysis A: General* 295:89–94.

Gray, M. R. 1994. *Upgrading Petroleum Residues and Heavy Oils*. New York: Marcel Dekker.

Gray, M. R., N. Srinivasan, and J. H. Masliyah. 2002. Pressure buildup in gas-liquid flow through packed beds due to deposition of fine particles. *Canadian Journal of Chemical Engineering* 80:346–354.

Gruesbeck C. 1982. Entrainment and deposition of fine particles in porous media. *Society of Petroleum Engineers Journal* 22:847–856.

Haraguchi, L., W. Loh, and R. S. Mohamed. 2001. Interfacial and colloidal behavior of asphaltenes obtained from Brazilian crude oils. *Journal of Petroleum Science and Engineering* 32:201–216.

Iliuta, I. and F. Larachi. 2003b. Fines deposition dynamics in packed-bed bubble reactors. *Industrial & Engineering Chemistry Research* 42:2441–2449.

Iliuta, I. and F. Larachi. 2005. Stretching operational life of trickle-bed filters by liquid-induced pulse flow. *AIChE Journal* 51:2034–2047.

Iliuta, I., F. Larachi, and B. P. A. Grandjean. 2003a. Fines deposition dynamics in gas–liquid trickle-flow reactors. *AIChE Journal* 49:485–495.

Jechura, J. 2016. Refinery feedstocks and products-properties and specifications. Colorado School of Mines. https://inside.mines.edu/~jjechura/Refining/02_Feedstocks_&_Products.pdf (accessed 20.05.2017).

Jones, D. S., and P. P. Pujado. 2006. *Handbook of Petroleum Processing*. Amsterdam, The Netherlands: Springer.

Knudsen, K. G., B. H. Cooper, and H. Topsøe. 1999. Catalyst and process technologies for ultra low sulfur diesel. *Applied Catalysis A: General* 189:205–215.

Leliveld, R. G., W. C. A. Huyben, A. J. van Dillen, J. W. Geus, and D. C. Koningsberger. 1997. Novel hydrotreating catalysts based on synthetic clay minerals. *Studies in Surface Science and Catalysis* 106:137–146.

Kotlyar, L. S., J. A. Ripmeester, B. D. Sparks, and D. S. Montgomery. 1987b. Characterisation of organic rich solids fractions isolated from Athabasca oil sands using a cold water agitation test. *Fuel* 67:221–226.

Kotlyar, L. S., B. D. Sparks, C. E. Capes, and R. Schutte. 1993b. Gel-forming attributes of colloidal solids from fine tails formed during extraction of bitumen from Athabasca oil sands by the hot water process. *AOSTRA Journal of Research* 8:55–61.

Kotlyar, L. S., B. D. Sparks, J. R. Woods, and K. H. Chung. 1999. Solids associated with the asphaltene fraction of oil sands bitumen. *Energy and Fuels* 13:346–350.
Kotlyar, L. S., H. Kodama, B. D. Sparks, and P. E. Grattan-Bellew. 1987a. Non-crystalline inorganic matter-humic complexes in Athabasca oil sands and their relationship to bitumen recovery. *Applied Clay Science* 2:253–271.
Kotlyar, L. S., J. A. Ripmeester, B. D. Sparks, and D. S. Montgomery. 1988. Characterisation of oil sands solids closely associated with Athabasca bitumen. *Fuel* 67:808–814.
Kotlyar, L. S., Y. Deslandes, B. D. Sparks, H. Kodama, and R. Schutte. 1993a. Characterization of colloidal solids from Athabasca fine tails. *Clays and Clay Minerals* 41:341–345.
Larachi, F., J. Chaouki, and G. Kennedy. 1995. 3-D mapping of solids flow fields in multiphase reactors with RPT. *AIChE Journal* 41:439–443.
Leffler, W. 2000. *Petroleum Refining in Nontechnical Language*, third ed. Tulsa, Oklahoma: Pennwell Corporation.
Liu, J., Z. Xu, and J. Masliyah. 2003. Interaction between bitumen and fines in oil sands extraction system: Implication to bitumen recovery. *Canadian Journal of Chemical Engineering* 82:655–666.
Liu, J., Z. Xu, J. Masliyah, and C. Tg. 2005. Processability of oil sand ores in Alberta. *Energy and Fuels* 19:2056–2063.
Luo, P., and Y. Gu. 2007. Effects of asphaltene content on the heavy oil viscosity at different temperatures. *Fuel* 86:1069–1078.
Mendoza de la Cruz, J. L., I. V. Castellanos-Ramírez, A. Ortiz-Tapia, E. Buenrostro-González, C. D. L. A. Durán-Valencia, and S. López-Ramírez. 2009. Study of monolayer to multilayer adsorption of asphaltenes on reservoir rock minerals. *Colloids and Surfaces A: Physicochemical and Engineering Aspects* 340:149–154.
Mochida, I., and K. Choi. 2006. Current progress in catalysts and catalysis for hydrotreating. In *Practical Advances in Petroleum Processing*, ed. C. S. Hsu and P. R. Robinson, P. R., pp. 257–296. New York: Springer.
Murgich, J. 2002. Intermolecular forces in aggregates of asphaltenes and resins. *Petroleum Science and Technology* 20:983–997.
Narayan, R., R. Coury, J. H. Masliyah, and M. R. Gray. 1997. Particle capture and plugging in packed-bed reactors. *Industrial & Engineering Chemistry Research* 36:4620–4627.
Newson, E. 1975. Catalyst deactivation due to pore-plugging by reaction products. *Industrial & Engineering Chemistry Research* 14:27–33.
Nourozieh, H., M. Kariznovi, and J. Abedi. 2011. Physical properties and extraction measurements for the Athabasca bitumen + light hydrocarbon system: Evaluation of the pressure effect, solvent-to-bitumen ratio, and solvent type. *Journal of Chemical & Engineering Data* 56:4261–4267.
Papayannakos, N., and G. Georgiou. 1988. Kinetics of hydrogen consumption during catalytic hydrodesulphurization of a residue in a trickle-bed reactor. *Journal of Chemical Engineering of Japan* 21:244–249.
Rajagopalan, S. 2010. Study of Bitumen Liberation from Oil Sands Ores. Thesis. University of Alberta.
Ramachandran, R., and R. K. Menon. 1998. An overview of industrial uses of hydrogen. *International Journal of Hydrogen Energy* 23:593–598.
Rana, M. S., V. Sámano, J. Ancheyta, and J. A. I. Diaz. 2007. A review of recent advances on process technologies for upgrading of heavy oils and residua. *Fuel* 86:1216–1231.
Rana, R., B. Badoga, A. K. Dalai, and J. Adjaye. 2017. The impact of process parameters on the deposition of fines present in bitumen-derived gas oil on hydrotreating catalyst. *Energy and Fuels* 31:5969–5981.
Rana, R., S. Nanda, A. Maclennan, Y. Hu, J. A. Kozinski, and A. K. Dalai. 2018a. Comparative evaluation for catalytic gasification of petroleum coke and asphaltene in subcritical and supercritical water. *Journal of Energy Chemistry*. doi:10.1016/j.jechem.2018.05.012.
Rana, R., S. Nanda, J. A. Kozinski, and A. K. Dalai. 2018b. Investigating the applicability of Athabasca bitumen as a feedstock for hydrogen production through catalytic supercritical water gasification. *Journal of Environmental Chemical Engineering* 6:182–189.
Ranade, V. V., R. Chaudhari, P. R. Gunjal. 2011. *Trickle Bed Reactors – Reactor Engineering and Applications*. Elsevier, Spain.
Satterfield, C. N. 1996. *Heterogeneous Catalysis in Industrial Practice*, second edition. Malabar, FL: Krieger.
Selucky, M. L., Y. Chu, T. Ruo, and O. P. Strausz. 1977. Chemical composition of Athabasca bitumen. *Fuel* 56:369–381.
Sparks, B., L. Kotlyar, J. O'Carroll, and K. Chung. 2003. Athabasca oil sands: Effect of organic coated solids on bitumen recovery and quality. *Journal of Petroleum Science and Engineering* 39:417–430.

Strausz, O. P., A. Morales-Izquierdo, N. Kazmi, D. S. Montgomery, J. D. Payzant, I. Safarik, and J. Murgich. 2010. Chemical composition of Athabasca bitumen: The saturate fraction. *Energy and Fuels* 24:5053–5072.

Strausz, O. P., E. M. Lown, A. Morales-Izquierdo, N. Kazmi, D. S. Montgomery, J. D. Payzant, and J. Murgich. 2011. Chemical composition of Athabasca bitumen: The distillable aromatic fraction. *Energy and Fuels* 25:4552–4579.

Sun, M., A. E. Nelson, and J. Adjaye. 2005. Adsorption and hydrogenation of pyridine and pyrrole on NiMoS: An *ab initio* density-functional theory study. *Journal of Catalysis* 231:223–231.

The Engineering ToolBox. Classification of Gas Oil. http://www.engineeringtoolbox.com/classification-gas-oil-d_165.html (accessed 26.03.2017).

The Wall Street Journal. 2016. Barrel Breakdown. http://graphics.wsj.com/oil-barrel-breakdown/ (accessed 05.01.2017).

Topsøe, H. 2007. The role of Co-Mo-S type structures in hydrotreating catalysts. *Applied Catalysis A: General* 322: 3–8.

U.S. Energy Information Administration (USEIA). 2016. Short-term Energy Outlook. https://www.eia.gov/forecasts/steo/report/global_oil.cfm (accessed 10.05.2017).

U.S. Energy Information Administration (USEIA). 2017. International Energy Statistics. Crude Oil Proved Reserves 2016. https://www.eia.gov/beta/international/data/browser (accessed 05.02.2017).

Vaezi, G. F., R. S. Sanders, and J. H. Masliyah. 2011. Flocculation kinetics and aggregate structure of kaolinite mixtures in laminar tube flow. *Journal of Colloid and Interface Science* 355:96–105.

van Santen, M. N. 2006. Concepts in theoretical heterogeneous catalytic reactivity. *Catalysis Reviews* 37:557–698.

Vosoughi, V., B. Badoga, A. K. Dalai, and N. Abatzoglou. 2016. Effect of pretreatment on physicochemical properties and performance of multiwalled carbon nanotube supported cobalt catalyst for Fischer-Tropsch synthesis. *Industrial & Engineering Chemistry Research* 55:6049–6059.

Wang, S. 2000. Chemistry of fine particles in hydrotreator reactor. Thesis, University of Alberta.

Wang, S., K. Chung, and M. R. Gray. 2001. Role of hydrotreating products in deposition of fine particles in reactors. *Fuel* 80:1079–1085.

Wang, S., K. H. Chung, J. H. Masliyah, and M. R. Gray. 1999. Deposition of fine particles in packed beds at hydrotreating conditions: Role of surface chemistry. *Industrial & Engineering Chemistry Research* 38:4878–4888.

Wardag, A. R. K., M. Hamidipour, M. Schubert, D. Edouard, and F. Larachi. 2012. Filtration and catalytic reaction in trickle beds: The use of solid foam guard beds to mitigate fines plugging. *Industrial & Engineering Chemistry Research* 51:1729–1740.

Zamani, A., and B. Maini. 2009. Flow of dispersed particles through porous media —Deep bed filtration. *Journal of Petroleum Science and Engineering* 69:71–88.

Zhao, S., L. S. Kotlyar, J. R. Woods, B. D. Sparks, J. Gao, J. Kung, and K. H. Chung. 2002. A benchmark assessment of residues: Comparison of Athabasca bitumen with conventional and heavy crudes. *Fuel* 81:737–746.

Zhou, Z. A., Z. Xu, J. A. Finch, J. H. Masliyah, and R. S. Chow. 2009. On the role of cavitation in particle collection in flotation – A critical review. II. *Mineral Engineering* 22:419–433.

Index

acetone-butanol-ethanol (ABE) fermentation 149–50
acidification 178–80, 182
acidogenesis 149
acridine 193
adenosine diphosphate (ADP) 164
adenosine triphosphate (ATP) 161–2, 164
adsorption 2, 38, 82, 89, 96, 162, 190–1, 196–8
Agave tequilana fructans 148
alcohol synthesis 115
algae 8–9, 25–6, 32–3, 55, 57–8, 112, 160–70, 172, 177, 179, 180–4, 197
algal high-rate pond 164
algal metabolism 159, 161
alkaline treatment 9, 47
alkyl levulinate 47
α-olefins 116–18
alumino-silicate 197
American Society of Testing and Materials (ASTM) 20, 54–5, 66
ammonia fiber explosion/expansion (AFEX) 10, 145
anaerobic digestion 7, 16, 34, 54, 160, 162, 168
Anderson-Schultz-Flory (ASF) hydrocarbon chain length 116
anhydride 123, 128–9, 131, 134
animal fat 7, 54–5, 57, 62, 64, 124, 135
animal manure 9
aprotic solvent 31, 40–1
arabinose 33, 145
Arachis hypogaea 56
Aspergillus aculeatus 148
Aspergillus oryzae 147
asphaltene 159, 187–201
asphaltene-coated kaolin 188, 197–8, 200
ATP-binding cassette 145
Attalea speciosa 57
autoclave reactor 24–5, 40, 172
Azadirachta indica 57

Bacillariophyta 171
bacteria 8, 10–11, 34, 144–5, 147, 149, 168, 171
Barrett-Joyner-Halenda (BJH) analysis 96
batch reactor 40, 171–2, 201
benzothiophenes 193
1, 4-β-D-glucosidase 142
β-glucosidase (BGL) 147–9
β-1, 4-glucosidic bond 142
β-xylosidase 145
biochar 9, 21–2, 160, 168–70
biodiesel 1, 6–8, 16, 33, 53–67, 109–10, 159, 162, 164, 168–9, 172, 182
biolubricant 123–34, 136
biomass pellet 16, 20
biomass-to-gas (BTG) 109–15, 119, 160
biomass-to-liquid (BTL) 110, 115, 119, 168, 170
bio-methanation 34
bio-oil 1, 7, 20–4, 54, 110, 160, 164, 168–72, 177–8
biopolymer 143, 167, 172
bi-reforming of methane (BOM) 71–4, 78, 88
bitumen 159, 187–92, 196–200, 203
bitumen-derived gas oil 187–8, 190–1
Borago officinalis 57
Botryococcus braunii 171
Boudouard reaction 77, 115
Brassica napus 56
Brønsted acid 34–6, 38–41, 43–4, 47–8, 64, 196
Brunauer-Emmett-Teller (BET) analysis 96
butanol 1, 6–7, 39–40, 110, 129, 131, 133, 141–2, 144, 149–1, 160, 168–9
1-butanol 129
2-butanol 39, 129
butyl levulinate 47

cabazole 193
Caldicellulosiruptor bescii 145
Caldicellulosiruptor obsidiansis 145
Caldicellulosiruptor saccharolyticus 146
calorific value 16, 20–2
Calvin cycle 161–2
Camelina sativa 57
Canadian Athabasca bitumen 187, 190
Canadian oil sands 203
Cannabis sativa 57
carbohydrate 2, 39–41, 43, 142–3, 145, 162, 172, 182–3
carbonization 168, 170
carbon sequestration 159, 162
catalyst fouling 187–9, 195–6, 199
cattle manure 110, 159
cellobiohydrolase 142
cellobiose 46, 142–3, 145–8, 150
cellulose 16, 22–3, 33–47, 111, 141–50, 170
cellulose-binding domain (CBD) 149
cellulosome 141, 148–9
cell wall binding domains (CWBD) 149
ceramic structured monolithic catalyst 119
cetane number 55, 58
chemical looping combustion (CLC) 3
chemical looping reforming (CLR) 3
chemo-enzymatic epoxidation 127
Chlorella vulgaris 171, 182–3
chloroplastic genome 162
circulating fluidized bed gasifier 111–13
Clavispora 148
Clostridium acetobutylicum 144, 149–50
Clostridium beijerinckii 149–51
Clostridium cellulolyticum 144, 147–51
Clostridium cellulovorans 147, 151
Clostridium phytofermentans 144–5
Clostridium saccharoperbutylacetonicum 150
Clostridium thermocellum 144–5, 147–50
Clostridium thermosaccharolyticum 150
Clostridium tyrobutyricum 150
cloud point 47, 55, 58
coagulation 166–7
coal-fired gasifier 112
co-combustion 15, 17–18
Cocos nucifera 56

co-firing 15, 17–20, 33
co-gasification 15, 17–18, 26–7
cohesin dockerin interaction 149
coker heavy gas oil (KHGO) 190
coker light gas oil (KLGO) 190
co-liquefaction 15, 17–18, 24–6
column photo-bioreactor 166
combustion 15–19, 21, 33, 53–4, 72, 142, 160, 168
co-milling 19
consolidated bioprocessing 141–51
continuous stirred tank reactor (CSTR) 172
conventional crude 187, 190, 202–3
coordinatively unsaturated sites (CUS) 196
co-pyrolysis 15, 17–18, 20, 21–7
cradle to grave approach 164, 178–80
critical micelle concentration (CMC) 198
Cuphea viscosissima 57
cyanobacteria 171
Cyanophyta 171
cyclohexanol 129
cyclone: boiler 18–19; feeder 19; separator 112

1-decanol 129
deep eutectic solvent (DES) 36
dehydration 34–5, 38, 40–1, 44, 48
dehydrogenase 145
delayed coking 192
densification 20
depolymerization 9, 24, 34–5, 44, 172
deprotonation 41
dibenzothiophenes 193
dibenzothiophine 196
di-ester 123, 129, 131, 133
differential scanning calorimetry analysis (DSC) 96
dimerization 124
2, 2-dimethyl di-propanol 129
dimethylformamide (DMF) 40–1
dissolved air flotation 166–7
dry reforming of methane (DRM) 72–3, 83–8

ecotoxicity 180
Elaeis guineensis 56
electro-coagulation 166
endo-1, 4-β-D-glucanase 142
endoglucanase 147, 149
endoxylanase 145
energy crops 7, 9–10, 16, 27, 32–3, 115, 142, 159–60, 177–8
Entner-Doudoroff pathway 145
entrained flow reactor 21, 112, 118
Environmental Impact Statement Assessment (EISA) 177
enzymatic epoxidation 127
enzymatic hydrolysis 142, 150, 168
epoxidation 123–36
Escherichia coli 149, 151
esterification 41, 44, 54–5, 61–5, 123, 129, 130–4, 168
ethanol 1, 6–8, 11, 16, 33–4, 40–4, 47, 54, 72, 94–5, 109–10, 128–9, 142–9, 160, 168–82
5-ethoxymethylfurfural (EMF) 7, 31, 41–5, 48
2-ethyl hexanol 129, 131, 133
ethyl levulinate (EL) 31, 34, 45, 47
eutrophication 178–80, 182
exo-1, 4-β-D-glucanase 142
exoglucanase 147, 149
extractives 143

fast pyrolysis 21–2, 171
fatty acid 34, 54–5, 58–66, 124–6, 131, 133, 160–2, 181
fatty acid methyl ester (FAME) 55–6, 60–1, 63, 65–7
fermentation 7, 10–11, 16, 34, 110, 142–51, 160, 168
fine particles 187–90, 196–203
Fischer-Tropsch process 3, 5, 72, 109–19
fixed bed reactor 23, 26, 114, 118–19, 172
flame ionization detector (FID) 66–7, 97
Flammulina velutipes 147
flash point 55, 58, 66, 123, 127, 131
flash pyrolysis 21
flat-plate photo-bioreactor 166
flexicoking 192
flocculation 162, 166–7, 198, 200
flotation 166–7
fluid catalytic cracking (FCC) 192
fluid coking 192
food waste 47, 74, 110, 115, 159
Fourier Transform Infrared Spectroscopy (FTIR) 53, 66, 96
fractional distillation 2, 190
free fatty acid (FFA) 54–65, 160
fructanase 14
fructose 35–48, 148, 170
fuel cell 3, 7, 72, 94
fungi 10, 141, 147
fungicides 35
furfural 10, 144, 170
furfuryl alcohol 46–7
Fusarium oxysporum 147
Fusarium verticillioides 147

γ-valerolactone 44
gas chromatography (GC) 53, 66, 97
gas hourly space velocity (GHSV) 71, 79–80, 83–5, 88, 94, 101–5
gasification 2, 5, 7, 9, 11, 15–27, 33, 54, 84, 94, 109–19, 160, 168, 172
gas-to-liquid (GTL) 109–10, 114–15, 119
genetic engineering 144, 147
Geobacillus 145
global warming 5, 18, 27, 31, 53, 141, 159, 179–80, 182
glucose 33–48, 142–8, 170
glutamine synthetase-glutamate synthase pathway 164
glycerol steam reforming (GSR) 93–7, 101, 104–6
glycolysis 145
glycoside hydrolase 145, 150
glycosyl hydrolase 149
Gossypium hirsutum 56
gravity sedimentation 167
greenhouse gas (GHG) 1–7, 15–19, 24, 27, 33, 71–3, 109, 141–2, 159, 163, 178, 180, 182, 184

Haematococcus pluvialis 165
haloids 44
hard wax 116
heating value 19–20, 183
heavy gas oil (HGO) 187, 190–1
heavy petroleum 116
heavy vacuum gas oil (HVGO) 190
Helianthus annuus 56
hemicellulose 33–5, 47, 111, 143–5, 147
hetero-atom contamination 195
heterogeneous catalyst 38, 43, 46–7, 53–4, 56, 61–7, 72, 74, 81, 128, 172

heterogeneous solid acid catalyst 44, 57, 61–2, 64
heterogeneous solid base catalyst 53, 65
heterogenized heteropoly acid catalyst 43
heterologous cellulolytic system 144
heteropoly acid 40, 43
1-hexanol 129
hexose 33, 142, 147
high-frequency reciprocating rig (HFRR) 131
high-temperature Fischer-Tropsch (HTFT) 118
homogeneous base-catalyzed transesterification 63
homogeneous catalyst 38, 46–7, 53–7, 61–7, 172
homogeneous liquid catalyst 61–2
hydrodearomatization (HDA) 195
hydrodemetallization (HDM) 193, 198
hydrodenitrogenation (HDN) 193–4, 196, 198
hydrodeoxygenation (HDO) 188, 193, 198
hydrodesulfurization (HDS) 193, 196
hydrofluorocarbons (HFC) 93
hydrogen 5, 7–8, 22, 24, 26–7, 33–4, 37, 44, 55, 71–2, 74, 94–5, 110, 112, 117, 124–5, 145, 160, 162, 178, 182, 184, 191–3, 196, 201
hydrogenation 94, 117, 124, 126, 192–3, 196
hydrogenolysis 193
hydrogen partial pressure 195
hydrolase 151
hydrolysis 34, 43–4, 47, 54, 57–8, 63, 125, 142–5, 147–8, 150–1, 168, 170
hydroprocessing 190–2, 196, 201
hydrothermal carbonization 168, 170
hydrothermal gasification 112, 119, 168
hydrothermal liquefaction 24, 159, 168, 170, 172
hydrotreating 181, 187–203
hydroxylation 123, 128
5-hydroxymethylfurfural (HMF) 10, 31, 34–7, 39–40, 45

immobilization 163, 166
indole 193
inductively coupled plasma optical emission spectroscopy (ICP-OES) 96
integrated gasification combined cycle (IGCC) 5
Intergovernmental Panel on Climate Change (IPCC) 93
inulin 36–8, 42–3
ion exchange resin 37, 39, 65, 127–8
ionic liquid 34, 38, 43–4, 46, 48, 168
isomerization 35, 43, 48
isopropanol 39–40

Jatropha 54–5, 57–9, 66, 129, 180

kaolin 188–9, 196–201
kaolinite 197, 200
kerosene 2, 115–16, 181
Khuyveromyces marxianusis 148
kinematic viscosity 47, 55, 66, 129, 134, 170
Klebsiella pneumonia 151
Kluyvera 150

Laminaria digitata 182
lauric acid 66
levulinic acid 31, 34, 44–6
Lewis acid 34–6, 38–40, 43–4, 47–8, 64, 196
life-cycle assessment/analysis (LCA) 10, 119, 177–84
light gas oil (LGO) 187, 190–2
light petroleum 116
light vacuum gas oil (LVGO) 190
lignin 9–10, 16, 22, 24–5, 33–5, 46–7, 111, 142–4, 170
lignite 2, 22–5
lignocellulosic biomass/feedstocks 7, 9, 11, 31, 33–5, 41, 44, 47, 111, 115, 141–4, 147, 150–1, 159, 177–8, 182
linoleic acid 59, 66, 127, 136
Linum usitatissimum 57
lipid 10, 54, 124–5, 143, 160, 162–3, 165, 168, 171–2, 182–3
liquefaction 2, 7, 15, 17–18, 21, 24–7, 33, 54, 110, 159–60, 168, 170–2
liquefied petroleum gas 116
liquid hourly space velocity (LHSV) 195
lower heating value (LHV) 183
low-temperature Fischer-Tropsch (LTFT) 118

maltene 190–1
mechanical extraction 16, 168
membrane separation 2
metabolic engineering 10, 142, 144, 151, 160, 162
metal organic framework (MOF) 38
metal-organic framework carbons (MDC) 37–8
metathesis 128
methanation 110
methanesulfonates 44
Methanol-to-Gasoline (MTG) 113
methyl levulinate 47
microalgae 9, 25–6, 55, 57–8, 160, 162–4, 166, 170, 172, 182, 184
microreactor 172
middle distillates 116
molecular sieve 40, 191
Moringa oleifera 55
Mrakia blollopis 148
Mrakiella 148
Mucor circinelloides 147
multi-tubular fixed bed reactor 118
municipal solid waste (MSW) 9, 15, 110, 112, 115
mystric acid 66

Nannochloropsis salina 171
naphthalene 195
natural gas combined cycle (NGCC) 20
Neochloris oleoabundans 171
nicotineamide adenine dinucleotide (NAD) 145
nicotineamide adenine dinucleotide phosphate (NADPH) 161–2, 164
nitrate reductase 164
non-local density functional theory (NLDFT) 96
non-polar solvent 40

oil seed 56–8, 160
olefin(s) 114–15, 193; epoxidation 127; synthesis 115
oleic acid 59, 63, 66, 127–8, 131–3
open pond system 8, 160, 164–6
organic coated solids 187, 196
oxidation onset temperature 131
oxidative cleavage 124
ozone layer depletion 178, 180

Paecilomyces variotii 147
palm fatty acid distillate (PFAD) 54, 66
palmitic acid 59, 66
palmitoleic acid 66